Author	MOORE
Title	Prostanoids: pharmacological, physiological and clinical relevance
Acc. No.	M41181 d

DATE OF RETURN AS BELOW

HERIOT—WATT UNIVERSITY LIBRARY

M41181 d

Prostanoids: pharmacological, physiological and clinical relevance

P. K. MOORE

Department of Pharmacology, Chelsea College, University of London

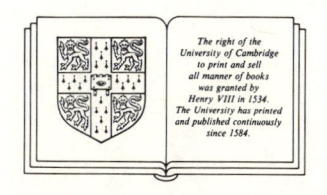

CAMBRIDGE UNIVERSITY PRESS
Cambridge
London New York New Rochelle
Melbourne Sydney

Published by the Press Syndicate of the University of Cambridge
The Pitt Building, Trumpington Street, Cambridge CB2 1RP
32 East 57th Street, New York, NY 10022, USA
10 Stamford Road, Oakleigh, Melbourne 3166, Australia

© Cambridge University Press 1985

First published 1985

Printed in Great Britain by the
University Press, Cambridge

Library of Congress catalogue card number: 85–4232

British Library cataloguing in publication data

Moore, P. K.
Prostanoids: pharmacological, physiological
and clinical relevance.
1. Prostaglandins 2. Thromboxanes
I. Title
612'.405   QP801.P68

ISBN 0 521 260817 hard covers
ISBN 0 521 278279 paperback

Contents

	Preface	vi
	Acknowledgements	vii
1	Biosynthesis and catabolism of prostaglandins, thromboxanes and leukotrienes	1
2	Effect of drugs on arachidonic acid metabolism	41
3	The cardiovascular system	68
4	Prostanoids and platelets	79
5	The reproductive system	100
6	Prostanoids and the gut	123
7	The respiratory system	148
8	Prostanoids and the kidney	173
9	Prostaglandins and the nervous system	200
10	Inflammation	220
	References	236
	Index	261

Preface

Few areas of medical research have progressed as rapidly during the last few years as has the study of the prostanoids. A good index of the enthusiasm for prostanoids from researchers world-wide is the steadily growing number of publications on the subject over the last 30 years. In 1962, for example, the total number of relevant published reports was less than 30 and the majority of these originated from just two laboratories. Ten years later this figure had risen to a little under 1000. In 1982 alone, scrutiny of abstracting journals such as *Index Medicus* revealed that 1800 publications on prostanoids from laboratories in over 50 different countries had found their way into print. This mass of literature has proved difficult to assimilate, even for active researchers in the field; it is thus no surprise that newcomers to prostanoid research are frequently confused and bewildered by the whole subject. The reason for writing this book is to provide a single source of readily understood but up-to-date information on the various physiological and clinical roles and uses of prostanoids. Only time will tell to what extent this goal has been achieved.

October 1984   P.K.M.

Acknowledgements

I am indebted to all of my family for their support over the last few months and particularly my father who typed the last of many draft versions of the manuscript. My thanks also go to friends and colleagues for their help during the writing of this book.

1

Biosynthesis and catabolism of prostaglandins, thromboxanes and leukotrienes

Introduction

Prostanoids are unsaturated lipids made from fatty acid precursors such as arachidonic acid. Unlike other local hormones (e.g. histamine, 5-hydroxytryptamine), prostanoids are not a single substance but a family of chemically related compounds comprising prostaglandins and thromboxanes. Mammalian tissues are able to synthesise up to 20 separate prostanoids each with subtly different biological activity. Even a single organ such as the lung or kidney can form several different prostanoids which may have similar or opposite biological actions in that organ or (much less frequently) no action at all. Recently, another family of biologically active substances (leukotrienes), synthesised from the same precursors as prostaglandins, have been discovered. Structurally, leukotrienes differ from prostanoids. The collective name for prostaglandins, thromboxanes and leukotrienes is eicosanoids.

A great deal of effort has been directed towards understanding how eicosanoids are made in the body. Progress has been slow and it has only been in the last few years that the final pieces in what has turned out to be a complicated jigsaw have been put together. The remainder of this chapter deals with the formation, breakdown and mechanism of action of eicosanoids.

Early history

In the first 30 years of this century several researchers, albeit unknowingly, observed some of the biological effects of prostanoids. Simple saline extracts of the human prostate gland were shown to reduce dog blood pressure following intravenous injection. It is likely that at least part of this vasodepressor effect was due to the presence of prostanoids. However, it is known that such prostate-gland extracts would also contain other biologically active substances (such as histamine and adenine nucleotides) which could also influence blood pressure. For this reason, the discovery of prostanoids is usually attributed to two New York gynaecologists, Kurzrok & Lieb (1930), who observed that freshly obtained human seminal fluid either contracted or relaxed human uterine strips set up in an isolated organ bath. Unlike extracted prostate glands, human semen does not contain histamine or other substances which are known to contract the uterus. Later, Goldblatt (1933) reported that human semen also contracted the rabbit intestine and von Euler (1934) showed that semen lowered blood pressure in the rabbit, cat and dog. von Euler was also responsible for the first attempted extraction of the active principle from semen which he showed was soluble in both water and organic solvents, was destroyed at extremes of pH, lacked nitrogen and was an unsaturated fatty acid. A similar vasodepressor spasmogen on gastro-intestinal and uterine smooth muscle was also detected in human and sheep prostate gland; this prompted von Euler to suggest the name, prostaglandin (von Euler, 1934).

The simple chemical techniques which were available in the 1930s for lipid analysis were inadequate to identify the chemical structure of prostaglandin. However, it was realised that prostaglandin formed a soluble barium salt; use was made of this property to extract relatively large amounts of prostaglandin from kilograms of seminal vesicle glands from Icelandic sheep. This was the first attempt at a large-scale extraction of prostaglandin from a biological source.

Identification of prostaglandins E and F

Further progress in prostaglandin research had to await the development of sufficiently sensitive chemical techniques to separate and determine the structure of the small amounts of prostaglandins present in biological tissues. Some 20 years after the first demonstration of biological activity in human semen, Bergström & Sjövall (1957) published the structure of a prostaglandin 'factor' which had been extracted from the seminal vesicle glands of sheep. Prostaglandin 'factor' later became shortened to prostaglandin F (PGF). These pioneering experiments (reviewed by von Euler & Eliasson, 1967) involved the development, from scratch, of new paper- and thin-layer chromatographic methods to separate PGF from other components of semen and the application of what was then the new process of gas-chromatography–mass-spectrometry (GC–MS). PGF contracted the rabbit intestine preparation but had only very weak vasodepressor activity in the anaesthetised rabbit; this suggested that semen contained a second biologically active substance, possibly another prostaglandin. Bergström & Sjövall (1960) later extracted a second prostaglandin from human semen which they named prostaglandin E (PGE). When injected into the blood stream of anaesthetised animals, PGE was a potent vasodepressor and also contracted the rabbit intestine preparation.

The PGE and PGF isolated by Bergström & Sjövall in the 1950s are now called PGE_2 and $PGF_{2\alpha}$ because each has two olefinic bonds linking C-5 and C-6 (*cis*) and a *trans* double bond between C-13 and C-14. The subscript α in $PGF_{2\alpha}$ indicates the stereochemical configuration of the hydrogen at position C-9. $PGF_{2\beta}$ can be synthesised chemically but does not occur naturally in the body. The majority of prostaglandins found in mammalian tissues contain 2 double bonds. However, smaller amounts of prostaglandins containing a single double bond (e.g. PGE_1) are made in

Fig. 1.1. Chemical structure of some naturally occurring prostaglandins. All prostaglandins have 20 carbon atoms numbered from the carbon attached to the carboxyl (-COOH) group (which is C-1). The 5-membered (cyclopentane) ring spans carbons 8–12. The hydroxyl group at position C-15 is required for biological activity. Prostanoic acid is a hypothetical substance which does not occur naturally in the body.

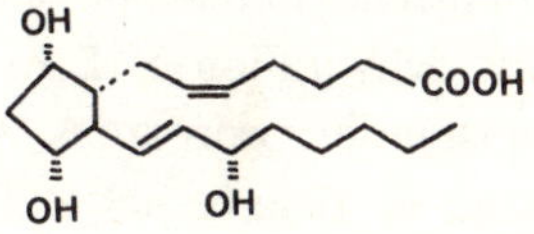

Prostaglandin $F_{2\alpha}$ ($PGF_{2\alpha}$)

Prostaglandin E_2 (PGE_2)

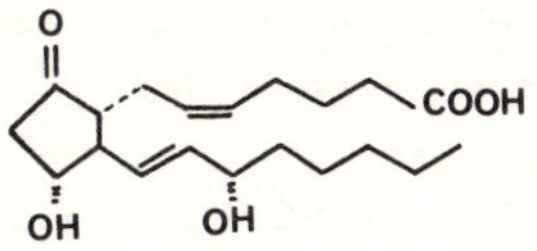

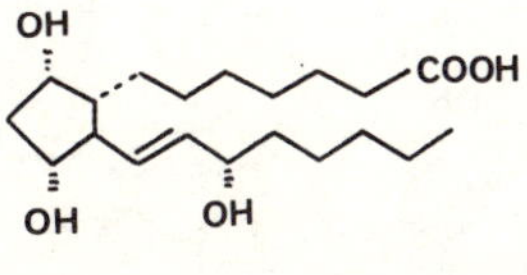

Prostaglandin $F_{1\alpha}$ ($PGF_{1\alpha}$)

Prostaglandin E_1 (PGE_1)

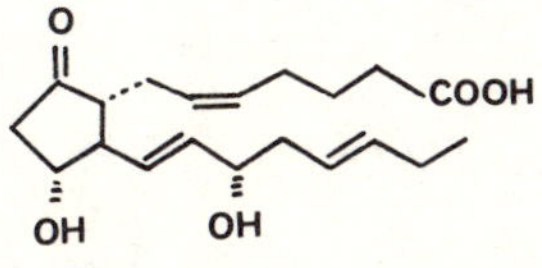

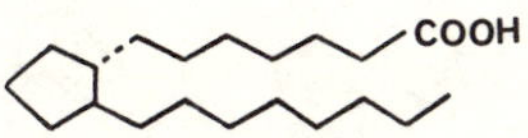

Prostanoic acid

some organs. Prostaglandins with an additional double bond linking C-17 and C-18 (e.g. $PGF_{3\alpha}$) may also occur naturally, for example in human semen. Like the C-5–C-6 double bond, that between C-17 and C-18 is in the *cis* transfiguration. The structures of some naturally occurring prostaglandins are shown in Fig. 1.1. The chemistry and stereochemistry of prostaglandins has been reviewed by Caton (1972).

Prostaglandins A, B, C and D

Several other prostaglandins have been described some of which have interesting biological actions and may occur naturally in the body. These prostaglandins are referred to by the letters A, B, C and D. Structurally these prostaglandins differ from those of the E and F series only by virtue of having different chemical groups attached to the cyclopentane ring. Although prostaglandins A, B, C and D (containing 1 and 2 double bonds) are shown in Fig. 1.2, there is certainly no chemical reason why such prostaglandins containing 3 double bonds should not exist. However, it is very unlikely that such prostaglandins occur naturally in the body. Prostaglandins A_2, B_2 and C_2 are dehydration products of PGE_2, having lost the hydroxyl group at C-11 with the formation of another double bond in the ring. This dehydration step occurs readily at acid pH and it is likely that the reported occurrence of these prostaglandins in extracts of biological tissues represents dehydration of E to A series prostaglandins during the extraction and isolation procedures. A prostaglandins are unstable and readily rearrange at alkaline pH to B prostaglandins. Some years ago enzymes were described in plasma which interconvert prostaglandins of the A, B and C series (Fig. 1.3.). These

Fig. 1.2. Chemical structure of PGD_2 and PGD_1. The cyclopentane ring region of A, B, and C prostaglandins are also shown. Notice the extra double bond in the ring.

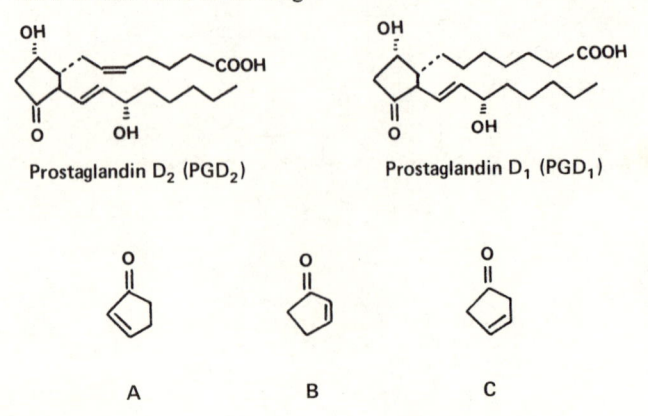

Prostaglandin D_2 (PGD_2)

Prostaglandin D_1 (PGD_1)

A

B

C

enzymes are probably of little or no importance physiologically. Prostaglandins of the D series (e.g. PGD_2) occur naturally and have potent biological effects on the cardiovascular system (Chapter 3), on platelet aggregation (Chapter 4) and in the brain (Chapter 9).

Fig. 1.3. Interconversion of E, A, B and C prostaglandins. The first step (E to A) also occurs chemically at acid pH.

$$PGE_2 \xrightarrow{\text{dehydrase}} PGA_2 \xrightarrow{\text{isomerase}} PGB_2 \xrightarrow{\text{isomerase}} PGC_2$$

Biosynthesis of prostaglandins in the body

After the elucidation of the structure of PGE_1, PGE_2 and PGE_3 and the corresponding F series prostaglandins, mainly as a result of the studies of Bergström and his colleagues in the period between 1960 and 1965, research was concentrated on identifying the pathway for the formation of prostaglandins in the body. Looking at the structure of prostaglandins, it was reasoned that the precursor could be an unsaturated fatty acid containing the same number of carbon atoms as prostaglandins and 3, 4 or 5 double bonds. The nomenclature of such fatty acids is shown in Table 1.1.

These fatty acids are usually referred to by their trivial names since the chemical names are long and unwieldy. It can now be seen why metabolites of these fatty acids are called eicosanoids. Some authors do use the 'short hand' notation shown in the last column of Table 1.1, where arachidonic acid is $20:4\ \omega\ 6$ (i.e. 20 carbon atoms, 4 double bonds, first double bond situated 6 carbon atoms away from the carboxy terminal).

The most abundant of these fatty acids in biological tissues is arachidonic acid (AA) which gives rise to prostaglandins containing 2 double bonds. Similarly eicosatri- and eicosapentaenoic acids are converted to prostaglandins containing 1 and 3 double bonds respectively.

Experimental confirmation of the role of fatty acids as precursors to

Table 1.1. *Prostanoid precursor nomenclature*

Trivial name	Chemical name	'Short hand'
Dihomo-γ-linolenic acid	*all-cis*-5,8,11-Eiocosatrienoic acid	20:3
Arachidonic acid	*all-cis*-5,8,11,14-Eicosatetraenoic acid	20:4
—	*all-cis*-5,8,11,14,17-Eicosapentaenoic acid	20:5

prostaglandins came with the simultaneous publication of results by Bergström and his colleagues (Bergström, Carlsson & Oro, 1964) and van Dorp and his colleagues (van Dorp *et al.*, 1964) working independently in laboratories in Sweden and the Netherlands respectively. Both groups described the conversion of radioactive arachidonic acid to radioactive PGE_2 by sheep seminal vesicle homogenates. Later the same enzyme source was shown to convert dihomo-γ-linolenic acid to PGE_1 and other organ homogenates were shown to synthesise preferentially prostaglandins of the F series. Guinea pig lung homogenate converted arachidonic acid mainly to $PGF_{2\alpha}$ (Anggard & Samuelsson, 1965). Subcellular fractionation of tissue homogenates revealed that the enzyme(s) responsible for prostaglandin formation occurred in the microsomal fraction of the cell. Microsomes are sealed spheres of endoplasmic reticular membrane formed during homogenisation and centrifugation of an organ homogenate. Little or no prostaglandins were synthesised when arachidonic acid was incubated with the cytoplasmic (100000 g) fraction of the cell.

Release of arachidonic acid from membrane phospholipids

Fatty acids which act as precursors to prostaglandins are present in the normal diet. Large amounts of arachidonic acid are found in meat and even larger amounts in seaweed. However, with most diets the intake of arachidonic acid, dihomo-γ-linolenic acid and eicosapentaenoic acid (found mainly in fish) is inadequate and thus the requirement for these fatty acids must be met by intake of linoleic acid. Unlike the fatty acids mentioned to date, linoleic acid has only 18 carbon atoms and 2 double bonds ($18:2\ \omega\ 6$) and occurs in many foods. Linoleic acid is enzymatically converted to prostaglandin precursors.

Although most mammals have sufficient amounts of the desaturase enzymes necessary for conversion of linoleic acid to arachidonic acid, cats are deficient in this enzyme and must take in all the arachidonic acid they require in a meat diet. Clearly, for this reason, cats cannot afford to be vegetarians.

Arachidonic acid and the other prostaglandin precursors are not normally found in the free state in the cell but occur bound in phospholipids, neutral lipids and with cholesterol. Since esterified fatty acids are not substrates for the enzymes which make prostaglandins, the first step in prostaglandin biosynthesis must be the release of these fatty acids from their combination in lipids. The most important source of fatty acids for prostaglandin formation is phospholipids. The structure of a typical phospholipid is shown in Fig. 1.4.

The backbone of the phospholipid structure is a 3-carbon compound

Fig. 1.4. Structure of a typical phospholipid. The fatty acid attached to 'hook' 1 is normally saturated while that attached at 'hook' 2 is normally unsaturated. Phospholipase A_2 cleaves specifically the unsaturated fatty acid (e.g. arachidonic acid) from the second hook. Release of arachidonic acid from phospholipids also follows the consecutive action of phospholipase C and diglyceride lipase.

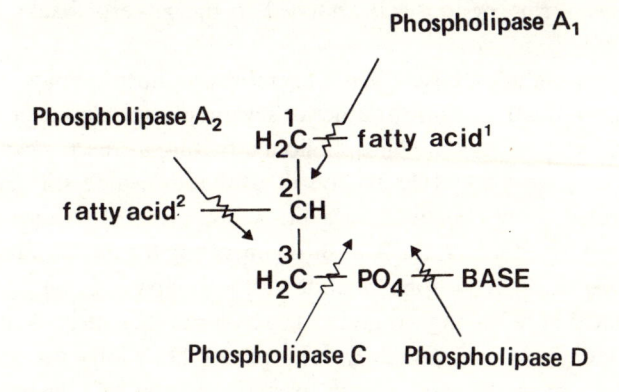

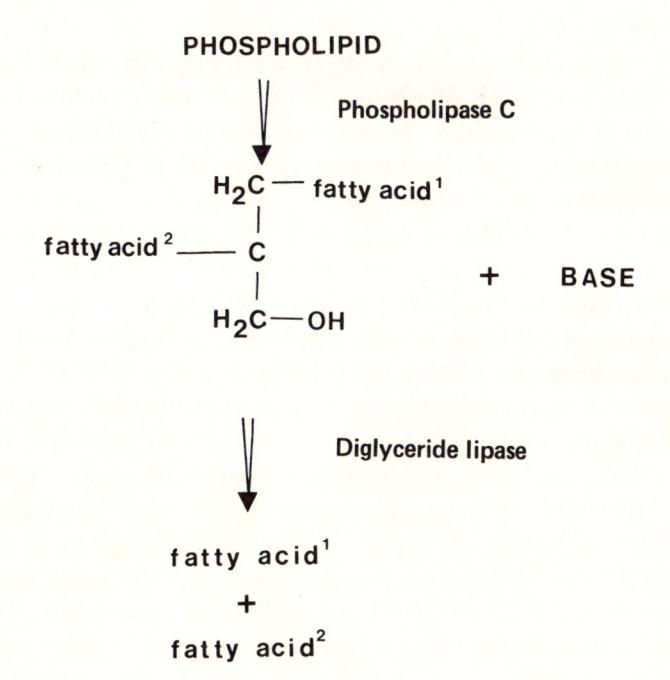

Phospholipase A_1

Phospholipase A_2

1
H_2C — fatty acid^1

fatty acid^2 — 2
CH

3
H_2C — PO_4 — BASE

Phospholipase C   Phospholipase D

PHOSPHOLIPID

⇓ Phospholipase C

H_2C — fatty acid^1
|
fatty acid^2 — C   + BASE
|
H_2C — OH

⇓ Diglyceride lipase

fatty acid^1
+
fatty acid^2

called glycerol which provides 3 'hooks' upon which the other components of the phospholipid may be 'hung'. The 1′ position is usually occupied by a saturated fatty acid (i.e. does not have any double bonds, palmitic acid is an example) and the 2′ position by an unsaturated fatty acid (such as arachidonic acid). The 3′ position is taken up with a base (usually choline, ethanolamine, serine or inositol) attached to the glycerol backbone by phosphate linkage.

Release of arachidonic acid from membrane phospholipids can be achieved enzymatically by one or other of several phospholipase enzymes. Phospholipase A_1 attacks the phospholipid structure at the 1′ position to release the fatty acid and give a 1′ lysophosphatide which still contains the unsaturated fatty acid and base bound to glycerol. Alternatively, phospholipase A_2 attacks at the 2′ position liberating the unsaturated fatty acid and leaving a 2′ lysophosphatide. Phospholipase C removes the phosphorylated base leaving a diglyceride which can then be broken down by a specific diglyceride lipase enzyme to release both fatty acids in free form. The lysophosphatides which are formed as a result of phospholipase action are either enzymatically converted to other products or reincorporated into membrane phospholipids. These various reactions are shown in Fig. 1.4.

Although there has been lively debate about which of these enzymes is normally responsible for liberating fatty acids for prostanoid biosynthesis it now seems that the most important is probably phospholipase A_2. In some cells, e.g. platelets, the concerted action of phospholipase C and diglyceride lipase is necessary for releasing arachidonic acid (Rittenhouse-Simmons, 1979). Phospholipase A_2 is a membrane-bound enzyme found in virtually every cell type and organ in the body. It is likely that the critical factor regulating the activity of phospholipase A_2 is the availability of calcium ions. Selective calcium ionophore agents such as A23187 stimulate phospholipase activity although calcium chelation (with EDTA or EGTA) does not inhibit phospholipase activity which strongly suggests that the calcium ions necessary come from an intracellular store. One thing which is clear is that phospholipase A_2 is activated with release of arachidonic acid following a wide variety of chemical, hormonal, immunological and mechanical stimuli. Probably the best way of ensuring prostaglandin biosynthesis in a particular organ is simply to homogenise that organ in aqueous buffer. When this was realised it soon became apparent that the very high concentrations of prostaglandins previously found in homogenates of organs such as lung, kidney and gut did not reflect the 'basal' content as was first envisaged but rather prostaglandins made during homogenisation due to activation of phospholipase A_2. As a general rule, any stimulus which disrupts or deranges membranes releases intracellular calcium ions and activates phospholipase enzyme activity.

Role of oxygen in prostaglandin biosynthesis

The role of molecular oxygen in the formation of prostanoids was studied in a series of elegant experiments by Samuelsson (1965). Bovine seminal vesicle microsomes were incubated with dihomo-γ-linolenic acid in an atmosphere containing radioactive oxygen. 50% of the oxygen molecules were $^{16}O^{16}O$ and the remaining 50% were $^{18}O^{18}O$. Analysis of the PGE_1 and $PGF_{1\alpha}$ formed revealed that equal quantities of prostaglandins with two ^{16}O at C-9 and C-11 and two ^{18}O atoms at these positions were formed (Fig. 1.5.). Only very rarely were prostaglandin molecules with mixed oxygen isotopes at C-9 and C-11 identified. Clearly both oxygen functions attached to the cyclopentane ring of PGE_1 and $PGF_{1\alpha}$ were derived from the same molecule of oxygen. This result was interpreted to suggest that a cyclic peroxide intermediate was formed from dihomo-γ-linolenic acid and then converted to prostaglandins.

Fig. 1.5. Incorporation of radiolabelled oxygen into PGE_1 and $PGF_{1\alpha}$.

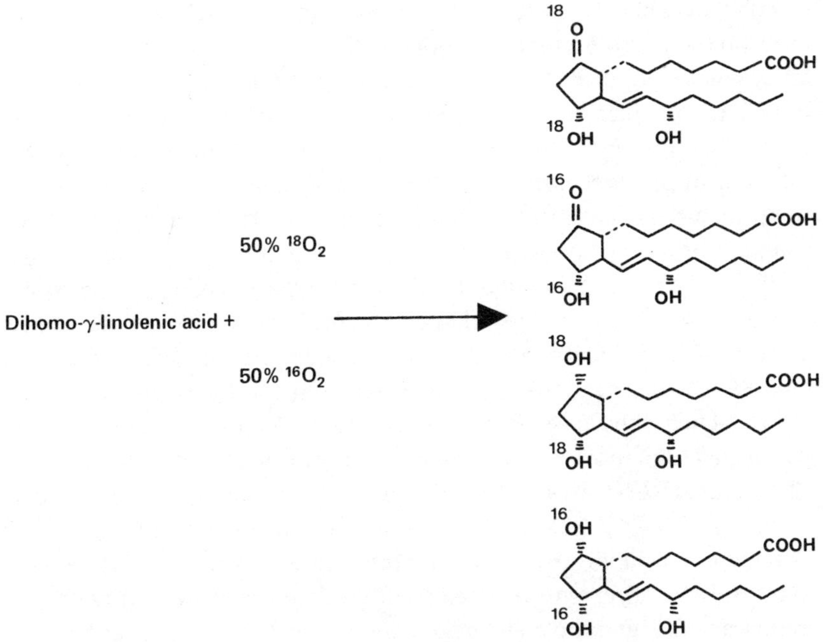

Discovery of rabbit aorta contracting substance

Direct evidence for the formation of an unstable cyclic peroxide intermediate in the biosynthesis of prostaglandins was obtained in 1969 by Piper and Vane. These workers used a specialised bioassay technique called 'multiple organ' or 'cascade' superfusion bioassay to detect the release of biologically active substances in the perfusate of isolated guinea pig lungs. In this method up to six isolated pharmacological preparations are suspended in air jackets, one above the other, and constantly superfused with physiological salt solution (usually Krebs' solution) which has been warmed and oxygenated. In this way the tissues can be kept alive and will respond to drugs for several hours. Lengthening or shortening of the tissues may be determined by attaching each preparation with cotton threads to force transducers. Contractions or relaxations of the tissues may be shown on a pen recorder.

In the experiments of Piper & Vane, lungs were removed from guinea pigs which had been immunologically 'sensitised' by injection 2 weeks previously with ovalbumin. The lungs were perfused with Krebs' solution via a cannula inserted into the pulmonary artery. In this way the Krebs' solution passed through the pulmonary vasculature before superfusing the bioassay preparations.

Injection through the lungs of ovalbumin produced an anaphylactic reaction and the release of a substance which contracted the isolated, spirally cut rabbit aorta strip but had disproportionately less contractile effect on the gastrointestinal smooth muscle preparations (e.g. rat stomach strip, guinea pig ileum) in the cascade (Fig. 1.6.). The unknown active principle was called 'rabbit aorta contracting substance' or RCS and its release was prevented by aspirin which was later shown to be a potent inhibitor of prostaglandin biosynthesis (see Chapter 2). Furthermore, by inserting a delay coil between the lungs and the rabbit aorta, RCS was found to be a very unstable substance with a half life ($t_{\frac{1}{2}}$) of about 2 min at 37 °C. Piper and Vane suggested that RCS was a cyclic endoperoxide, probably identical to that postulated by Samuelsson some years previously. Release of RCS from perfused, antigen-challenged guinea pig lungs was later shown to occur following injection into the lungs of bradykinin, slow-reacting substance of anaphylaxis (SRS-A) and arachidonic acid. Mechanical agitation simply by manipulating the lungs between the fingers also released RCS activity (Piper & Vane, 1971). Other organs were quickly found to release RCS. Rabbit spleen slices or microsomes incubated with arachidonic acid or the perfused rabbit spleen stimulated either with bradykinin or adrenaline released a labile substance which contracted the rabbit aorta. Human platelet aggregation *in vitro* was also accompanied by release of RCS (Vargaftig & Dao, 1972) as was addition of arachidonic acid to dog whole blood (Vargaftig & Zirinis, 1973). The pathway of arachidonic acid, as it was known at the end of 1969, is shown in Fig. 1.7.

The chemical structure of RCS was determined independently by Hamberg & Samuelsson (1973) and Nugteren & Hazelhof (1973). Each group isolated a 15-hydroxy cyclic endoperoxide from incubations containing arachidonic acid and bovine seminal vesicle microsomes. This substance was called PGH_2 by Hamberg and Samuelsson and PGR_2 by Nugteren and Hazelhof. A similar prostaglandin endoperoxide with a 15-hydroperoxy group was later identified and called PGG_2 and PGS_2 by the two research groups. The nomenclature of Hamberg and Samuelsson has been adopted. Both PGG_2 and PGH_2 are unstable in aqueous solution, undergoing chemical decomposition to a mixture of PGE_2, $PGF_{2\alpha}$, PGD_2, malondialdehyde (MDA) and a 17-carbon compound, 12L hydroxy-5,8,10-heptadecatrienoic acid (HHT). The half-lives of both prostaglandin endoperoxides are approximately 5 min in aqueous solution at 37 °C. Since

Fig. 1.6. Discovery of rabbit aorta contracting substance (RCS). Lungs from ovalbumin-sensitised guinea pigs were removed, superfused with Krebs' solution via a cannula inserted into the pulmonary artery and allowed to drip over four isolated pharmacological preparations in cascade. Challenging the lungs with ovalbumin causes the release of a substance which contracts the rabbit aorta. When lung effluent was collected and injected over the tissues 23 min later, no contraction of the rabbit aorta was observed, showing that RCS is chemically unstable. Responses of the preparations to PGE_2 are also shown. Reprinted by permission of *Nature*, **223**, 29–35, Copyright © 1969 Macmillan Journals Limited, and by permission of the authors.

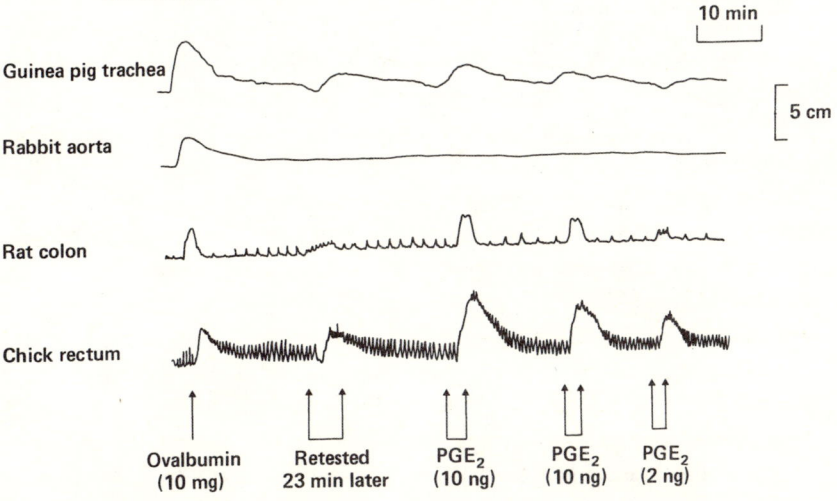

PGG$_2$ and PGH$_2$ contracted the rabbit aorta preparation and induced platelet aggregation, Nugteren and Hazelhof suggested that PGG$_2$ or PGH$_2$ (or both) were RCS. This conclusion was supported by Willis and his colleagues (1974) who demonstrated the formation of an unstable principle that induced platelet aggregation and contracted the rabbit aorta in short-term incubations of arachidonic acid and sheep seminal vesicle gland homogenates. The product(s) were named 'labile aggregation stimulating substance(s)' (LASS). Subsequent isolation and purification revealed that LASS was chemically and biologically indistinguishable from PGH$_2$. Prostaglandin endoperoxides are synthesised from arachidonic acid by a two-step enzyme reaction, catalysed by the enzyme prostaglandin endoperoxide synthase. Cyclisation of arachidonic acid by the cyclo-oxygenase enzyme component forms PGG$_2$ which is converted by the peroxidase enzyme component to PGH$_2$ by reduction of the C-15 hydroperoxy group. The pathway of prostaglandin biosynthesis, as understood up to the end of 1973, is shown in Fig. 1.8.

Discovery of thromboxane A_2 (TxA_2)

Two observations suggested that RCS and prostaglandin endoperoxides were not identical. Firstly, the $t_{\frac{1}{2}}$ of RCS (about 2 min) and the $t_{\frac{1}{2}}$ of PGG$_2$/PGH$_2$ (about 5 min) were significantly different. Secondly, the amount of prostaglandin endoperoxides released from aggregating platelets or from perfused guinea pig lungs challenged with arachidonic acid was insufficient to account entirely for the observed rabbit aorta contracting

Fig. 1.7. Pathway of arachidonic acid metabolism (1969).

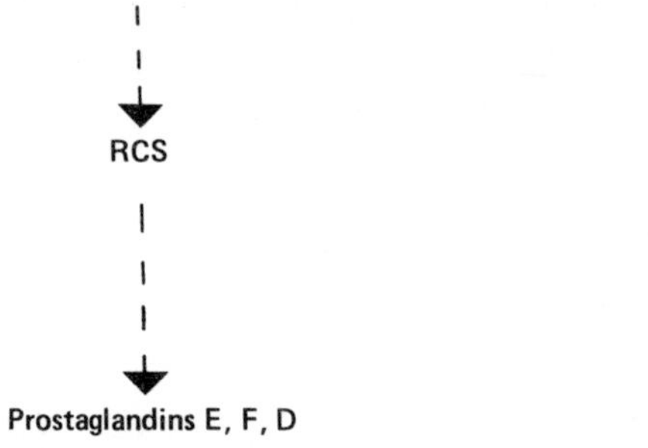

Phospholipid
↓ ← Phospholipase A$_2$ or C
↓
Arachidonic acid
↓
↓
↓
↓
RCS
↓
↓
↓
↓
Prostaglandins E, F, D

ability. These discrepancies initiated a search for additional metabolites of arachidonic acid with RCS-like biological properties, but with a shorter $t_{\frac{1}{2}}$ than the prostaglandin (PG) endoperoxides. The result of this search was the publication of the structure of a novel metabolite of arachidonic acid called thromboxane A_2 (TxA$_2$) which was intermediate in the conversion of PGG$_2$ to PHD (a polar compound, the hemiacetal derivative of 8(1-hydroxy-3-oxopropl-9,12L-dihydroxy-5,10-heptadecaenoic acid), now called thromboxane B_2 (TxB$_2$). These experiments were originally carried out by incubating platelet microsomes with PGH$_2$ for very short

Fig. 1.8. Pathway of arachidonic acid metabolism (1973). Chemical structure of prostaglandin endoperoxides, PGG$_2$ and PGH$_2$.

Membrane phospholipids

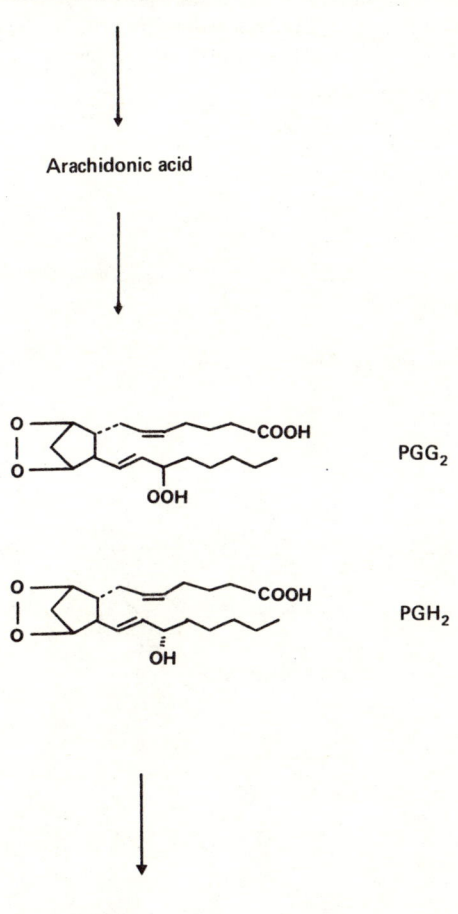

↓

Arachidonic acid

↓

[structure] PGG$_2$

[structure] PGH$_2$

↓

Prostaglandins

periods of time and then trapping the extremely labile TxA$_2$ by addition of a large excess of very cold dry acetone (Needleman *et al.*, 1976). TxA$_2$ contracted the rabbit aorta powerfully and aggregated platelets but was even less stable than the PG endoperoxides. The $t_{\frac{1}{2}}$ of TxA$_2$ at physiological temperature and pH, is only about 30 s, it breaks down into the chemically stable TxB$_2$. The enzyme responsible for converting PGH$_2$ to TxA$_2$ has been named thromboxane synthetase. The pathway of prostaglandin and thromboxane synthesis, as it stood at the end of 1976, is shown in Fig. 1.9. It should be noticed that TxA$_2$ and TxB$_2$ (so-called because of their oxane ring structure and potent platelet or thrombocyte aggregating activity) have no cyclopentane ring, are not derivatives of prostanoic acid

Fig. 1.9. Pathway of arachidonic acid metabolism (1976). Chemical structure of thromboxane A$_2$ (TxA$_2$) and thromboxane B$_2$ (TxB$_2$). TxA$_2$ is chemically labile and spontaneously decomposes to TxB$_2$.

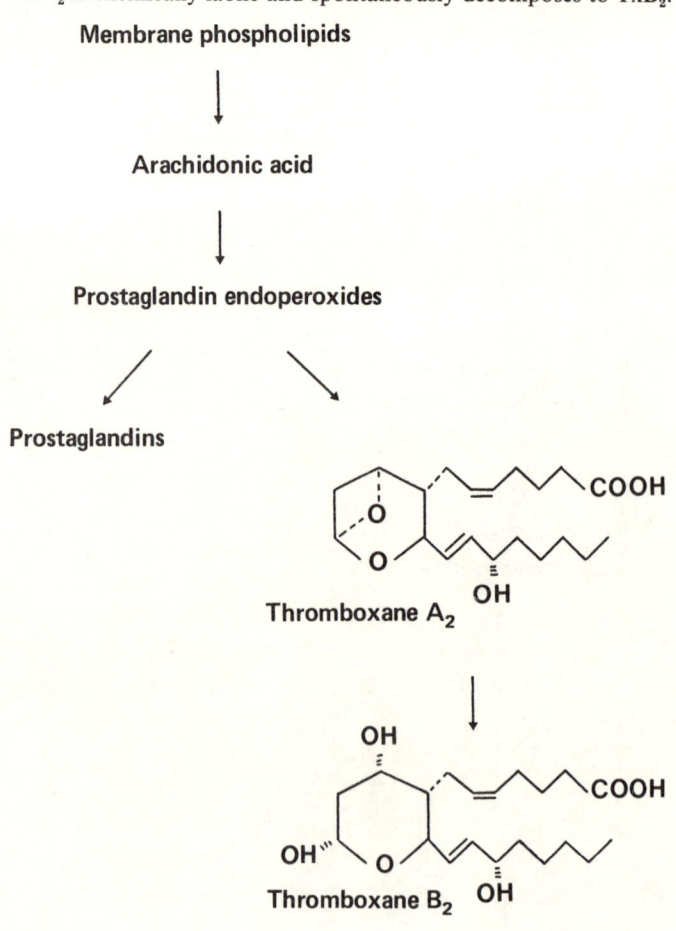

Membrane phospholipids

↓

Arachidonic acid

↓

Prostaglandin endoperoxides

↙ ↘

Prostaglandins

Thromboxane A$_2$

↓

Thromboxane B$_2$

and are thus not prostaglandins. Prostaglandins and thromboxanes are collectively called prostanoids.

Prostacyclin (PGI_2)

In an attempt to determine whether PG endoperoxides were converted to TxA_2 in tissues other than platelets, Moncada et al. (1976) incubated arachidonic acid or PGG_2 with microsomal preparations of several organs and followed the appearance of TxA_2 by monitoring the contractile effect of the incubate on the superfused rabbit aorta preparation in a cascade bioassay set up. If PGG_2 was incubated in aqueous buffer for a few minutes at 37 °C, then the endoperoxide underwent chemical decomposition to a mixture of PGE_2 and $PGF_{2\alpha}$ which contracted the rat colon which was positioned beneath the rabbit aorta in the cascade. However, when PGG_2 was incubated with resuspended pig or rabbit aorta microsomes, no contractions whatsoever were observed on either the rabbit aorta or the rat colon preparations (Fig. 1.10.). This result suggested that aortic tissue did not convert PGG_2 either to TxA_2 (would have contracted the rabbit aorta) or $PGE_2/PGF_{2\alpha}$ (would have contracted the rat colon); but, nevertheless, a metabolite of PGG_2 had been formed since otherwise chemical breakdown to E, F and D prostaglandins would have ensured a large contraction of the rat colon preparation.

In order to determine the biological activity of what they believed could be a new prostaglandin, other pharmacological preparations were included in the superfusion cascade. The new metabolite of PG endoperoxide breakdown (by this time called PGX) was shown to relax the spirally cut rabbit mesenteric and coeliac arteries, in contrast to TxA_2 which causes a contraction of these preparations. In addition PGX was also shown to inhibit platelet aggregation. PGX has now been renamed prostacyclin (PGI_2) or epoprostenol. Although PGI_2 and TxA_2 have little or no biological actions in common, each is chemically unstable in aqueous solution. The $t_{\frac{1}{2}}$ of PGI_2 in aqueous buffer (pH 7.4, 37 °C) is about 3 min although its stability can be dramatically improved in alkaline media (e.g. $t_{\frac{1}{2}}$ in glycine buffer pH 10.0 is about 100 h). PGI_2 undergoes spontaneous chemical hydrolysis to yield a stable metabolite, 6-oxo-$PGF_{1\alpha}$. The pathway of arachidonic acid metabolism following the discovery of prostacyclin is shown in Fig. 1.11. The enzyme responsible for converting PGH_2 to PGI_2 is called prostacyclin synthetase.

It is of some interest, albeit possibly only historical, that PGI_2 was probably discovered several years before it was shown to be the major metabolite of arachidonic acid in the aorta. Pace-Asciak & Wolfe (1971) described the conversion by rat stomach homogenates of arachidonic acid

Fig. 1.10. Discovery of prostacyclin (PGI_2). Contractions of the rabbit aorta spiral (upper trace) and rat colon (lower trace) to (a) PGG_2 alone, (b) PGG_2 incubated in Tris buffer, (c) PGG_2 incubated with pig aortic microsomes (AM) and (d) PGG_2 incubated with boiled pig AM. All incubations were carried out at 22 °C (room temperature). When incubated in Tris buffer for up to 30 min, PGG_2 contractions of the aorta gradually declined while contractions of the rat colon gradually increased as PGG_2 chemically decomposes to PGE_2. In the presence of pig AM PGG_2 is enzymatically converted to a substance with little spasmogenic activity on either bioassay preparation. Reprinted by permission from *Nature*, **263**, pp. 663–5. Copyright © 1976 Macmillan Journals Limited, and by permission of the authors.

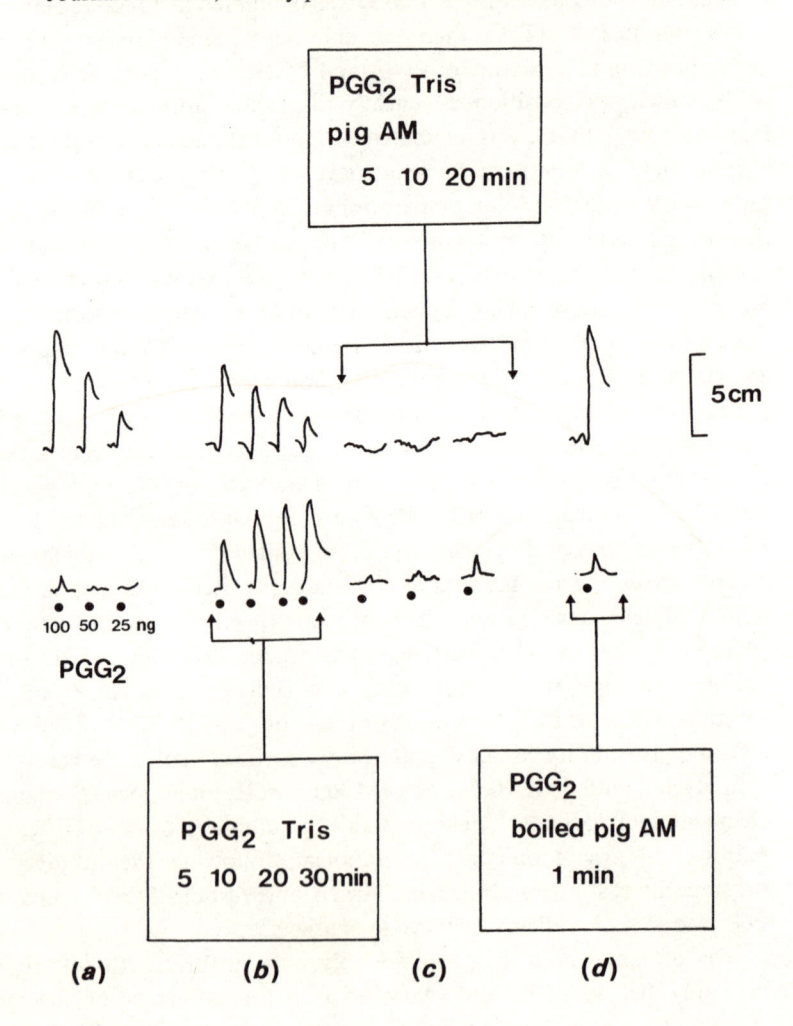

into two products with chromatographic mobility similar to that of PGE_2. Mass spectrometric analysis of the two structures revealed that one was a prostaglandin with an ether bridge between carbon atoms 6 and 9 and the other was later identified as 6-oxo-$PGF_{1\alpha}$. It now seems very likely that the first compound was PGI_2, which we know to be the major prostanoid formed in the rat stomach. Unfortunately, Pace-Asciak did not study the biological activity of these substances and thus did not appreciate the importance of his findings.

Characterisation of the enzymes involved in prostanoid biosynthesis

Prostaglandin endoperoxide synthase (EC 1.14.99.1;1)

Prostaglandin endoperoxide synthase has been purified to homogeneity from bovine and sheep seminal vesicle glands. The first successful

Fig. 1.11. Pathway of arachidonic acid metabolism (1979), showing the chemical structure of PGI_2 and its chemical hydrolysis product 6-oxo-$PGF_{1\alpha}$.

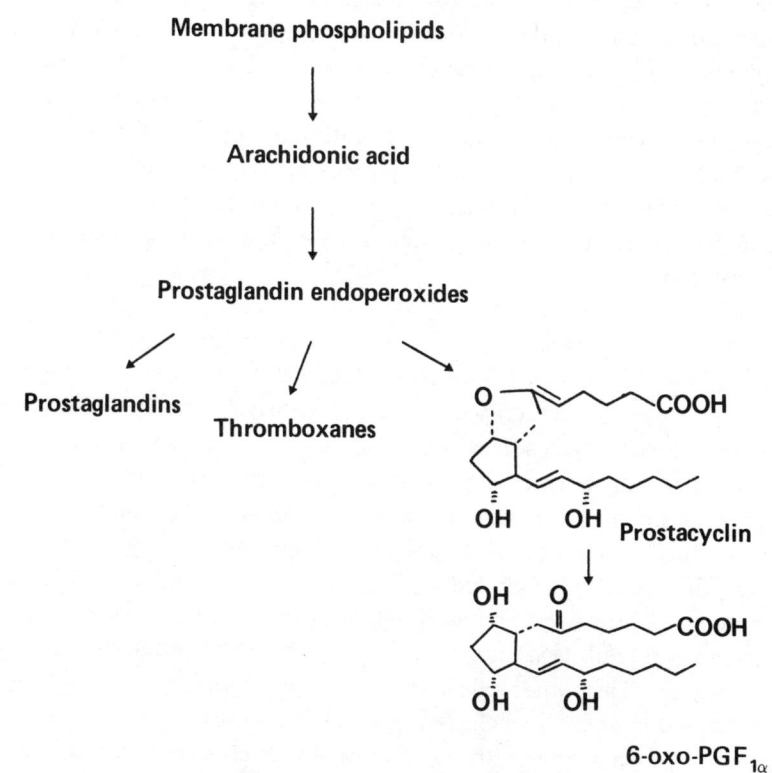

purification involved solubilisation of bovine seminal vesicle microsomes followed by DEAE cellulose chromatography and isoelectric focussing (Miyamoto et al., 1976). The material so obtained was pure as judged by polyacrylamide gel electrophoresis (PAGE) and required haem, either in the form of haemoglobin or as myoglobin, for activity. The enzyme is a glycoprotein containing two subunits of approximate molecular weight 70 000. The molecular weight of the whole dimer is between 124 000 and 129 000. Seminal vesicle PG endoperoxide synthase has been purified sufficiently to allow the raising of antibodies and, more recently, monoclonal antibodies in rabbits. This in turn has resulted in the development of a sensitive radioimmunoassay for this enzyme. The amino acid composition of ram seminal vesicle PG endoperoxide synthase has also been reported (van Ouderas & Buytenhek, 1982). To date, it has proved impossible to separate two individual proteins responsible for the cyclooxygenase (i.e. AA $\rightarrow$ PGG$_2$) and peroxidase (PGG$_2 \rightarrow$ PGH$_2$) steps. The peroxidase activity is probably an inherent part of the cyclooxygenase enzyme. However, differences do exist between the two enzymes. The cyclooxygenase component enzyme, for example, is inhibited by aspirin whilst the peroxidase enzyme is not (Miyamoto et al., 1976). In contrast the peroxidase (but not the cyclooxygenase) enzyme is stimulated by several compounds, including adrenaline and phenol. End-product regulation of PG endoperoxide synthase has also been reported. PGG$_2$ stimulates cyclooxygenase activity whilst co-oxidation of normal cell constituents as a result of the peroxidase activity produces oxygen-containing radicals which potently inhibit cyclooxygenase (Lands, 1979). Whether such end-product regulation is important for the control of this enzyme *in vivo* is not known.

Prostaglandin endoperoxide E isomerase (EC 5.3.99.3;3)

Responsible for converting PGG$_2$/PGH$_2$ to PGE$_2$, prostaglandin endoperoxide E isomerase has been extensively purified from sheep seminal vesicle microsomes (Ogino et al., 1977). These authors initially solubilised the enzyme with Tween 80 detergent and then purified it by DEAE cellulose chromatography and hydrophobic chromatography on aminobutyl Sepharose 4B. Unfortunately, enzyme activity is rapidly lost once the enzyme is purified; this process may be prevented by thiol compounds such as reduced glutathione (GSH), cysteine or 2-mercaptoethanol. GSH not only stabilises the enzyme but also serves as a cofactor. This latter effect does not occur with either cysteine or mercaptoethanol. The inclusion of GSH in incubations of organ homogenates or microsomes with arachidonic acid directs the breakdown of PG endoperoxides towards PGE_2 and PGD_2. PGG_2 may act as substrate for this enzyme although PGH_2 is probably the preferred substrate. From PGG_2 the product formed is 15-hydroperoxy-PGE_2, which is chemically unstable and rapidly breaks down to form PGE_2. However, the major route of biosynthesis of E series prostaglandins in the body is probably arachidonic acid (AA) → PGH_2 → PGE_2.

Prostaglandin endoperoxide D isomerase

The study of PGD_2 biosynthesis is complicated by the non-enzymatic formation of both PGD_2 and PGE_2, following the chemical decomposition of PG endoperoxides. Non-enzymatic breakdown of PGG_2/PGH_2 to PGD_2 is greatly enhanced in the presence of serum albumin (Hamberg & Fredholm, 1976). A specific PG endoperoxide D isomerase enzyme does occur in several tissues and has been purified from rat spleen, brain and blood vessels. This enzyme is found in the cytoplasmic fraction of the cell and thus differs from all other enzymes of arachidonic acid metabolism which are in the particulate (microsomal) fraction of the cell. The rat spleen enzyme, like PG endoperoxide E isomerase, is unstable once purified; it has a molecular weight of about 30 000 and requires GSH as cofactor. Other non-selective enzymes such as glutathione-S-transferase will also convert PGH_2 to PGD_2.

Prostaglandin F reductase

There is some difference of opinion as to whether $PGF_{2\alpha}$ biosynthesis occurs non-enzymatically by breakdown of PGH_2 or enzymatically by a prostaglandin endoperoxide F reductase. Several proteins with this enzyme activity have been identified. Guinea pig uterine microsomes contain a reducing factor which converts PGH_2 and has several characteristics of an enzyme i.e. saturable at high substrate concentrations, temperature dependent etc. This factor was not destroyed by boiling or activated by GSH, which argues against it being an enzyme. No such enzyme activity was detected in guinea pig lung, liver or kidney microsomes; this is surprising, since all three of these organs are known to form relatively large amounts of $PGF_{2\alpha}$.

Thromboxane synthetase

The microsomal enzyme, thromboxane synthetase, has been purified from human and bovine platelets (Yoshimoto & Yamamoto, 1982) where it occurs mainly in membranes of the intracellular dense tubular system (see Chapter 4). Thromboxane synthetase converts PGH_2 to equal amounts of TxB_2 and HHT. PGH_3 is also converted to

TxA_3. Although there has been one report of the conversion of PGH_1 to TxA_1, subsequent studies have been unable to confirm this finding. The C-5–C-6 double bond is probably essential for TxA_2 biosynthesis. PGG_2 is also converted by thromboxane synthetase to yield equal amounts of 15-hydroperoxy-TxA_2 and 12-hydroperoxy-HHT. Although PGG_2 is a substrate for thromboxane synthetase *in vitro* and yields equal amounts of 15-hydroperoxy-TxA_2 and 12-hydroperoxy-HHT, this reaction probably does not occur naturally in the body.

Prostacyclin synthetase

Prostacyclin synthetase has been partially purified from pig and rabbit aorta and converts PGH_2 and PGH_3 to PGI_2 and PGI_3 respectively. No conversion of PGH_1 to PGI_1 has been reported. The generation of monoclonal antibodies to prostacyclin synthetase has been reported (DeWitt & Smith, 1982). No doubt, these antibodies will prove to be very helpful in determining the sites of prostacyclin formation in the body.

Catabolism of prostanoids
'Classical' prostaglandins

Classical prostaglandins are defined as those of the A, B, C, D, E and F series. Of these, the inactivation of E, F and D series prostaglandins will be discussed here. Prostaglandins A, B and C do not occur naturally and are therefore of little interest. The initial step in the catabolism of these prostaglandins is their uptake from the extracellular space across the plasma membrane and into the cytoplasm of the cell which contains the degradative enzymes. Surprisingly, considering that they are lipids, prostanoids do not pass freely across biological membranes but are transported by a poorly understood uptake carrier mechanism (Bito, Baroody & Reitz, 1977). Uptake of prostanoids into cells has been studied in several tissues but most experiments have concentrated upon the perfused rat, rabbit and guinea pig lung. The features of the carrier mechanism include:

(1) *Selectivity*. The rate of uptake of prostanoids from the perfusing fluid into the pulmonary tissue is greatest for prostaglandins of the E, F and D series. Little or no uptake of PG metabolites (e.g. 15-oxo-PGE_2), PGI_2 or PGA_2 occurs in perfused lungs. Uptake is stereospecific since $PGF_{2\beta}$ is not a substrate.

(2) *Energy-requiring*. Uptake is slowed at low temperatures and prevented altogether by probenacid. Oddly, metabolic poisons such as cyanide do not affect PG uptake.

(3) *Blocked* by high concentrations of non-steroid anti-inflammatory

drugs such as indomethacin and by certain dyestuffs such as bromocresol blue.

The enzyme responsible for the biological inactivation of classical prostaglandins is a specific 15-hydroxyprostaglandin dehydrogenase (15-PGDH, EC 1.1.1.131) which has been purified to homogeneity from various sources (reviewed by Hansen, 1976). 15-PGDH is a cytoplasmic enzyme occurring in the 100 000 g supernatant fraction of organ homogenates. Two types of 15-PGDH enzyme have been identified. Type I enzyme requires oxidised nicotinamide adenosine diphosphate (NAD^+) for activity and occurs in most organs of the body including lung, spleen, parts of the gastrointestinal tract, kidney and heart. Type II 15-PGDH is found in the kidney, brain, erythrocytes and utilises $NADP^+$ as cofactor. Of the classical prostaglandins, the best substrate for 15-PGDH (types I and II) is PGE_2, followed by $PGF_{2\alpha}$.

Prostaglandins of the A, B, C and D series are, in general, poor substrates for this enzyme although a specific PGA_2 15-PGDH has been reported in rabbit kidney and a specific PGD_2 15-PGDH in pig brain. Both enzymes are NAD^+-dependent. 15-PGDH inactivates classical prostaglandins since the 15-oxo-prostaglandins so formed have only minimal biological activity in most assays, normally less than 1% of the parent compound (Crutchley & Piper, 1975). However, 15-oxo-$PGF_{2\alpha}$ has been reported to display bronchoconstrictor activity similar to $PGF_{2\alpha}$ in the anaesthetised dog. The enzyme prostaglandin $\Delta 13$ reductase (PG-$\Delta 13$R) reduces the C13-C14 double bond of 15-oxo-prostaglandins (but not of the parent PGs themselves) to yield the corresponding 13,14-dihydro-15-oxo-prostaglandin metabolites. PG-$\Delta 13$R has been partially purified from chicken heart, human placenta and bovine lung and requires NADH (or, more rarely, NADPH) for activity. Specific antibodies to bovine lung PG-$\Delta 13$R have been raised in rabbits. The $t_{\frac{1}{2}}$ of 13,14-dihydro-15-oxo-PGE_2 is approximately 8 min in circulation in man (Hamberg & Samuelsson, 1971) compared with about 30 s for PGE_2 itself. Like 15-oxo-PG metabolites, the 13,14-dihydro-15-oxo metabolites of PGE_2 and $PGF_{2\alpha}$ have very little biological activity.

The liver, kidney and intestine contain β- and/or ω-oxidising enzymes, which remove one or more methylene groups from the side chains of prostanoids. These enzymes are not specific for prostaglandins but are common to many fatty acids. In man, the major plasma metabolite of $PGF_{2\alpha}$ is the corresponding 13,14-dihydro-15-oxo-$PGF_{2\alpha}$ compound whilst 7-α-hydroxy-5,11-dioxotetranorprostane-1,16-dioic acid is the main urinary metabolite. Similar plasma and urinary metabolites of PGE_2 occur. The profile of plasma and urinary metabolites of prostaglandins is very complex and has been extensively studied in man and experimental animals (reviewed in Hamberg & Wilson, 1973; Granström, 1981).

Another enzyme found in the liver and kidney of pregnant rabbits and in lower amounts in lung and uterus of non-pregnant rabbits converts PGE_2 and $PGF_{2\alpha}$ to 19-hydroxy- and 20-hydroxyprostaglandins. These ω-hydroxylase enzymes may represent activity of the microsomal cytochrome P450 enzyme system since formation of 19-hydroxyprostaglandins *in vitro* is blocked by proadifen, a selective inhibitor of the mixed-function oxidase system (of which cytochrome P450 is an integral part). Unlike other metabolites of classical prostaglandins the 19- and 20-hydroxyPGs retain some biological activity.

An as-yet unidentified enzyme found in guinea pig kidney reoxidises the C-15 ketone function of 13,14-dihydro-15-oxo-PGs to yield 13,14-dihydro-PGs (Hoult & Moore, 1977). This reaction may be of interest since 13,14-dihydro-PGs retain considerable biological activity. For example, 13,14-dihydro-PGE_2 has about 30% of the potency of PGE_2 on the isolated rat stomach strip. Whether or not this reaction occurs *in vivo* is not known. The catabolism of classical prostaglandins is shown in Fig. 1.12.

Fig. 1.12. Catabolism of PGE_2 by prostaglandin 9-keto-reductase (PG-9KR), 15-hydroxyprostaglandin dehydrogenase (15-PGDH) and prostaglandin-$\Delta13$ reductase (PG-$\Delta13$R). Biological inactivation occurs when the C15 hydroxyl group is oxidised to a ketone by 15-PGDH.

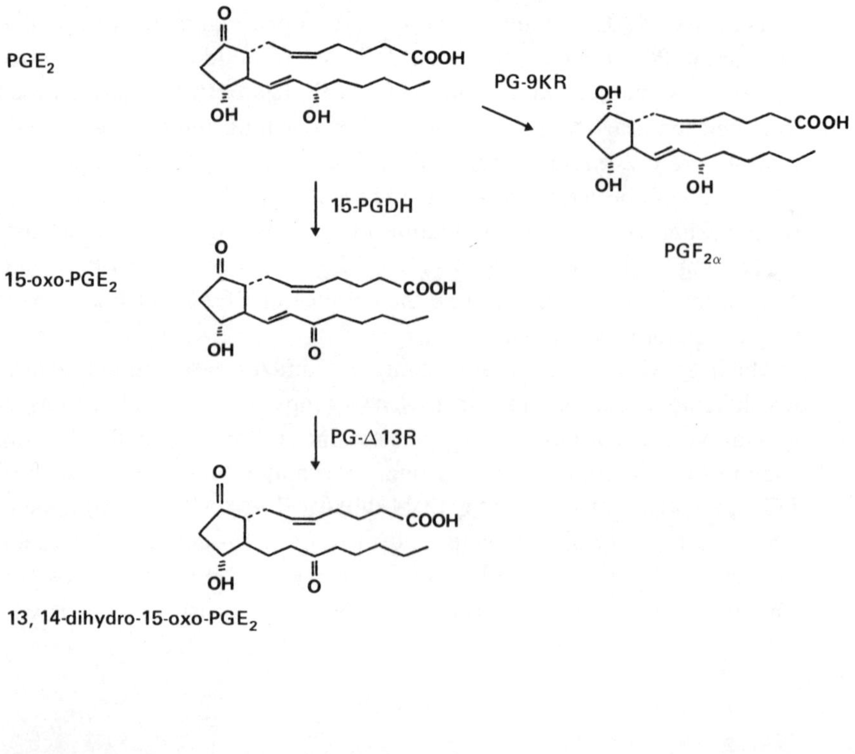

Prostaglandins of the E, F and D series may also be enzymatically interconverted. Prostaglandin 9-ketoreductase (PG-9KR) reduces the C9 ketone function of PGE_1 and PGE_2 (and their metabolites) to produce the corresponding PGF compounds (Lee & Levine, 1974). This enzyme requires NADH or NADPH for activity and large amounts are found in mammalian kidneys and parts of the gut. The reverse reaction catalysed by prostaglandin 9-hydroxydehydrogenase (PG-9HDH) requires NAD^+ and occurs in liver, kidney and platelets. The activity of these enzymes may be of physiological importance since there are marked differences in the biological activity (both qualitative and quantitative) between PGE_2 and $PGF_{2\alpha}$. The importance of these enzymes for the regulation of some of the renal actions of prostaglandins, for example, is discussed in Chapter 8. The conversion of PGD_2 to $PGF_{2\alpha}$ is catalysed by an NAD^+-dependent prostaglandin 11-keto-reductase (PG-11KR) enzyme which is not identical to the PG-9KR described above. PGD_2, infused intravenously into anaesthetised cats, results in the appearance of large amounts of $PGF_{2\alpha}$ in the animals' blood streams. Rabbit liver PG-11KR has been purified to homogeneity but, so far, has received only scant attention by prostaglandin researchers since its physiological relevance, if any, is not known.

Prostacyclin

Prostacyclin undergoes chemical hydrolysis to yield 6-oxo-$PGF_{1\alpha}$. A significant route of catabolism of endogenous PGI_2 is likely to be oxidation by 15-PGDH to yield 15-oxo-PGI_2 which may be further enzymatically catabolised as described above for PGE_2 and $PGF_{2\alpha}$ or alternatively hydrolysed spontaneously to the corresponding 6-oxo-$PGF_{1\alpha}$ derivative. PGI_2 (but not 6-oxo-$PGF_{1\alpha}$) is a good substrate for 15-PGDH *in vitro*. For this reason, 6,15-dioxo-$PGF_{1\alpha}$ may well be the major circulating metabolite of PGI_2. Several other plasma and urinary metabolites of PGI_2 have been described in different species. In rats, for example, the major metabolites of PGI_2 in the urine are dinor-6-oxo-$PGF_{1\alpha}$ and dinor-6,15-diketo-13,14-dihydro-20-carboxyl-$PGF_{1\alpha}$.

A second enzymatic pathway for PGI_2 catabolism is by oxidation of the hydroxyl function at position C9 to yield 6-oxo-PGE_1 (Moore & Griffiths, 1983a). The PG-9HDH enzyme necessary is particularly abundant in human platelets. Unlike other metabolites of PGI_2, 6-oxo-PGE_1 has considerable biological activity. To date, conversion of PGI_2 to 6-oxo-PGE_1 has been demonstrated *in vivo* in the dog but not in human subjects.

Thromboxane A_2

Like prostacyclin, TxA_2 undergoes spontaneous hydrolysis to TxB_2, which is almost certainly the major route of catabolism in the body. TxB_2 may then be further broken down by 15-PGDH, PG-Δ13R and β/ω oxidation. Dawson *et al.*, (1976) identified 13,14-dihydro-15-oxo-TxB_2 in the perfusate of arachidonic-acid-challenged guinea pig lungs. The major urinary metabolite of TxB_2 in man is 2,3-dinor-TxB_2.

Occurrence, distribution and release of prostanoids

Prostanoids are not stored in a pre-formed state in mammalian organs but may be extracted from most animal tissues and some plants. The prostanoid concentration of tissue homogenates therefore reflects biosynthesis from endogenous arachidonic acid during homogenisation and work-up of the sample. In order to measure the basal prostanoid concentration in a particular tissue certain precautions must be taken as follows:

(1) To prevent biosynthesis of prostanoids during tissue preparation, homogenisation can be carried out in organic solvents (e.g. ethanol/acid mixture) to destroy enzyme activity, or in the presence of a high concentration of indomethacin (an inhibitor of cyclooxygenase activity).

(2) Organ prostanoid levels increase rapidly in the first few moments after death, necessitating either very rapid freezing of the tissue or alternatively a method of sacrifice which destroys enzyme activity at the same time as causing death.

(3) Barbiturate anaesthetics increase prostanoid concentration. Thus experiments should ideally be performed on non-anaesthetised animals thereby adding to the practical and ethical problems of prostanoid measurement in man.

Some of these problems can be avoided by measuring prostanoid levels in body fluids which obviously do not require homogenisation or other work-up procedures. However, other problems arise in these studies. Plasma prostanoid concentrations, for example, vary with the presence or absence and type of anticoagulant used, the sort of syringe and even the length and gauge of the needle. The trauma of venepuncture is probably sufficient to cause platelet aggregation locally, thus generating cyclo-oxygenase products and distorting the plasma prostanoid profile. There have been numerous studies of the prostanoid composition of human body fluids. Table 1.2 shows the body fluids in which prostanoids have been detected. Human seminal fluid, for example, contains the highest concentration and number of different prostanoids yet isolated from a single source. Menstrual fluid is also rich in prostaglandins (see Chapter 5 for further details).

Most estimates of the prostanoid content of animal tissues have come from experiments in which little or no attempt has been made to prevent endogenous prostanoid biosynthesis during tissue preparation. In the few instances where such precautions have been taken, basal levels have been very low (in the pg/g tissue range) and generally insufficient to produce any significant biological action in that tissue. These so called 'basal' levels obviously provide us with very little information about the amount and type of prostanoid formed in a particular organ. A more worthwhile measure is the prostanoid-synthesising capacity. In such experiments, organ homogenates or microsomal fractions are incubated either with arachidonic acid, PGG_2 or PGH_2 for a fixed period of time (usually at

Table 1.2. *Prostanoids in human body fluids*

Blood	Menstrual fluid
Plasma	Urine
Serum	Pleural fluid
Cerebrospinal fluid	Lymph
Follicular fluid	Milk
Amniotic fluid	Aqueous humour
Inflammatory exudate	Nasal fluid (during hay fever)
Semen	Saliva

Table 1.3. *Major prostanoids synthesised in mammalian organs*

Tissue	Prostanoid	Comments
Lung	6-oxo-$PGF_{1\alpha}$ > TxB_2 > PGE_2 > $PGF_{2\alpha}$	
Kidney	PGE_2 > $PGF_{2\alpha}$	TxB_2 in ureter-obstructed kidney
Spleen	TxB_2 > 6-oxo-$PGF_{1\alpha}$ > PGE_2	
Platelets	TxB_2 > PGD_2, PGE_2 > $PGF_{2\alpha}$	
Uterus	6-oxo-$PGF_{1\alpha}$ > PGE_2 > $PGF_{2\alpha}$	
Corpus luteum	6-oxo-$PGF_{1\alpha}$ > PGE_2	
Stomach	6-oxo-$PGF_{1\alpha}$ > PGE_2 > $PGF_{2\alpha}$	Very species-dependent
Colon	6-oxo-$PGF_{1\alpha}$ = PGD_2 = PGE_2	Very species-dependent
Brain	PGD_2 > TxB_2	
Blood vessels	6-oxo-$PGF_{1\alpha}$	PGE_2 in placental vessels
Adipocytes	6-oxo-$PGF_{1\alpha}$ > PGE_2 = $PGF_{2\alpha}$	
Gall bladder	6-oxo-$PGF_{1\alpha}$ $\gg$ PGE_2 = $PGF_{2\alpha}$	

37 °C); the different prostanoids synthesised can be separated and identified using one or other of the methods discussed later in this chapter. The most abundant prostanoids formed in several mammalian tissues are shown in Table 1.3.

Prostaglandins of the E and F series are also formed in non-mammalian vertebrates, invertebrates, fish and insects (Christ & van Dorp, 1972). Relatively large amounts of a prostaglandin A derivative are found in a Mediterranean species of coral called *Plexaura homomalla*. Indeed extrac-

Table 1.4. *Release of prostanoids from intact organs*

Organ	Stimulus for release[a]
Lung	Chemical (ATP, ADP, histamine, 5-HT, arachidonic acid, BK, AGT II, leukotrienes, nicotine)
	Mechanical (injection of microspheres into the pulmonary artery, gentle massage)
	Others (hyperventilation, hypoxia)
	Pathological (bronchial asthma, anaphylactic shock, oedema, pulmonary hypertension, pulmonary embolism)
Kidney	Chemical (arachidonic acid, AGT II, diuretics e.g. frusemide, ethacrynic acid, leukotrienes)
	Pathological (renal hypertension, renal ischaemia)
Spleen	Chemical (arachidonic acid, BK, AGT II)
	Mechanical (injection of microspheres into the splenic artery)
Brain	Pathological (pyrogenic endotoxins, subarachnoid haemorrhage, cerebral embolism)
Gut	Chemical (arachidonic acid, histamine, 5-HT, ACh, catecholamines, clonidine)
	Mechanical (stretch during peristalsis)
	Pathological (diarrhoea, gastritis, intestinal obstruction)
Uterus	Chemical (AGT II, BK, oxytocin)
	Mechanical (contractions during parturition)
	Pathological (dysmenorrhoea, endometriosis, presence of intrauterine devices)
Blood platelets	Chemical (platelet aggregating agents e.g. collagen, thrombin)
	Pathological (some thrombotic diseases e.g. diabetes mellitus, sickle cell anaemia, polycythaemia)
Muscles	Pathological (arthritis)
Skin	Pathological (allergy, eczema, sunburn)
Eye	Pathological (uveitis, glaucoma)
Teeth	Pathological (gingivitis)

[a] Abbreviations: ATP, adenosine triphosphate; ADP, adenosine diphosphate; BK, bradykinin; AGT II, angiotensin II; ACh, acetylcholine; 5-HT, 5-hydroxytryptamine.

tion of this prostaglandin from *Plexaura* was at one time a commercially viable way of making prostaglandins.

The fact that prostanoids can be extracted from a particular tissue when supplied with precursors does not necessarily indicate a physiological function for those prostanoids in that tissue *in vivo*. Almost all cell types (except erythrocytes) have the capacity to form prostanoids but how many of these do so naturally in the body is another question. In order to demonstrate a possible physiological role of a particular prostanoid(s) in an organ it is necessary also to demonstrate that the organ releases that prostanoid when stimulated either naturally or with hormones or local hormones. Table 1.4 lists some of the organs from which both basal and stimulated prostanoid release has been shown. Notice that the stimuli required to elicit prostanoid release from intact organs range from the physiological (exposure to angiotensin II, kinins, catecholamines, stretching of membranes following peristaltic movement of the intestine or the normal breathing movements of the lungs) to the pathophysiological (ischaemia, lack of oxygen, disease states like angina pectoris etc.).

Assay of prostaglandins and thromboxanes

Several methods are currently used to assay the small amounts of prostanoids present in biological tissues. Each method has its advantages and its disadvantages; care must be taken when interpreting the results of a particular paper that the assay used is appropriate to the problem under study. The various assay techniques for prostanoids are compared in Table 1.5 and discussed below.

Bioassay

Biological assay techniques are inexpensive, relatively easy to perform and sensitive to small amounts of prostaglandins. The rat stomach fundus strip and rabbit duodenum, for example, will contract to as little as 100 pg of PGE_2; the same amount of PGI_2 will produce a significant inhibition of human platelet aggregation *in vitro*. Bioassay preparations show different sensitivity to different prostanoids. The rat and gerbil colon are much more sensitive to prostaglandins of the F than the E series whilst the guinea pig ileum, rat and hamster stomach strips show the reverse sensitivity to prostaglandins. The response to different prostanoids of 14 commonly used smooth muscle bioassays is shown in Table 1.6.

The major problem with the bioassay of prostaglandins is lack of selectivity. The rat stomach strip, for example, which contracts to as little as 100 pg PGE_2 and 200 pg $PGF_{2\alpha}$ will also respond to sub-picogram amounts of 5-hydroxytryptamine (5HT). The problem with lack of selectivity may be overcome in several ways. Firstly, by the use of extraction procedures (usually acidify to pH 3.0–4.0 and extract into ethyl acetate, ether or chloroform) which removes prostanoids but does not extract other biologically active substances such as acetylcholine (ACh), histamine, noradrenaline (NA) etc. Secondly, bioassay may be carried out

Table 1.5. *Assay methods for prostanoids*

Method	Threshold sensitivity	Capacity	Selectivity	Comments[a]
Bioassay				
PGE_2, $PGF_{2\alpha}$	Rat stomach strip (about 1 ng)	> 50	Poor	Contracts to 5HT, ACh, histamine
PGD_2	Inhibition of platelet aggregation, 10 ng	> 20	Average	Platelet aggregation also inhibited by PGE_1, PGI_2, adenosine
PGI_2	Inhibition of platelet aggregation, 0.5 ng	> 20	Average	Platelet aggregation also inhibited by PGE_1, PGD_2, and adenosine
TxA_2	Rabbit aorta	> 10	Poor	Contracted by NA, ADR, 5HT, AGT II, PGG_2, PGH_2
Gas-chromatography–mass-spectrometry				
All prostanoids	< 1 ng	5+	Excellent	Allows separation and identification
Gas-chromatography–electron-capture				
All prostanoids	0.1 ng	5+	Excellent	Allows separation and identification
Radioimmunoassay				
All prostanoids	0.05 ng	> 200	Usually good	Cross-reactivity to other PGs < 5%
Spectrophotometry				
PGEs, PGAs	5 μg	10	Poor	Insensitive

[a] Abbreviations: NA, noradrenaline; ADR, adrenaline.

in the presence of antagonists to these substances to decrease or remove their effect on the assay tissues. For this purpose most authors advocate the use of a 'cocktail' of antagonists including atropine, mepyramine, methysergide, pronethalol and phentolamine. Biological assay of stable prostaglandins can be performed using classical organ-bath techniques by bracketing doses of unknown between standard doses of prostaglandin in a Latin square design. For the chemically unstable prostanoids, this approach is less useful and a superfusion-cascade-type of bioassay is indicated. This method combines the sensitivity to prostanoids of classical bioassay with the capacity to study the action on five or more assay tissues virtually simultaneously. Additionally, by incorporating a perfused organ (e.g. guinea pig lung) at the top of the cascade, prostaglandin endoperoxides and TxA_2 may be generated *in situ* and the effect on the bioassay preparations determined. PGI_2 and PGD_2 are conveniently bioassayed for the ability to prevent ADP-induced rabbit or human platelet aggregation. Prostanoid bioassay has been reviewed by Moncada, Ferreira & Vane (1978).

Gas chromatography methods

For gas chromatography, samples must first be extensively purified by one or more steps of thin-layer/column/high-pressure-liquid chromatography. Prostanoids must be derivatised to provide an analogue which is volatile at high temperatures. Derivatisation can be achieved by

Table 1.6. *Some commonly used smooth muscle bioassay preparations*

Preparation	Sensitivity to prostanoids[a]
Bovine coronary artery	$I_2 > E_2 (\downarrow), TxA_2 > G_2 (\uparrow)$
Rat stomach strip	$E_2 > E_1 > F_{2\alpha}$
Rat colon	$F_{2\alpha} > E_2$
Rat uterus	$E_2 > E_1 > F_{2\alpha} > I_2$
Gerbil colon	$E_2 > F_{2\alpha}$
Hamster colon	$F_{2\alpha} > E_2$
Hamster stomach pouch	$E_2 > F_{2\alpha}$
Guinea pig ileum	$E_2 > E_1 > F_{2\alpha}$
Guinea pig uterus	$E_2 > F_{2\alpha}$
Rabbit jejunum	$F_{2\alpha} > E_2$
Rabbit aorta spiral	$TxA_2 > G_2$
Rabbit coeliac artery	$G_2 > E_2 (\downarrow), TxA_2 (\uparrow)$
Rabbit mesenteric artery	$G_2 > E_2 (\downarrow), TxA_2 (\uparrow)$

[a] Arrow pointing upward indicates contraction while arrow pointing downward indicates relaxation.

converting the carboxyl group of the molecule to the methyl ester with diazomethane, any carbonyl functions to the methoxime and the hydroxyl groups to the trimethylsilylether. An excellent review of the application of gas chromatographic techniques to the assay of prostaglandins, thromboxanes and their metabolites can be found in Salmon & Flower (1979).

Separation by gas chromatography depends upon the fact that derivatised prostaglandins partition differently between a gas phase (usually N_2 at about 230 °C) and a stationary phase (a siloxane polymer coated onto a support medium and packed into glass tubes) thus emerging from the column at different rates. Detection of the derivatised prostaglandins in the column effluent requires a mass spectrometer, an electron-capture device or a flame-ionisation detector. For mass spectrometry, derivatised prostaglandins are bombarded with a beam of electrons causing them to be split into their component ions which then deviate towards or away from a magnetic field. The extent of this deviation may be measured and represents the mass of the ion. From this sort of experiment, a mass spectrum of different derivatives of prostaglandins can be constructed. No two prostaglandins have exactly the same mass spectrum. This method thus allows separation and identification of a single prostaglandin in a mixture. GC/MS can easily be used as a quantitative assay for prostaglandins simply by adding a small quantity of the appropriate deuterated prostaglandin to the sample. After purification and derivatisation, the ratio between the deuterated and non-deuterated molecules can be measured with the mass spectrometer. If the amount of deuterated compound added is known, then the amount of non-deuterated sample can be calculated easily. The principal advantages of GC/MS are its very high selectivity and sensitivity. However, the lengthy derivatisation steps add to the difficulty of the procedure and increases the risk of incurring losses of the sample. Furthermore, very sophisticated (and expensive) equipment is needed.

The electron-capture device is probably the most sensitive detector for use with gas chromatography. The apparatus consists of a cell in which β-particles from a radioactive source (e.g. ^{63}Ni) ionise molecules of the carrier gas, forming electrons which migrate continuously to an anode within the cell. The resultant current may be measured. The more electron-attracting the substance which emerges from the column, the greater will be the fall in current registered. Greatest sensitivity is attained with compounds having high electron affinities. The method is particularly useful for prostaglandins of the E series but, by appropriate choice of derivatising agent (e.g. to form halogenated analogues which are highly electron-attracting), the method can be applied to other prostaglandins.

Radioimmunoassay

Radioimmunoassay (RIA) is a technique based upon the competition between labelled and unlabelled molecules of a prostaglandin or thromboxane for binding sites of an antibody directed against that particular prostanoid (reviewed by Granström & Kindahl, 1978). In the presence of only a low concentration of unlabelled substance, a large proportion of the labelled molecules are bound to the antibody. When the concentration of unlabelled substance is high, the binding of labelled prostaglandin to antibody is correspondingly reduced. In order to assay the amount of prostanoid present all that need be done is to separate free from antibody-bound radioactivity, either by immunoprecipitation with an antibody directed to the first antibody or (more simply) with dextran-coated charcoal.

Radioimmunoassay is a highly sensitive technique capable of detecting very small amounts (limits of detection of most RIAs is 10–50 pg) of prostaglandins, thromboxanes and their metabolites and analogues. The technique is also selective in that antibody cross-reactivity to other prostaglandins is normally less than 2–3%. RIA allows the assay of many samples (over 150) at the same time and, apart from the purchase of radiolabelled tracers and the time-consuming raising of antibodies, the technique is relatively inexpensive and easy to perform. The main drawback of RIA is that, unlike bioassay, it is not an innovative technique capable of identifying new prostanoids since an antibody to a specific (and therefore, obviously, a known) prostanoid must be used.

Spectrophotometry

Spectrophotometry technique depends upon the conversion of prostaglandins of the E or A series to the corresponding PGB derivatives at alkaline pH. PGB_2 has a characteristic ultraviolet absorbance $\lambda_{max}^{ethanol}$ (extinction coefficient = 28 600) at 278 nm. This property may be used to assay PGB_2 (and thus PGE_2) although large amounts (greater than 500 ng) are required. The sensitivity of this method can be increased several fold by coupling the prostaglandin with a suitable chromophore, e.g. p-bromophenacyl PG ester. In general, spectrophotometry lacks selectivity since several other substances which do occur naturally also exhibit an absorbance peak at or close to 278 nm. Although spectrophotometry is simple and inexpensive, the lack of sensitivity and selectivity of the method (and the fact that it can only be applied to E and A type prostaglandins) has meant that it has been used only very rarely in the last few years. A slightly more sensitive spectrofluorimetric assay of PGE_1, PGE_2 and PGE_3 has been described but has not been widely used.

Prostanoid receptors

There is now little doubt that prostanoids, like other biologically active substances in the body, exert their many effects by combining with selective prostanoid receptors on the cell membrane. Compared with our extensive knowledge of receptors for substances such as histamine, acetylcholine and noradrenaline, surprisingly little is known about receptors for prostaglandins of the E and F series, and even less of PGI_2 and TxA_2 receptors. The reasons for this are many but include technical problems of measuring binding of chemically unstable prostanoids and also the lack of tried and tested selective prostanoid antagonists. Evidence for the existence of prostanoid receptors has come from two separate lines of research:

(1) radioligand binding studies, in which the capacity of radiolabelled prostanoid to bind with high affinity to intact cells or particulate fractions thereof is investigated;
(2) characterisation of the biological actions of a range of prostanoids and chemically similar analogues.

Binding studies for PGE_1/PGE_2, $PGF_{2\alpha}$, PGI_2 and PGD_2 have been demonstrated using the radioligand binding technique in biological tissues from at least six species (Table 1.7).

Care must be exercised in the interpretation of these studies since 'binding sites' are not necessarily synonymous with receptor sites. High affinity binding of prostanoids may, for example, reflect interaction with an uptake carrier protein or even a catabolising enzyme. Three criteria

Table 1.7. *Organs in which prostanoid receptors have been identified*

PGI_2	Human: platelets, uterus, corpus luteum
	Cow: coronary artery
	Guinea pig: pulmonary artery
PGE_1/PGE_2	Human: platelets, uterus, corpus luteum, sperm
	Cow: platelets, pineal gland
	Guinea pig: uterus, corpus luteum, skeletal muscle
	Rat: platelets, uterus, adipocytes, kidney, thymocytes, stomach, adrenal gland, fibroblasts, liver, skin, neuroblastoma cells
$PGF_{2\alpha}$	Human: uterus, corpus luteum, sperm
	Hamster: uterus
	Rabbit: oviduct
	Rat: corpus luteum
	Horse: corpus luteum
	Cow: corpus luteum
PGD_2	Human: platelets
PGA_2	Rabbit: kidney

should be satisfied before 'binding sites' can properly be equated with 'receptor sites'. Binding must be shown to:
(1) be saturable at high concentrations of prostanoid;
(2) be specific (i.e. not mimicked by structurally similar but biologically inactive stereoisomers, analogues or metabolites) and reversible;
(3) parallel the biological effect produced by that prostanoid in that tissue.

Of these criteria (1) and (2) have been satisfied in all the studies shown in Table 1.7. Technically, the satisfaction of the final criterion is more difficult since ligand-binding studies are normally performed in tissue homogenates or subcellular fractions in which it is difficult, if not impossible, to elicit a biological response. However, since many of the effects of prostanoids are mediated by changes in intracellular levels of cyclic $3'5'$-adenosine monophosphate (cAMP) and prostanoid receptors in some cells at least are known to be linked to a membrane adenylate cyclase enzyme, it has been possible in some cases to correlate the binding of a prostanoid to its receptor with its effect on adenylate cyclase. The majority of such studies have been carried out using either human platelets or corpus luteum of several species; examples from these two tissue sources will be discussed here.

PGE_1, PGD_2 and PGI_2 inhibit platelet aggregation in species such as rat, rabbit, dog, guinea pig and man (see Chapter 4). This effect is mediated by activation of a membrane adenylate cyclase enzyme resulting in a rise in cAMP concentration within the platelet. Many studies point to the existence on the platelet membrane of at least two different types of prostanoid receptor, each linked to adenylate cyclase. One of these receptors interacts selectively with PGD_2 and the other with PGI_2 and PGE_1. Both radioligand and biological studies confirm this subdivision of platelet prostanoid receptors. For example:

(1) The increase in platelet cAMP by PGI_2 is rapidly increased by PGE_1 but unaffected by PGD_2 (Bonne et al., 1981).
(2) Cross-desensitisation occurs between prostanoids of the E and I series but not between PGE_1/PGI_2 and PGD_2. Exposure of human platelets to PGE_1 reduced the anti-aggregatory effect and elevation of platelet cAMP induced by PGI_2, but not that induced by PGD_2.
(3) The prostaglandin antagonist NO1644 blocks PGD_2-induced inhibition of platelet aggregation (and increase in cAMP) but has no effect on either PGE_1 or PGI_2 platelet inhibition (MacIntyre, 1977).
(4) Platelets from several species show differential sensitivity to the anti-aggregatory effect of PGE_1, PGD_2 and PGI_2, e.g. rat platelets

are exquisitively sensitive to the anti-aggregatory action of PGE_1 and PGI_2 but insensitive to PGD_2 (Whittle, Moncada & Vane, 1978b).

There is a relatively good agreement between the K_d values for low-affinity binding sites for PGE_1, PGD_2 and PGI_2 and the EC_{50} values for stimulation of platelet adenylate cyclase (Table 1.8). The nature of the high affinity binding sites for PGE_1 and PGI_2 on platelets is not known but it is unlikely that these binding sites represent true receptor sites. A similar good agreement between binding data and adenylate cyclase activation has been reported for PGE receptors on cultured lymphoma cells, neuroblastoma cells and adrenal glands. With these exceptions, the K_d values for PGE binding in other tissues shown in Table 1.7 are frequently 1–2 orders of magnitude lower than the concentration required for half-maximal stimulation of adenylate cyclase in the same tissue. Thus binding sites in these tissues may not necessarily represent the receptor sites. It should be pointed out that these binding and adenylate cyclase assays were normally performed in different laboratories.

The possibility that separate receptors exist for PGE_1 and PGE_2 has also been put forward. Corpus luteum and thymocyte receptors have been shown to bind PGE_1 and PGE_2 with equal affinity. In contrast, hamster and monkey uterine myometrial receptors showed greater affinity for PGE_2 compared with PGE_1, while neuroblastoma and lymphoma cells exhibited binding sites with greater affinity for PGE_1. In all tissues prostaglandins of the D and F series and their metabolites were only very poorly bound (Samuelsson et al., 1978).

$PGF_{2\alpha}$ receptors have been demonstrated in several tissues. Most attention has centred upon the corpus luteum since a physiological role

Table 1.8. [^3H]PG binding to human platelets and effect on adenylate cyclase

PG	System	Affinity constant	EC_{50} for inhibition of adenylate cyclase in vitro
PGD_2	Intact platelets	(i) 4.1×10^{-7} M (ii) 5.4×10^{-8} M	5×10^{-7} M
PGI_2	Intact platelets	(i) 4.1×10^{-9} M (ii) 9.9×10^{-7} M	1.8×10^{-7} M
PGE_1	Intact platelets	(i) 9.5×10^{-9} M (ii) 1.6×10^{-6} M	1×10^{-6} M

for $PGF_{2\alpha}$ as a luteolytic hormone has been suggested (Chapter 5). Specific binding of $PGF_{2\alpha}$ has been demonstrated in sheep and bovine corpus luteum ($K_d = 50$–100 nM). Structural modification of $PGF_{2\alpha}$ (e.g. by oxidation of the hydroxyl groups at C-9 or C-15) greatly reduced the affinity for the corpus luteum binding sites; a reasonable correlation between binding affinities to these receptor sites and the luteolytic potencies of the analogues in heifers has been demonstrated. Subcellular distribution studies have revealed that these binding sites are localised to the plasma membrane.

Other authors have classified prostanoid receptors not by their capacity to bind to organ homogenates or membranes but by their relative agonist potencies in a variety of isolated pharmacological preparations. Kennedy et al. (1982) have proposed a classification of prostanoid (P) receptors where tissues contain DP, EP, FP, IP and/or TP receptors, depending upon whether they show greatest sensitivity to prostaglandins of the D, E, F and I series (respectively) or to TxA_2. According to this classification, for example, guinea pig ileum and cat trachea contain mainly EP receptors whilst rat aorta contains TP receptors and cat iris sphincter muscle mainly FP receptors. The classification is strengthened by the finding that prostanoid antagonists appear to interact selectively with the different receptor groups. The experimental drug SC19220 blocks responses following EP receptor stimulation but has little or no effect on other receptors. In contrast, AH19437 selectively blocks TP receptors. In an earlier, separate study Gardiner & Collier (1980) suggested that airway smooth muscle contained at least three separate receptors for E series prostaglandins. It is to be hoped that a pharmacological classification of prostanoid receptors, possibly along the lines of that suggested by Kennedy et al. (1982), is quickly accepted and adopted by prostanoid researchers. Only when this has happened will it be possible to consider a sensible approach to the development of prostanoid receptor antagonists, which may be of potential clinical use.

Leukotrienes

Leukotrienes are a family of local hormones formed from the same precursor fatty acids as the prostanoids. Leukotrienes have a conjugated triene structure (three adjacent C=C double bonds) and a peptide chain containing 1, 2 or 3 amino acid residues linked via a sulphur atom (thioether bond) to carbon atom 5. Unlike prostanoids, leukotrienes have no cyclopentane ring.

Several different leukotrienes occur naturally and each is identified by a letter of the alphabet. To date, leukotrienes of the A, B, C, D and E series have been detected in biological tissues and their chemical structures are shown in Fig. 1.13. The most abundant naturally occurring leukotrienes contain 4 double bonds and are synthesised enzymatically from arachidonic acid. Leukotrienes are normally abbreviated to LtB_4, LtC_4 etc. Leukotrienes which contain 3 or 5 double bonds may also be biosynthesised from dihomo-γ-linolenic acid and eicosapentaenoic acid respectively. Arachidonic acid is quantitatively the most important precursor fatty acid for both leukotriene and prostanoid biosynthesis. Note that cyclisation of arachidonic acid produces prostanoids with 2 double bonds fewer than the

Fig. 1.13. Chemical structure of leukotrienes.

Leukotriene A_4

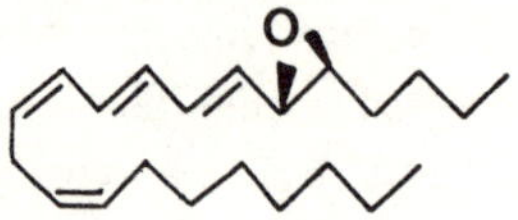

Leukotriene B_4

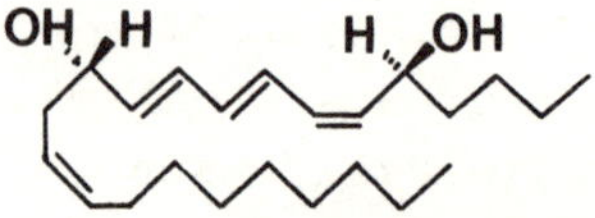

Leukotriene C_4

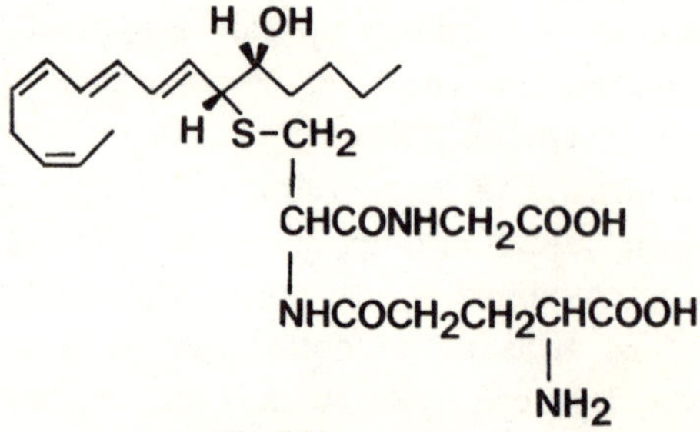

Leukotriene D_4

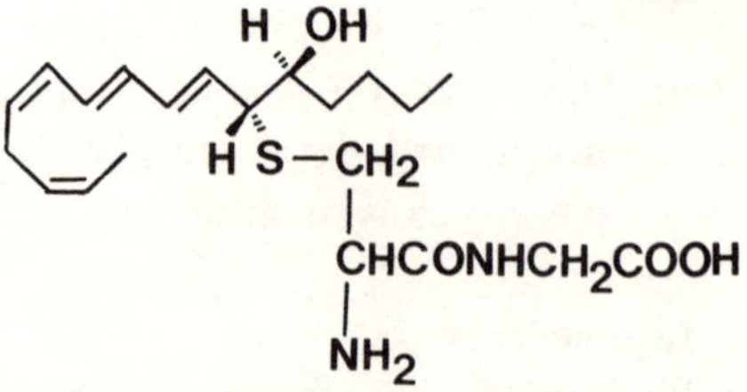

Leukotriene E_4

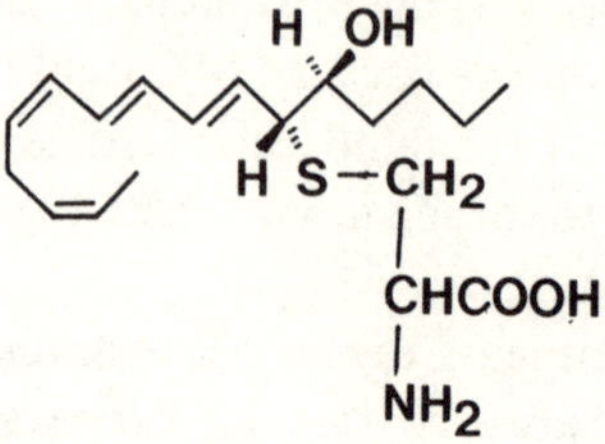

parent fatty acid but leukotrienes with the same number of double bonds as precursor arachidonic acid. Thus, arachidonic acid is converted to prostanoids of the 2 series (e.g. PGE_2, PGI_2) but leukotrienes of the 4 series (e.g. LtA_4, LtD_4).

The chemical structure of LtC_4 was announced by Borgeat & Samuelsson (1979) at the Fourth International Conference on Prostaglandins in Washington D.C., USA in May 1979. However the fact that biologically active substances other than prostanoids could be formed from arachidonic acid has been known for many years. Feldberg & Kellaway (1938) observed release from perfused guinea pig lungs challenged with cobra venom of a smooth muscle contracting factor which had none of the chemical or pharmacological properties of histamine, or any other spasmogen, on smooth muscle known at the time. The factor released from lungs produced a very characteristic slow contraction of the isolated guinea pig ileum preparation and was consequently called 'slow-reacting substance' or SRS. Later workers demonstrated the release of a substance with similar biological activity to SRS from immunologically sensitised guinea pig lungs challenged with antigen. This substance, renamed slow-reacting substance of anaphylaxis (SRS-A) to distinguish it from the non-immunologically released SRS-C (slow-reacting substance released by cobra venom) was later shown to be made in human lung. Similar slow-reacting substances are released from a variety of tissues by a number of different stimuli, but most attention has been paid to SRS-A because of its possible pathophysiological significance in hypersensitivity reactions such as asthma. Evidence has steadily accumulated during the last 30 years that SRS-A is a non-prostanoid metabolite of arachidonic acid synthesised by the action of a lipoxygenase enzyme. Some of this evidence is listed below:

(1) Synthesis of SRS-A by human and guinea pig lung can be increased by injecting arachidonic acid (Walker, 1973).
(2) Rat basophilic leukaemic cells (RBL cells) convert radiolabelled arachidonic acid to radiolabelled SRS purified by two-dimensional chromatography.
(3) 5,8,11,14 eicosatetryanoic acid (ETYA), an inhibitor both of cyclooxygenase and lipoxygenase, markedly inhibited SRS-A biosynthesis.
(4) Indomethacin, which inhibits cyclooxygenase but not lipoxygenase, enhances SRS-A production in perfused lungs. Most likely, this is because indomethacin blocks prostanoid formation in the lungs, thereby allowing more arachidonic acid to be converted to SRS-A.
(5) In rat basophilic leukaemic cells, stimulated with the calcium

ionophore A23187, generation of SRS was increased up to 5 times in the presence of exogenous arachidonic acid whereas other long chain fatty acids (e.g. linoleic acid, oleic acid) either failed to enhance SRS biosynthesis or actually inhibited its formation.

Early attempts to purify and characterise SRS-A were largely unsuccessful because of the small amounts of this substance present in lung perfusate and its chemical instability. SRS-A was found to contain sulphur and had an ultraviolet absorption spectrum with an absorbance peak at a wavelength of 280 nm. Further progress to elucidate the identity of SRS-A required a biological source capable of synthesising much larger amounts of this substance than could be extracted from guinea pig lung perfusate. Borgeat, Hamberg & Samuelsson (1976) studying the metabolism of arachidonic acid by rabbit polymorphonuclear leucocytes (PMNL) found that the major metabolite was not a prostanoid but 5-hydroxyeicosatetraenoic acid (5-HETE), a stable substance made following the chemical breakdown of the corresponding 5-hydroperoxyeicosatetraenoic acid (5-HPETE). The enzyme responsible for 5-HPETE biosynthesis is called lipoxygenase.

Later work by the same authors showed that rabbit PMNL incubated with arachidonic acid made an even more polar metabolite identified as $5(S)$ $12(R)$ dihydroxy-6,8,10,14-eicosatetraenoic acid and later named leukotriene B_4. Using isotopic oxygen (either ^{18}O or ^{16}O), it was shown that the oxygen of the alcohol group at C-5 of LtB_4 originated in molecular oxygen whereas the oxygen of the alcohol group at C-12 was derived from water. This result strongly suggested that the first step in the formation of LtB_4 was an attack by molecular oxygen on carbon atom 5, forming an unstable intermediate. The proposed structure of this intermediate which has now been isolated from human PMNL is 5,6-oxido-7,9,11,14-eicosatetraenoic acid or leukotriene A_4. At physiological temperature and pH LtA_4 has a half-life of only 3 min. The UV spectrum of LtB_4 was similar to that of SRS-A purified from cultured mouse mastocytoma cells stimulated with A23187. The major difference was that the peak absorbance of SRS-A was at a wavelength 10 nm higher than that of LtB_4. This difference was, in fact, consistent with the structure of a sulphur substituent linked to a conjugated triene structure.

Previous research using the relatively crude technique of spark-source spectrometry had revealed the presence of sulphur in SRS-A. Experiments in which rat basophilic leucocytes were incubated with numerous radio-labelled sulphur-containing molecules showed that the source of the sulphur in SRS was the amino acid cysteine. Degradation of SRS by a variety of chemical reagents yielded breakdown products which indicated that the arachidonic acid derivative and cysteine were linked by a thioether bond. Further chemical studies revealed more information about the cysteine part of the molecule. Chemical extraction of the sulphur from cysteine would be expected to yield alinine. However, desulphuration of SRS prepared biologically from mouse mastocytoma cells was not accompanied by the release of free alinine, which suggested that the cysteine present in SRS was actually attached to one or more different amino acids. The structure of the peptide part of SRS-A was determined by classical biochemical protein sequencing techniques as γ-glutamyl-cysteinylglycine otherwise called glutathione which is present in millimolar amounts in most cells.

Thus, these experiments confirmed that the structure of SRS from mouse mastocytoma cells (which had all the biological properties of immunologically formed SRS-A) is 5-hydroxy-6(S)-glutathionyl-7,9,11,14-eicosatetraenoic acid (LtC_4). Morris *et al.* (1980) subsequently extracted SRS-A from the perfusate of 800 sensitised, allergen-challenged guinea pig lungs and confirmed its structure as that of leukotriene C_4. The pathway for the biosynthesis of the various leukotrienes from arachidonic acid is shown in Fig. 1.14. If rat basophilic cells are stimulated with A23187, the first leukotriene to be formed is LtC_4 which is detected in the first few minutes of incubation and is then degraded by the enzyme γ-glutamyltranspeptidase to cysteinylglycyl SRS (LtD_4) which is the major product after 15 min incubation. LtD_4 is in turn enzymatically degraded

Fig. 1.14. Pathway of leukotriene biosynthesis from arachidonic acid.

```
                    Phospholipids
                         |
                         v
                  Arachidonic acid
     Cyclooxygenase  /        \  Lipoxygenase
                    /          \
                   v            v
             Prostanoids      5 HPETE
                                 |
                                 v
          Glutathione-s-transferase   Leukotriene A₄
                        \           /       |
                         v         v        v
                       Leukotriene C₄   Leukotriene B₄
   Glutamyl transpeptidase    |
                              v
                       Leukotriene D₄
   Cysteinyl glycinase        |
                              v
                       Leukotriene E₄
```

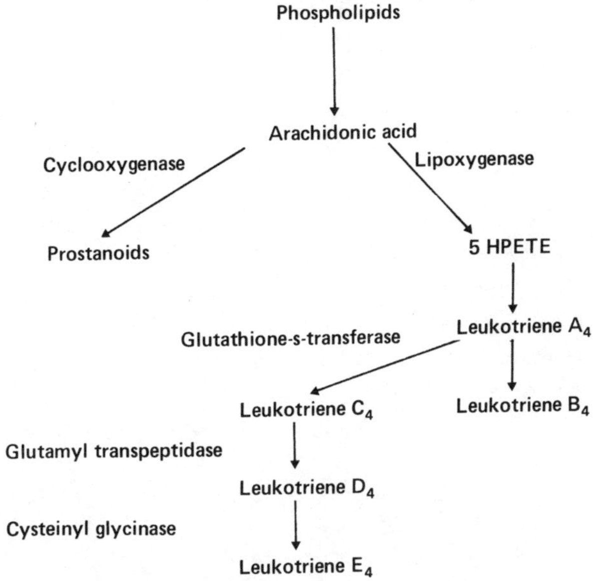

to LtE_4. It is now believed that the numerous SRSs which have been described over the last 40 years are mixtures of biologically active leukotrienes C_4 and D_4. The preparation and properties of both chemically synthesised and biologically derived leukotrienes are now known to match almost exactly thus finally confirming the structures of leukotrienes as those shown in Fig. 1.13. A great deal of interest has already been shown in the formation and biological properties of leukotrienes, particularly since they may be intimately involved in the aetiology of asthma and perhaps other inflammatory conditions. These aspects of leukotriene pharmacology are discussed in some detail in Chapters 7 and 10.

2

Effect of drugs on arachidonic acid metabolism

Introduction

The experiments which led to the discovery of classical prostaglandins, prostacyclin, thromboxanes and leukotrienes have been described in Chapter 1. The aim of this chapter is to review the vast literature which

Fig. 2.1. Effect of drugs on eicosanoid (prostanoid and leukotriene) biosynthesis.

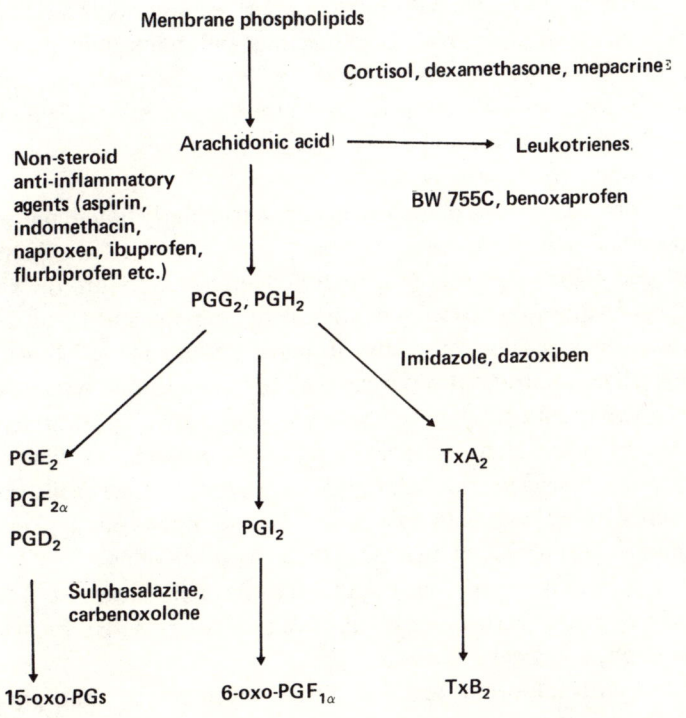

```
                    Membrane phospholipids
                              │
                              │   Cortisol, dexamethasone, mepacrine
                              ▼
                      Arachidonic acid ──────────►  Leukotrienes
Non-steroid                   │
anti-inflammatory             │        BW 755C, benoxaprofen
agents (aspirin,              │
indomethacin,                 │
naproxen, ibuprofen,          │
flurbiprofen etc.)            ▼
                         PGG₂, PGH₂
                       ╱      │      ╲   Imidazole, dazoxiben
                      ╱       │       ╲
                     ╱        │        ▼
        PGE₂                  │        TxA₂
        PGF_{2α}              │
        PGD₂                 PGI₂
          │                   │         │
          │ Sulphasalazine,   │         │
          │ carbenoxolone     │         │
          ▼                   ▼         ▼
      15-oxo-PGs         6-oxo-PGF_{1α}   TxB₂
```

has accumulated over the last 15–20 years dealing with the way in which drugs interfere with prostanoid and leukotriene biosynthesis and catabolism. The major groups of drugs which inhibit one or other enzymes of the arachidonic acid cascade are shown in Fig. 2.1.

Drugs which inhibit the phospholipase step in eicosanoid formation

The first step in the biosynthesis of both prostanoids and leukotrienes is the release of free arachidonic acid from membrane-bound phospholipids. The enzyme responsible, phospholipase A_2 in most cells, is activated by a variety of chemical, mechanical or immunological stimuli which have in common disruption of or trauma to the cell membrane. Inhibition of the phospholipase step in the arachidonic acid cascade effectively 'turns off' the formation of all cyclooxygenase and lipoxygenase metabolites by starving these enzymes of their precursors.

Several phospholipase A_2 enzyme inhibitors have been identified. Of these one of the first drugs to be used experimentally was the anti-malarial agent mepacrine. This drug complexes with phospholipids and thereby prevents their attack by phospholipase enzymes. Other substances which block phospholipase in this way include cationic polyamines like putrescine and spermidine, and a hotch-potch of pharmacologically active agents like propranolol, chlorpromazine, dibucaine and imipramine. In contrast, p-bromophenacyl bromide (pBPB) inhibits phospholipase activity by binding not to the phospholipid but to the enzyme itself. pBPB chemically reacts with a histidine amino acid residue at, or close to, the active centre of the phospholipase enzyme.

Unfortunately none of these drugs are particularly potent phospholipase blockers. For example, mepacrine blocks the release of TxA_2 from isolated, perfused guinea pig lungs *in vitro*, following injection into the pulmonary artery of phospholipase activators like angiotensin II or histamine. However mepacrine is only active at concentrations above 500 μM. Not only are such concentrations never reached in human blood when this drug is given therapeutically but, at these levels, mepacrine also inhibits a wide range of other enzymes including the cyclooxygenase which converts arachidonic acid to prostaglandin endoperoxides. Like mepacrine, pBPB is similarly lacking both in potency and selectivity as a phospholipase inhibitor. This does not mean that these drugs are of no value. In contrast, mepacrine in particular has been used as an experimental tool for several years by researchers interested in deducing the role of prostaglandins in many physiological processes.

The most important group of drugs which prevent phospholipase A_2

The phospholipase step in eicosanoid formation 43

activity are the corticosteroids (Blackwell & Flower, 1983). The main naturally occurring corticosteroids are cortisol (hydrocortisone) and corticosterone. Many synthetic corticosteroids have been developed and are used clinically. These include dexamethasone, prednisolone, triamcinolone, betamethasone and paramethasone. The structures of some commonly used anti-inflammatory steroids are shown in Fig. 2.2. Synthetic corticosteroids are used chiefly as anti-inflammatory drugs in a variety of autoimmune and hypersensitivity states including arthritis, asthma, skin diseases like eczema and psoriasis, conjunctivitis and uveitis of the eye, ulcerative colitis and systemic lupus erythematosus. They are extremely valuable drugs but have the drawback of a number of potentially very harmful side effects including peptic ulcer, osteoporosis (brittleness of bones due to removal of calcium) and a generalised increase in susceptibility to infection. When these drugs are prescribed the clinician must therefore try to balance the potential benefits of steroid therapy with their side effects.

There have been many suggestions for the mechanism of action of the anti-inflammatory steroids. One theory, popular a few years ago, was that corticosteroids stabilised the membranes of cell lysosomes and thus prevented the release of proteolytic enzymes which could digest away the surrounding tissue and thus provoke an inflammatory reaction. When it became clear in the early 1970s that prostanoids were formed at the site

Fig. 2.2. Chemical structure of three widely used anti-inflammatory steroids.

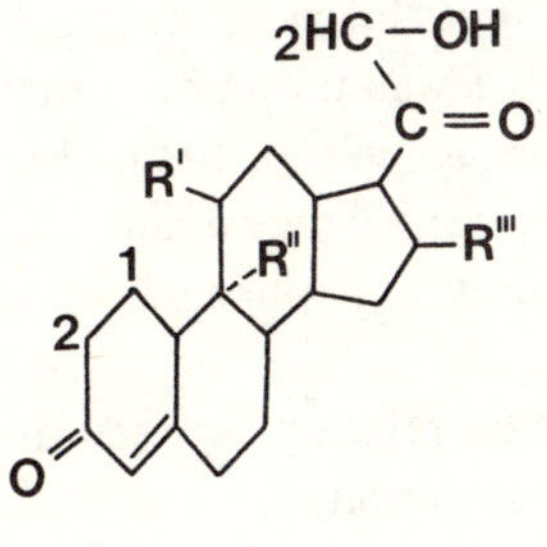

R^I	R^{II}	R^{III}	C-1–2	
=O	–H	–H	Saturated	Cortisol
–OH	–H	–H	Unsaturated	Prednisolone
–OH	–F	–CH$_3$	Unsaturated	Dexamethasone

of an inflammatory response and that they mimicked many of the symptoms of inflammation e.g. they cause vasodilatation, erythema, oedema and pain (see Chapter 10 for a full discussion of prostanoids as inflammatory mediators) it was proposed that steroids may reduce inflammation by blocking the biosynthesis of prostanoids. This possibility gained support from the observation that aspirin-like drugs, which are also used clinically to treat inflammation, potently blocked prostanoid formation.

Several research groups reported that corticosteroids prevented prostanoid biosynthesis and release by cells in culture (e.g. guinea pig macrophages, rat leucocytes and human fibroblasts) as well as intact tissues like the perfused mesenteric vascular bed of the rabbit constricted with noradrenaline and the antigen-challenged perfused guinea pig lungs (Gryglewski et al., 1975). Two pieces of evidence strongly suggested that steroids stopped prostanoid formation by an action on the phospholipase step in the arachidonic acid cascade. The first piece of evidence came from experiments using isolated perfused guinea pig lungs challenged with rabbit aorta contracting substance releasing factor (RCS-RF). This substance is a peptide which is released from sensitised guinea pig lungs challenged with the appropriate antigen. RCS-RF activates phospholipase enzymes. Blackwell et al. (1978) demonstrated that corticosteroids infused into guinea pig lungs in vitro prevented the release of TxA_2 (assayed on the rabbit aorta) induced by injection of RCS-RF but not that induced by injection of arachidonic acid. This result suggests that the site of action of these steroid drugs was the phospholipase and not the cyclooxygenase step in the pathway. The second piece of evidence was the finding that corticosteroids blocked prostanoid formation in intact cells or tissues but not in organ homogenates or broken-cell preparations. Once again, this localised the site of action of corticosteroids to inhibition of phospholipase enzymes.

It was soon realised that there was an excellent correlation between the ability of a series of corticosteroids to block phospholipase and their anti-inflammatory activity in a number of animal models of inflammation. In addition steroid hormones like aldosterone, oestradiol and progesterone, which are devoid of anti-inflammatory activity, were also without effect on phospholipase enzyme activity. A list of cells and tissues in which prostanoid biosynthesis is inhibited by corticosteroids is shown in Table 2.1.

Although it is now widely accepted that corticosteroids exert most of their anti-inflammatory action by preventing phospholipase activity, and thereby stopping the formation of inflammatory prostanoids and leukotrienes, the way in which they do this has been a focus of research throughout the 1970s and 1980s. The first steps in understanding their mechanism of action have only recently been taken. The experiments which underly our modern-day understanding of steroid action represents not only an intriguing pharmacological detective story but also an excellent

Table 2.1. *Inhibition of phospholipase by corticosteroids*

Tissue	Stimulus to prostanoid biosynthesis
Perfused guinea pig lung	RCS-RF, histamine, ATP, anaphylactic shock
Rat peritoneal leucocytes	
Rabbit neutrophils	F-met-leu-phenylalanine
Fibroblasts	Phorbol myristyl acetate
Macrophages	Zymosan
Human synovial tissue	

Fig. 2.3. General mechanisms of action of steroid hormones. The steroid crosses the cell membrane (*a*), combines with a steroid receptor (*b*), enters the nucleus (*c*), directs the biosynthesis of a mRNA (*d*) coding for a specific protein or peptide (*e*) which either acts within the same cell or is exported to adjacent cells to bring about the biological effect.

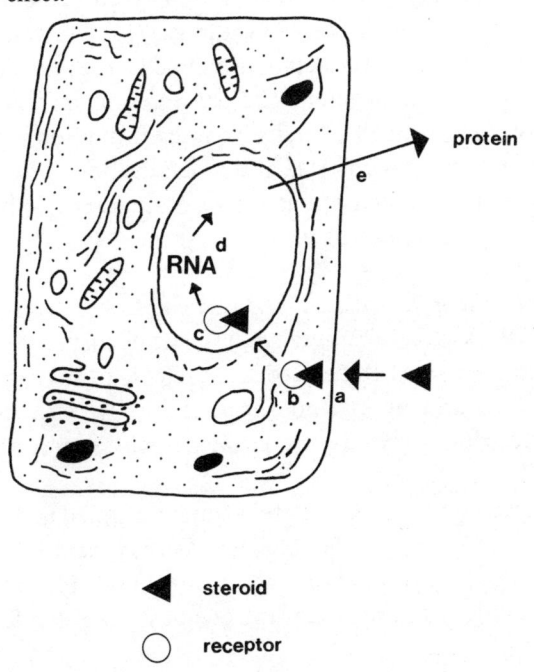

example of scientific investigation. For this reason these experiments will be discussed here in some detail.

An unexpected early finding was that corticosteroids do not inhibit phospholipase activity by a direct action on the enzyme since, unlike mepacrine, they have no effect on phospholipase in tissue homogenates. For steroid drugs to be effective, intact cells are required. Furthermore, in all test systems studied, inhibition of prostanoid biosynthesis by corticosteroids involves a time lag between steroid administration and prevention of prostanoid formation of between 15 and 30 min. Together these two observations raised the possibility that corticosteroids are simply hormonal messengers switching the cell on to produce its own phospholipase inhibitor.

This conclusion is entirely consistent with what is known of the general mechanism of action of all steroids. Current thoughts on the steroid/cell interaction are depicted diagrammatically in Fig. 2.3. Briefly, the steps involved are as follows: steroid enters the cell freely across the plasma membrane (a) and combines with a selective steroid receptor (b) which has a molecular weight of about 50000. The result is the exposure of steroid-binding sites in the nucleus. The steroid/receptor complex enters the nucleus and interacts with DNA to cause the transcription of a new messenger RNA (mRNA) which codes for a new protein (c). This protein may be an enzyme, cofactor for an enzyme or a modulator of an enzyme already present in the cell. It is the newly synthesised protein and not the steroid itself which changes the cell's metabolism. Each and every biological action of a steroid, be it a sex hormone like progesterone or a mineralocorticoid like aldosterone, probably results from the formation within the cell of a separate active protein. The partial characterisation of the protein responsible for the anti-inflammatory action of corticosteroids has been achieved by two groups of researchers.

Experiments using the isolated, perfused guinea pig lungs

Flower & Blackwell (1979) studied the effect of corticosteroids on RCS-RF and arachidonic-acid-challenged TxA_2 release from the perfused guinea pig lungs. They used an ingenious technique in which two guinea pig lungs were perfused with warmed, oxygenated Krebs' solution in series as shown in Fig. 2.4.

The first 'generator lung' received Krebs' solution warmed to 37 °C and oxygenated with 95% O_2 in CO_2. The effluent from the generator lung was reoxygenated and pumped through a second 'test lung'. Phospholipase activity in the test lung was measured using one or both of the following assays:

Fig. 2.4. Experimental design to demonstrate the effect of corticosteroids on prostanoid biosynthesis in superfused guinea pig lungs. Steroid is infused via the pulmonary artery into the 'generator lung' (A) and the emerging Krebs' solution effluent collected, reoxygenated and pumped into the 'test lung' (B). Prostanoid biosynthesis in the test lung is assessed by challenging the lung with RCS–RF and measuring TxA_2 release by monitoring the contraction of the rabbit aortic spiral preparation.

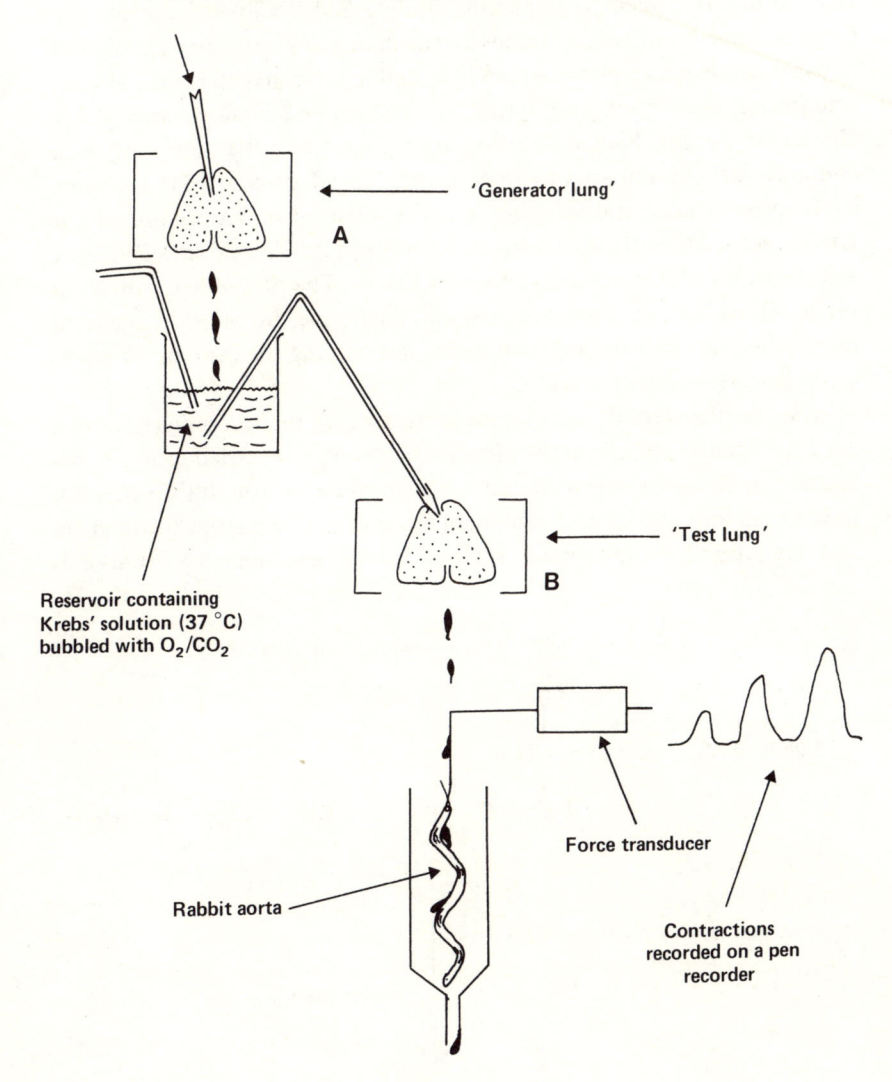

Bioassay. Injection of RCS-RF into the test lung stimulated phospholipase activity and released TxA_2 from the lung. A spirally cut rabbit aorta strip positioned beneath the test lung in cascade was used to bioassay the released TxA_2.

Radiochemical. Briefly, the radiochemical method involves injection of a mixture of $[^3H]2'$-oleoylphosphatidylcholine and $[^{14}C]$oleic acid through the test lung. As this mixture passes through the pulmonary circulation phospholipase enzymes present in lung cell membranes split off $[^3H]$oleic acid from $2'$-oleoylphosphatidylcholine. The reaction is shown in Fig. 2.5. Consequently the Krebs' solution emerging from the test lung now contains varying amounts of three radiolabelled species: $[^3H]$oleic acid, $[^3H]2'$-oleoylphosphatidylcholine and $[^{14}C]$oleic acid. Extraction of the Krebs' perfusate with hexane removes both $[^3H]$oleic acid and $[^{14}C]$oleic acid but not $[^3H]2'$-oleoylphosphatidylcholine. The percentage hydrolysis of the phospholipid, $2'$-oleoylphosphatidylcholine, by phospholipase A_2 in the lung may be measured simply by determining the ratio of ^3H to ^{14}C using a scintillation counter.

When corticosteroids were infused directly into the test lung (B, in Fig. 2.4.) no immediate effect on phospholipase was recorded using either method. In contrast, when steroid was first infused through the generator lung (A, in Fig. 2.4.) and the effluent from this lung passed through the test lung then phospholipase activity in the test lung was markedly

Fig. 2.5. Radiochemical assay for phospholipase activity.

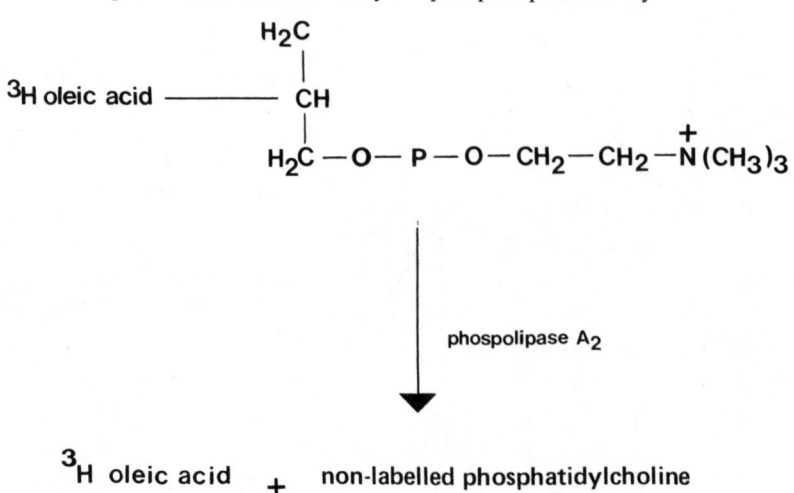

$$^3H \text{ oleic acid} - \begin{array}{c} H_2C \\ | \\ CH \\ | \\ H_2C \end{array} - O - P - O - CH_2 - CH_2 - \overset{+}{N}(CH_3)_3$$

$$\downarrow \text{phospolipase } A_2$$

^3H oleic acid + non-labelled phosphatidylcholine

inhibited. This was apparent both in the reduced hydrolysis of [^{3}H]2'-oleoylphosphatidylcholine and by a fall in the amount of TxA$_2$ released to contract the rabbit aorta.

These results have been interpreted to indicate that corticosteroids release a substance from the generator lung which inhibits phospholipase activity in the test lung. Not surprisingly other steroids like aldosterone, which have little or no anti-inflammatory activity, do not cause the release of this substance from the perfused guinea pig lung. Furthermore, if both generator and test lungs are perfused with Krebs' solution containing either an inhibitor of RNA transcription (e.g. actinomycin D) or of ribosomal protein biosynthesis (e.g. puromycin) then corticosteroids lose the ability to release the phospholipase inhibitor from the generator lungs. At more or less the same time that these experiments were being carried out in perfused lungs, broadly similar conclusions were being drawn from experiments using white blood cells.

Experiments using rat peritoneal leucocytes

Leucocytes are an excellent source of both prostanoids and leukotrienes and therefore make a good test system for assessing the effect of drugs on phospholipase activity (Glatt et al., 1977). Before they can be used for this purpose, cells must first be collected or harvested. This may be achieved by washing out the peritoneal cavity of a freshly killed rat, using heparinised saline and carefully removing the now cell-rich fluid from the carcass. Leucocytes may be spun down and resuspended in whatever buffer solution is required. In order to gather together sufficient leucocytes for the experiment, cells from several animals are normally pooled.

When leucocytes are incubated in salt solution at physiological temperature they spontaneously synthesise small amounts of both prostanoids and leukotrienes. The main prostanoid is PGE$_2$ which can be assayed either biologically or by radioimmunoassay. In the presence of killed bacteria, leucocytes synthesise very much larger amounts of prostanoids as they rapidly engulf and phagocytose the microorganisms.

DiRosa and his colleagues (DiRosa & Persico, 1979 a, b) incubated leucocytes with hydrocortisone (37 °C, 90 min), centrifuged them to bring down a leucocyte pellet, and tested aliquots of the supernatant fluid for their ability to prevent prostanoid biosynthesis by 'test' leucocytes (Fig. 2.6.). Leucocytes treated with the steroid released a substance into the supernate which blocked prostanoid biosynthesis in the test leucocytes. Thus, as in the perfused lung, steroids appear to interact with leucocytes to initiate the synthesis/release of a phospholipase blocker. Once again, experiments using agents which prevent RNA transcription or protein biosynthesis at the ribosome blocked the formation of this novel phospholipase inhibitor (suggesting that it is a protein).

Characterisation of macrocortin

The phospholipase inhibitor released from rat peritoneal leucocytes and perfused guinea pig lungs challenged with corticosteroids has been named macrocortin (macro, implying made in macrophages; cortin, suggesting that it mimics the action of corticosteroids). Macrocortin may be a mixture of several proteins. Chromatographic separation of the anti-phospholipase activity has revealed four major proteins with this biological activity. The most abundant has a molecular weight of 15 000 and is a carbohydrate-containing protein or glycoprotein. Studies using the partially purified macrocortin are still in their infancy. However, purified macrocortin has been shown to block leucocyte prostanoid formation and also decreases inflammation induced in rats by injecting carrageenan, a seaweed extract, into the paw.

The discovery of macrocortin has answered many questions about the way in which corticosteroids exert their anti-inflammatory effect. However, like all worthwhile discoveries, it has raised many other questions. For example, is there spontaneous formation of macrocortin by cells? Do all

Fig. 2.6. Experimental design to demonstrate the effect of corticosteroids on prostanoid biosynthesis in rat peritoneal leucocytes. Steroid is first incubated with leucocytes collected from several animals and pooled together. After centrifugation to bring down a leucocyte pellet aliquots of the supernate are tested for ability to prevent prostanoid formation by leucocytes in the presence of killed bacteria. Prostanoids so formed are bioassayed on the rat stomach strip preparation.

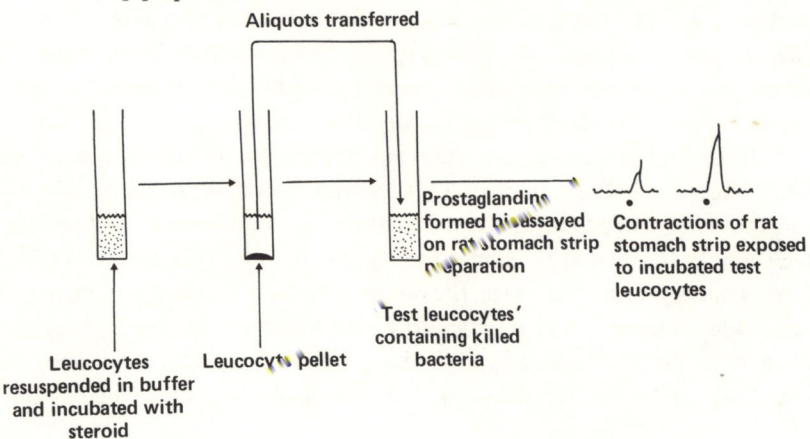

cells synthesise macrocortin or is this a property of only some cell types? What happens to the macrocortin once it has been made and does it cross biological membranes? Is macrocortin biosynthesis reduced in inflammatory disease states and, if so, does this deficiency of what is an endogenous anti-inflammatory molecule actually exacerbate the inflammation? Finally, can macrocortin, or (possibly at some future date) an analogue of macrocortin, be used to treat crippling inflammatory conditions such as arthritis? Since macrocortin is responsible only for the anti-inflammatory actions of corticosteroids and not for their side effects, which are presumably brought about by a separate messenger protein, it is conceivable that macrocortin would be a viable alternative to corticosteroids for the treatment of inflammation.

Cyclooxygenase inhibitors

Once released from membrane stores, arachidonic acid is enzymatically converted either by a cyclooxygenase enzyme to PG endoperoxides or by a lipoxygenase enzyme to leukotrienes. Drugs which block lipoxygenase activity are dealt with on pp. 62–4. What is normally referred to as cyclooxygenase is in fact two separate enzymes, as discussed in Chapter 1. The first component enzyme converts arachidonic acid to prostaglandin G_2 (this enzyme is properly named cyclooxygenase) and a second peroxidase enzyme converts PGG_2 to PGH_2. Both of these enzymes have been partially purified and require different cofactors for full activity. All of the drugs discussed in this section affect the cyclooxygenase enzyme but have little or no action on the peroxidase step.

Drugs which inhibit cyclooxygenase enzyme activity are usually referred to as non-steroid anti-inflammatory drugs (NSAID). This name is unsatisfactory since immunosuppressant drugs like methotrexate are certainly non-steroid and anti-inflammatory but are not classed as NSAID since they do not affect prostanoid biosynthesis. Other authors refer to NSAID as aspirin-like agents but again this title does not take into account all

Table 2.2. *Some clinically used NSAID*

Alclofenac	Indomethacin
Aspirin	Ketoprofen
Benorylate	Meclofenamic acid
Fenclofenac	Mefanamic acid
Fenoprofen	Naproxen
Flufenamic acid	Niflumic acid
Flurbiprofen	Paracetamol
Ibuprofen	Piroxicam

of those NSAID which are structurally unrelated to aspirin. Perhaps a better name for this class of drugs is simply 'cyclooxygenase inhibitors'. However, the term NSAID is in common use and will be retained throughout the rest of this volume. Clinically useful NSAID are shown in Table 2.2. and the structures of some of these drugs are shown in Fig. 2.7.

Aspirin (acetylsalicylate) is the prototype of all NSAID. The discovery of the curative power of salicylates is usually attributed to the Reverend Edmund Stone who treated patients suffering from fever and rheumatism with extracts of willow bark, now known to contain salicylic acid. Stone reported these findings to the Royal Society in 1763. The name 'aspirin' is actually derived from another plant source of salicylic acid, the meadowsweet. The letter 'a' in aspirin reflects the acetyl group added to salicylic acid and 'spir' represents the Latin name of the meadowsweet (*Spirea ulmaria*). Exactly how aspirin got its 'in' has remained a mystery.

Fig. 2.7. Chemical structure of three commonly used non-steroid anti-inflammatory drugs (NSAID).

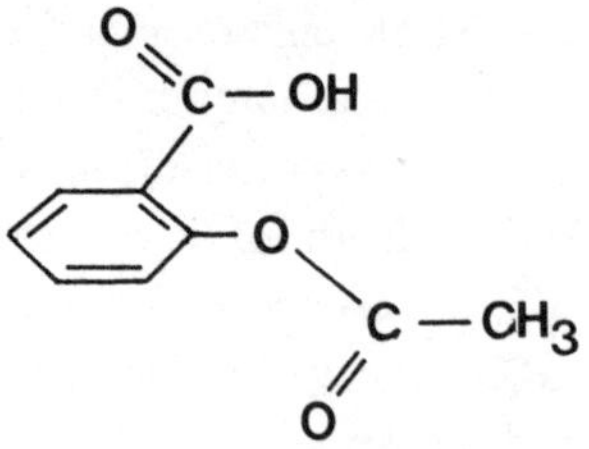

Aspirin

Mefanamic acid

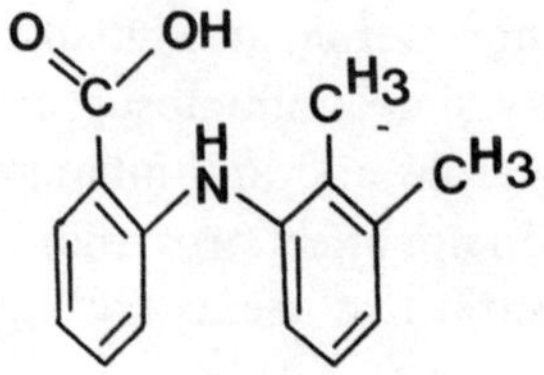

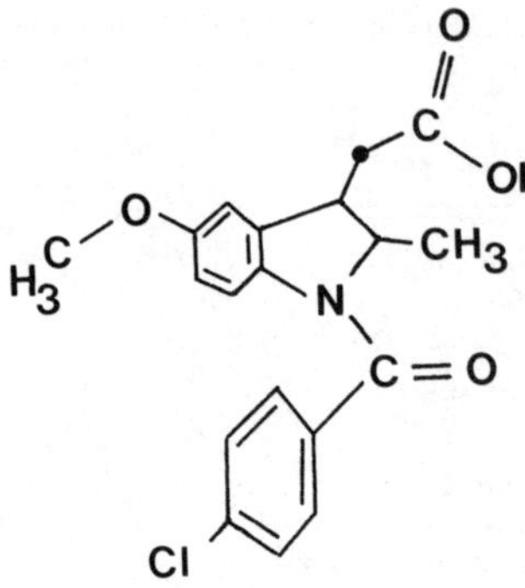

Indomethacin

Despite the fact that aspirin was first introduced onto the pharmacists' shelf over 80 years ago it still remains one of the most widely prescribed drugs today. World production of aspirin is over 100000 tons per year and, in the United Kingdom in 1978 alone, over 35 million prescriptions for aspirin-based preparations were written by clinicians. Despite the continuing popularity of aspirin, a number of other NSAID have been introduced onto the market. These include anthranilic acids like mefanamic acid and propionic acids like ibuprofen, flurbiprofen and naproxen.

Mechanism of action of NSAID

The possibility that NSAID interfere in some way with the biosynthesis of prostanoids was inferred in some early studies carried out by Collier and his colleagues in the early 1960s (see Berry & Collier, 1964). They observed that aspirin blocked the contractions of the isolated guinea pig ileum to bradykinin and also antagonised bronchoconstriction in anaesthetised guinea pigs following the intravenous injection of slow reacting substance of anaphylaxis (SRS-A) and ATP. Since it was most unlikely that aspirin could act as a receptor antagonist for each of these chemically very dissimilar substances it was suggested that aspirin blocked the formation of a bronchoconstrictor which mediated the effect of both SRS-A and ATP. At the time the identity of the proposed bronchoconstrictor was not known. Now it is clear that bradykinin, SRS-A and ATP all released prostaglandin endoperoxides and TxA_2 (potent bronchoconstrictors) from lung tissue. The efforts of Collier and his colleagues in the 1960s are, unfortunately, frequently overshadowed in the blaze of interest which was to follow.

The first positive finding which suggested that aspirin prevented biosynthesis of prostanoids came in 1969 when Piper & Vane observed that aspirin prevented release of RCS (rabbit aorta contracting substance, now known to consist mainly of TxA_2) from antigen- or arachidonic-acid-challenged guinea pig lungs. Proof that aspirin and other NSAID blocked prostanoid biosynthesis by an action on the cyclooxygenase enzyme came 2 years later, with the simultaneous publication in the journal *Nature* of three papers which revolutionised thinking on the pharmacology of these drugs. In the first paper, Vane (1971) demonstrated that aspirin and indomethacin reduced the conversion of arachidonic acid to bioassayable prostaglandins by cell-free homogenates of the guinea pig lung in a dose-dependent manner. This result was confirmed in the second paper by Smith & Willis (1971) who detected a similar inhibition of platelet prostaglandin formation by aspirin. Finally, in the third paper, Ferreira,

Moncada & Vane (1971) inhibited prostaglandin release from nerve-stimulated perfused dog spleen preparations, using indomethacin.

Since 1971, inhibition of prostanoid biosynthesis by NSAID has been examined in a number of different tissues, as shown in Table 2.3. This list is not exhaustive and is not intended to be. Rather it is intended to indicate the universality of this action of aspirin. Whichever tissues are chosen, however they are prepared and whatever the method used to assay the synthesised prostanoids, aspirin-like drugs invariably inhibit cyclooxygenase. Indeed the author is not aware of any experiments in which NSAID have failed to inhibit cyclooxygenase activity in tissue homogenates or subcellular fractions prepared therefrom. However, NSAID do vary in potency from one tissue to the next. There are several reasons for this. The degree of cyclooxygenase inhibition by a particular NSAID depends upon the circumstances of the experiment (e.g. whether a crude tissue homogenate, centrifuged supernate, microsomes or dried acetone powder

Table 2.3. *Tissues in which NSAID inhibit prostanoid biosynthesis*

Species	Tissue	Preparation	Assay[a]
Man, cat, dog, monkey, rat, guinea pig, mouse, gerbil	Lung	Homogenates/microsomes	B, RIA, GC/MS
Cat, monkey, rat, rabbit, guinea pig	Lung	Perfused via pulmonary artery and challenged with arachidonic acid	B, RIA, GC/MS
Ram, bull	Seminal vesicle	Homogenates/microsomes	B, RIA, Sp.
Several species	Kidney	Homogenates/microsomes/slices	B, RIA
Several species	Kidney	Perfused via renal artery	B, RIA
Guinea pig, human	Uterus	Homogenates/microsomes/pieces	B, RIA, GC/MS
Several species	Spleen, brain, gut	Homogenates/microsomes	B, RIA
Dog, cat	Spleen	Perfused via splenic artery	B
Several species	Inflammatory exudate	Aspirated from joint or air bleb	B, RIA, GC/MS
Several species	Semen	Collected and extracted	B, RIA, GC/MS
Several species	Platelets, leucocytes	Collected and extracted	B, RIA, GC/MS
Several species	Whole body	Measured in urine to determine body turnover	GC/MS

[a] Abbreviations: B, biological assay; RIA, radioimmunoassay; GC/MS, gas-chromatography – mass-spectrometry, Sp., spectrophotometry.

is used), the concentration of substrate arachidonic acid and the inclusion or non-inclusion of cofactors. For example, the concentration of indomethacin required for 50% inhibition of cyclooxygenase activity increases as the concentration of arachidonic acid is increased in incubations of sheep seminal vesicle microsomes. Furthermore, an analogue of indomethacin, fluoroindomethacin, at concentrations which completely inhibit microsomal prostaglandin biosynthesis in sheep seminal vesicles has little or no effect on the same enzyme from the same source which had been dried with acetone to a powder and then resuspended in buffer solution before incubation with arachidonic acid. Clearly the choice of incubation conditions will affect the NSAID potency and should therefore be made very carefully. The potencies of six different NSAID measured as the concentration required to bring about 50% inhibition of cyclooxygenase activity (IC_{50}) from five different tissues is shown in Table 2.4.

Even if the differences in incubation conditions are taken into account there are still unexplained variations in the sensitivity of different preparations to different NSAID. This is particularly apparent with paracetamol, which is a potent inhibitor of brain cyclooxygenase from several species but has only a weak effect against this enzyme in peripheral tissues.

Differential sensitivity to NSAID is presumably caused by subtle differences in the structure of the enzyme in various parts of the body. This finding is more than a simple academic curiosity since it raises the intriguing possibility that NSAID may be developed which preferentially block cyclooxygenase activity in inflamed tissues without affecting, for example, gastric mucosal cyclooxygenase. Such a drug, if it could be

Table 2.4. *Potency of NSAID as cyclooxygenase inhibitors in five different tissue preparations in vitro*

Drug	Tissue source of cyclooxygenase[a]				
	BSV	SSV	RbK	GP Lung	Dog spleen
Aspirin	9000	83	2500	35	37
Indomethacin	236	0.45	3.7	0.75	0.17
Naproxen	24	6.1	ND	ND	ND
Meclofenamic acid	ND	2.1	ND	ND	0.7
Ibuprofen	4.5	1.5	ND	ND	ND
Phenylbutazone	6.4	13	1.9	ND	7.2

[a] All figures show concentration of NSAID required to produce 50% inhibition (IC_{50}) of cyclooxygenase activity in μmol/l. Abbreviations: BSV, bovine seminal vesicle microsomes; SSV, sheep seminal vesicle microsomes; RbK, rabbit kidney homogenate; GP lung, guinea pig lung homogenate; ND, not determined.

developed, would be expected to combine a good anti-inflammatory effect without the risks of gastric ulceration and bleeding (which are common side effects of prolonged NSAID therapy).

The molecular mechanism of action of NSAID has also been studied. Aspirin acetylates an ε-amino group of a lysine amino acid residue at or close to the active centre of the cyclooxygenase enzyme producing an irreversible block (Roth & Majerus, 1975). This explains why a single oral dose of aspirin blocks platelet cyclooxygenase and prevents platelet aggregation for the lifespan of the platelet in the circulation (see Chapter 4 for a comprehensive review of the effect of NSAID on platelet function). Unlike aspirin, most other NSAID produce a reversible inhibition of cyclooxygenase activity.

The pharmacology of NSAID

Aspirin and other NSAID have three major pharmacological properties:
  (1) reduction of swelling (oedema) and redness (erythema) which is associated with inflammation (the anti-inflammatory effect);
  (2) the ability to lower elevated body temperature due to a fever caused by invasion of the body by foreign microorganisms (anti-pyretic effect);
  (3) the ability to reduce the sensation of pain (analgesic effect).

This combination of biological properties make the NSAID very useful in the treatment of a wide range of inflammatory diseases. NSAID are clinically indicated in the treatment of rheumatoid and osteoarthritis, bursitis, tendonitis, lupus erythematosus as well as gout, post-operative pain and the pain associated with dysmenorrhoea. Of course aspirin and its numerous preparations are used in vast quantities each year to treat infections of the upper respiratory tract such as influenza and the common cold. The major side effects of NSAID are gastrointestinal disturbances such as nausea and vomiting, irritation of the gastric mucosa with formation of ulcers and haematemesis (vomiting up of blood), renal disease and, occasionally, bronchospasm in susceptible individuals (like asthmatics).

Several attempts have been made to identify a single, unifying mechanism of action of NSAID which would explain all of their biological actions as well as their side effects. With the finding that all NSAID shared the ability to inhibit prostanoid biosynthesis, Ferreira & Vane (1974) were the first to propose that this underlied all the effects of NSAID. This proposal has met with general acceptance and the evidence on which it is based is discussed in the following sections. For further information the reader is advised to consult the reviews of Collier (1969, 1984); Ferreira & Vane (1974, 1978), Flower (1974); and Vane (1978).

Analgesia by NSAID – inhibition of prostanoid biosynthesis? If NSAID relieve pain by inhibiting cyclooxygenase then it should be possible to demonstrate that one or more prostanoids are capable of producing pain. In the last few years prostaglandins of the E, F and I series have been found both to cause pain (algesia) in their own right and to sensitise pain receptors to other inflammatory mediators (hyperalgesia). In general, prostaglandins are algesic only at high concentrations following, for example, intradermal or intramuscular injection into human volunteers. In contrast, concentrations which are likely to occur at the site of inflammation do not produce algesia but do cause a long-lasting hyperalgesia. This condition may be demonstrated experimentally in human volunteers by carefully applying prostaglandins, histamine and bradykinin and various combinations of the three onto blister bases raised on the skin of the forearm. Although low doses of bradykinin and histamine cause pain, PGEs do not. Combining PGE_1 with either bradykinin or histamine provokes a considerable increase in the pain produced. An animal model in which NSAID produce analgesia is the abdominal constriction and writhing which occurs in mice injected intraperitoneally with either arachidonic acid or irritant chemicals (Collier & Schneider, 1972). Pre-injection of these animals with aspirin prevents writhing and thus, presumably, stops the animals feeling pain due to arachidonic acid but not that due to intraperitoneal injection of prostaglandins themselves. Further experimental evidence that prostaglandins are algesic and hyperalgesic is discussed in Chapter 10.

These results satisfy the criterion that aspirin reduces the appreciation of pain by a peripheral action to block the biosynthesis of algesic and hyperalgesic prostaglandins. This fits well with previous research by Lim and colleagues (1964) who confirmed that, in man, aspirin is analgesic by an action not in the brain (like morphine) but at peripheral sites. This also explains why aspirin is effective in mild to moderate pain such as headache, dysmenorrhoea and toothache (in which some degree of tissue inflammation, and thus prostaglandin biosynthesis, would be expected) but is not a particularly powerful analgesic in pain associated with contraction of visceral muscle or with very severe trauma. Not all of the evidence accords with the hypothesis that NSAID are analgesic by preventing the formation of pain-producing prostaglandins. It has been pointed out that PGE_1 and PGE_2 sensitise pain receptors to cause hyperalgesia for several hours whereas aspirin, administered for example to relieve toothache, is normally effective within 15–30 mins. In this case, even if the dose of aspirin used prevented all prostaglandin formation in the region of the decaying tooth (which is unlikely), the pain would still be expected to persist for 3–4 h; this represents the approximate duration of the hyperalgesic action of PGEs. Ferreira, Nakamura & Abreu Castro (1978) have countered this objection by proposing that PGI_2 is the prostanoid responsible for promoting pain in these circumstances. PGI_2, unlike PGE_2 and $PGF_{2\alpha}$, produces a hyperalgesia which lasts for only about 30 min. Presumably this is because PGI_2 is unstable and rapidly hydrolysed to $6\text{-oxo-}PGF_{1\alpha}$, which is not hyperalgesic. In this case NSAID cause effective pain control by preventing PGI_2 formation.

However, this conclusion does presuppose that PGI_2 is the major pain-producing prostanoid at the site of inflammation. This is not necessarily the case since PGE_2 is also present in large amounts in damaged tissues.

Anti-pyresis by NSAID – inhibition of prostanoid biosynthesis? Assuming that aspirin and other NSAID lower body temperature during fever by preventing prostanoid biosynthesis, it should be possible to demonstrate that one or other of the prostanoid family are pyrogenic (increase body temperature) by an action somewhere in the brain. Prostaglandins certainly occur in the brain and have been linked with a number of central physiological processes including the regulation of hormone release from the hypothalamus and pituitary gland. That they may also be involved in temperature control was suggested by a series of studies in the early 1970s (Milton & Wendlandt, 1971; Potts & East, 1972) in which intraventricular injection of E-type prostaglandins in the brains of rabbits, rats, sheep or cats caused a rise in body temperature of 1–2 deg C. Using iontophoresis it was possible to map out which part of the brain was responding to the prostaglandins. The pyrogenic effect was localised to a small region of neurones close to the third ventricle referred to as the anterior hypothalamus/pre-optic area or AH/POA.

Subsequently, Feldberg & Gupta (1973) showed that injection of *Shigella dysenteriae* bacteria into anaesthetised cats caused (*a*) a fever which could be localised to the AH/POA and (*b*) the release of large amounts of a prostaglandin into the cerebrospinal fluid. Originally this prostaglandin was estimated biologically but later experiments using radioimmunoassay revealed it to be PGE_2. Both the fever and the appearance of PGE_2 in cerebrospinal fluid were abolished by pre-treating the animals with NSAID like paracetamol, indomethacin and aspirin. These observations led to the hypothesis that a pyrogen caused fever by

promoting PGE_2 biosynthesis locally in the AH/POA and that NSAID exerted their anti-pyretic action by blocking cyclooxygenase activity in this part of the brain. There is now a considerable amount of data to support this hypothesis (discussed in detail in Chapter 10). However, a few experimental observations have come to light during the last 10 years which are difficult to reconcile with this theory. This evidence includes:
(1) PG receptor antagonists, like the experimental compound SC19220, do not block bacterial pyrogen-induced fever in cats;
(2) PGs are not pyrogenic in all species (e.g. in the chicken, fever can be induced by intravenous injection of pyrogen in the normal way but PGE_1 and PGE_2 actually lower body temperature when injected intraventricularly);
(3) animals maintained on a diet deficient in essential fatty acids (and therefore capable of forming only very small amounts of prostaglandins) can still respond to bacterial pyrogens with a fever in the normal way.

Further evidence against the proposed mechanism has come from experiments using NSAID themselves. In anaesthetised rabbits, Cranston, Hellon & Mitchell (1975) administered a dose of sodium salicylate which blocked the rise in cerebrospinal fluid prostaglandin activity produced by infused bacterial pyrogen but nevertheless had no effect whatsoever on the fever. Thus it seems that blocking the formation of prostaglandins in the brain may only be a side effect of NSAID therapy possibly unrelated to their ability to lower fever. Obviously further experiments of this type are necessary to clarify the role of prostaglandins in fever.

Anti-inflammation by NSAID – inhibition of prostanoid biosynthesis? If inhibition of prostanoid formation explains why NSAID alleviate inflammation in man and experimental animals then one or more prostanoids may be assumed to mimic the biological features of the inflammatory response. Amongst other things, prostanoids should cause vasodilatation of the microcirculation, increase capillary permeability to cause oedema, produce pain and promote movement of white blood cells to the damaged tissues. There is good evidence that prostaglandins of the E and I series dilate small blood vessels, raise capillary permeability and (as we have seen) potentiate the pain-producing effect of other mediators likely to be found at the site of inflammation. However, prostanoids influence white cell migration *in vitro* only at concentrations which could not possibly occur in damaged tissue *in vivo*. On the other hand, leukotrienes have potent chemotactic and chemokinetic actions on leucocytes at concentrations which do occur naturally. Judged on this information, therefore, NSAID which do not block lipoxygenase activity would not appear to be ideal drugs for the treatment of inflammation. What is really required is a drug which inhibits both cyclooxygenase and lipoxygenase enzyme activity.

The evidence for and against inhibition of prostanoid biosynthesis as the mechanism of the anti-inflammatory activity of NSAID is listed below.

Evidence for
(1) The ability of a series of NSAID to prevent prostanoid biosynthesis by cell-free homogenates prepared from dog spleen correlates with their anti-inflammatory potency to prevent carrageenan-induced oedema in rat paws (Flower *et al.*, 1972).
(2) Isomers of naproxen and indomethacin, which have little or no anti-inflammatory activity, have only very feeble ability to inhibit cyclooxygenase *in vitro* (Tomlinson *et al.*, 1972).
(3) Anti-inflammatory doses of NSAID reduce prostanoid levels at the site of inflammation in experimental animals and in man *in vivo*. These results extend those obtained from experiments *in vitro*. NSAID block cyclooxygenase and thus reduce prostanoid levels following subcutaneous implantation of sponges soaked in carrageenan into the flank of rats (Walker, Smith & Ford-Hutchinson, 1976) and in synovial fluid from arthritic rats, rabbits and humans.

If NSAID do prevent inflammation by blocking prostanoid formation then it should be possible to demonstrate that normal therapeutic doses of these drugs produce plasma concentrations capable of inhibiting cyclooxygenase activity. This is undoubtedly the case for the majority of NSAID. The free plasma concentration of indomethacin in man, after taking into consideration plasma binding, is approximately 0.5 μg/ml which is 2–10 times the IC_{50} for this drug against cyclooxygenase from most tissues.

Some inconsistencies have arisen with the concept that all of the anti-inflammatory effect of NSAID can be attributed to an inhibition of prostanoid biosynthesis.

Evidence against
Sodium salicylate is a weak inhibitor of cyclooxygenase activity *in vitro* but nevertheless is equipotent with aspirin as an anti-inflammatory agent in animal models. In an attempt to reconcile these observations it has been suggested that sodium salicylate is an example of a pro-drug, which is enzymatically converted into an active metabolite when administered. This proposal was prompted by experiments which suggested that sodium

salicylate did inhibit prostanoid formation *in vivo*. For example, Hamberg (1972) reported that sodium salicylate was equieffective with aspirin in reducing the urinary excretion of prostaglandin metabolites in human volunteers. From studying the metabolites of sodium salicylate in man, a possible candidate for this 'active metabolite' was identified as gentisic acid (Fig. 2.8.). Later it was found that, in molar terms, gentisic acid was approximately equipotent with aspirin as a cyclooxygenase blocker. A major problem with the notion that gentisic acid is the active metabolite of sodium salicylate is that, if this were true, then sodium salicylate should be equipotent with aspirin as an analgesic and anti-pyretic. In reality, sodium salicylate is considerably less potent in both cases. Other enzymes involved in the arachidonic acid cascade are also inhibited by sodium salicylate and this should be taken into account when considering the anti-inflammatory effect of this substance. For example, sodium salicylate inhibits the peroxidase enzyme which converts 12-HPETE to 12-HETE. Whether this is of importance for its anti-inflammatory effect will only become clear when the precise roles of 12-HPETE and 12-HETE (if any) in the inflammatory process have been elucidated.

Paracetamol is a good analgesic and anti-pyretic NSAID but is virtually devoid of anti-inflammatory activity. As pointed out previously, cyclo-oxygenases from different tissues vary in their responsiveness to paracetamol. As a general rule, brain cyclooxygenase is inhibited by this drug whereas cyclooxygenase from peripheral tissue is not. An action localised

Fig. 2.8. Pathway for the catabolism of salicylic acid in the body.

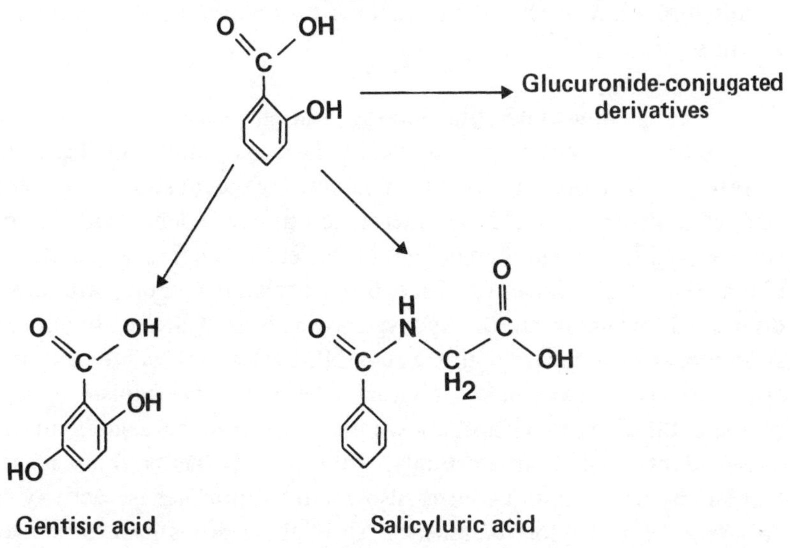

to the brain may account for the anti-pyretic action of paracetamol but not for its analgesic effect (which is known to be of peripheral and not central origin). Furthermore, according to some reports, paracetamol does reduce the concentration of prostanoids in rat inflammatory exudate even though it fails to affect the inflammation.

Side effects of NSAID – inhibition of prostanoid biosynthesis? The most frequently observed side effects of NSAID are ulceration and bleeding from the gut, renal damage and bronchoconstriction. Since prostanoids formed naturally in the body prevent formation of gastrointestinal ulcers, dilate the renal vasculature (thus preventing the kidney from being damaged by ischaemia) and dilate the airways, it has been proposed that all of these side effects may be due to inhibition of prostanoid biosynthesis (see Ferreira & Vane, 1974; Higgs & Vane, 1983).

In summary there is good evidence that many NSAID are therapeutically effective in reducing fever, relieving pain and reducing inflammation by virtue of inhibiting prostanoid biosynthesis. However, it is likely that this action alone does not account for all of the biological properties of all NSAID. In some cases alternative explanations which may not involve the arachidonic acid pathway at all must be sought. One possibility which has received some attention is that NSAID inhibit the production of the so-called superoxide anion (O_2^-) by phagocytosing leucocytes. This is a very reactive chemical species which, amongst other things, can break down hyaluronic acid which is an important component of joint synovial fluid. Excess formation of O_2^- may contribute to the chronic inflammation in arthritis.

Drugs which inhibit the lipoxygenase enzyme

Several seemingly selective lipoxygenase inhibitors have been reported. The list includes structural analogues of the substrate, arachidonic acid (15-hydroxy-5,8,11,13-eicosatetraenoic acid, 5,6-dehydroarachidonic acid) as well as structural analogues of the leukotrienes which are formed (5,6-methanoleukotriene A_4). These drugs probably compete with arachidonic acid for binding to the lipoxygenase enzyme. Chemically unrelated substances like norhydroguairetic acid (NDGA), propylgallate and proxicrimil are also lipoxygenase inhibitors. The precise mechanism of action of these three drugs is not known but may well be related to their anti-oxidant ability. Interestingly, FPL 55712 (better known as a leukotriene-receptor antagonist) also inhibits lipoxygenase activity (see Ashida *et al.*, 1983 for references). All of these drugs have little cyclooxygenase blocking activity and vary considerably in potency as inhibitors of lipoxygenase. The concentration of some of these agents required to cause 50% inhibition of cyclooxygenase and lipoxygenase enzyme activity in incubated rat basophilic leukaemic cells (RBL-1) are shown in Table 2.5. In addition to blocking leucocyte lipoxygenase activity these drugs also prevent the synthesis and release of leukotrienes and SRS-A from the perfused guinea pig lung. Although the lipoxygenase blockers have proved to be useful experimental tools for the researcher there are at present no reports of their clinical use in man.

Drugs which inhibit cyclooxygenase and lipoxygenase enzymes: the 'dual blockers'

Agents in this group include phenidone and its structural analogue BW755C (see Fig. 2.9), benoxaprofen and eicosatetraynoic acid, an analogue of arachidonic acid in which all the double bonds are replaced with triple bonds. These agents are frequently called 'dual blockers', highlighting the fact that they block both cyclooxygenase and lipoxygenase enzymes. The IC_{50} values for BW755C and benoxaprofen against rat basophilic leukaemic cell enzymes are included in Table 2.5.

The discovery of dual-blocking drugs has raised the exciting possibility of a completely new approach to the treatment of inflammation. There is considerable evidence that leukotrienes like prostanoids have many of the characteristics of inflammatory mediators. In addition to their vasodilator

Table 2.5. IC_{50} values for some lipoxygenase inhibitors[a]

Drug	Lipoxygenase	Cyclooxygenase
NDGA[b]	0.2	> 30
Propylgallate	3.3	> 30
FPL 55712	20.6	> 100
Proxicrimil	139.4	> 300
BW755C	3.6	7.0
Benoxaprofen	7.8	14.6
Indomethacin	> 100	0.08
Flurbiprofen	> 100	0.05

[a] Potency of some cyclooxygenase, lipoxygenase and dual-blocking drugs as inhibitors of rat basophilic leukaemic cell cyclooxygenase and lipoxygenase enzyme activity. Results show the concentration of each drug required to produce 50% inhibition of enzyme activity measured in μmol/l.
[b] NDGA is nordihydroguiaretic acid.

property several of the leukotriene family are powerful chemotactic agents i.e. they attract phagocytic white cells to the damaged tissues. Thus, dual blockers which reduce the levels of both prostanoids and leukotrienes in the inflamed region would be expected to be better anti-inflammatory agents than NSAID which block only prostanoid formation. In theory at least the anti-inflammatory effect of dual blockers should be on a par with glucocorticoid drugs which also prevent prostanoid and leukotriene biosynthesis but by inhibiting phospholipase activity. It is to be hoped that the dual blockers will have fewer harmful side effects than the steroid anti-inflammatory drugs.

Of the dual blockers so far developed, BW755C has been most thoroughly studied. This compound is anti-inflammatory *in vivo*. Oral administration of BW755C to rats significantly reduced oedema due to injection of carrageenan into the paw and also decreased the accumulation of leucocytes in inflammatory exudates (Higgs, Flower & Vane, 1979*b*). Additionally, BW755C prevents the release of leukotrienes from antigen-challenged perfused guinea pig lungs which suggests that this agent may be of use in the treatment of conditions like asthma which may be caused, at least in part, by the release of large amounts of bronchoconstrictor leukotrienes from the lungs.

It is still too early to predict whether dual blockers will rival glucocorticoid steroids in the treatment of inflammatory diseases. Over-enthusiastic claims for these drugs should be avoided. However, the possibility of developing anti-inflammatory drugs with the potency of steroids but without their toxic side effects is indeed a very attractive one and should ensure that these drugs remain a focus for research for some time to come.

Drugs which inhibit thromboxane biosynthesis

Thromboxane synthetase, the enzyme which catalyses the formation of TxA_2 from prostaglandin endoperoxides, is inhibited by several chemical substances as shown in Table 2.6.

Fig. 2.9. Structure of BW755C.

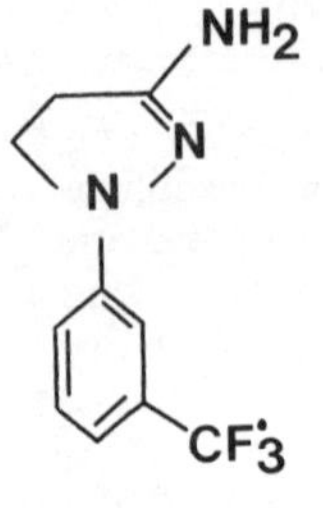

Of these drugs, dazoxiben has been most studied. An analogue of imidazole the full chemical name of dazoxiben is 4-(-2-(1H-imidazol-1-yl)ethoxy benzoic acid) and its structure is shown in Fig. 2.10. Pharmacologically, dazoxiben is a potent inhibitor of human platelet microsomal thromboxane synthetase (IC_{50} = about 3×10^9 M) whereas it has little or no effect on either cyclooxygenase or PGI_2 synthetase enzymes at concentrations as high as 10^{-4} M. Dazoxiben also prevents thromboxane formation *in vivo*. Intravenously injected dazoxiben in rabbits challenged with arachidonic acid prevents the build-up of TxB_2 in plasma and also blocked platelet aggregation. Similarly, dazoxiben given in a single oral dose reduced the plasma levels of TxB_2 in human volunteers (Tyler *et al.*, 1981).

Since dazoxiben blocks the formation of pro-aggregatory TxA_2 by platelets but leaves the biosynthesis of anti-aggregatory PGI_2 by the vascular endothelium unchanged, one obvious potential clinical use of this drug is in the treatment of diseases with thrombosis as the underlying cause. Some preliminary clinical experiments in patients with peripheral vascular disease have suggested that dazoxiben may have a promising future as an anti-thrombotic agent (see Chapter 4 for further details).

Drugs which prevent prostanoid catabolism

Classical prostaglandins of the E, F and D series are biologically inactivated by a sequence of enzyme steps. The first enzyme, which is also the rate-limiting enzyme, is 15-hydroxyprostaglandin dehydrogenase (15-PGDH) which converts these prostaglandins to their inactive 15-oxo metabolites. Inhibition of 15-PGDH would be expected to result in a build

Table 2.6. *Inhibitors of thromboxane synthetase*

9,11-azoprost-5,13-dienoic acid (a PG endoperoxide analogue)
N-0164 (a PG receptor blocker)
2-isopropyl-3-nicotinylindole
imidazole
pyridine
dazoxiben

Fig. 2.10. Structure of dazoxiben.

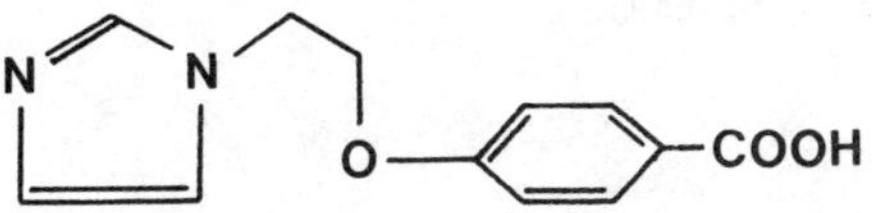

up of classical prostaglandins in the tissues. Several drugs have been identified which inhibit 15-PGDH (see Table 2.7. and Hansen, 1976).

Many NSAID have only a feeble ability to inhibit 15-PGDH. Concentrations 50–1000 times greater than those required to block cyclooxygenase are normally required. It is therefore most unlikely that any of the biological effects of NSAID can be attributed to inhibition of prostaglandin catabolism. Of the other 15-PGDH inhibitors, sulphasalazine and its analogues have received the most attention. Sulphasalazine represents a molecule of 5-aminosalicylic acid (5-ASA) linked to a molecule of sulphapyridine by an azo bond. Its structure is shown in Fig. 2.11. Sulphasalazine inhibits 15-PGDH in all broken-cell tissue preparations so far studied as well as in intact tissues like the perfused guinea pig lungs and in anaesthetised animals *in vivo* (Hoult & Moore, 1978). It is a fairly potent inhibitor of 15-PGDH with an IC_{50} value *in vitro* of between 30 and 60 μM. Sulphasalazine will inhibit both cyclooxygenase and lipoxygenase enzymes, although concentrations in excess of 10^{-3} M are required. Both sulphasalazine and carbenoxolone are used clinically to treat ulcerative disease affecting the colon and stomach respectively. Since several prostanoids exert an anti-ulcer effect in all parts of the gastrointestinal tract, it is tempting to speculate that these drugs may prevent ulcer formation in man by blocking gut 15-PGDH activity thereby increasing the concentration of endogenous protective prostaglandins in the gut mucosa.

PGI_2 and TxA_2 are almost certainly inactivated in the body by chemical hydrolysis to yield 6-oxo-$PGF_{1\alpha}$ and TxB_2 respectively. This chemical reaction cannot be influenced by drugs. However, both PGI_2 and TxA_2 bind to and are stabilised by plasma proteins which delay their chemical hydrolysis. The half-life of PGI_2, for example, in aqueous buffer at

Table 2.7. *Inhibitors of 15-hydroxyprostaglandin dehydrogenase (15-PGDH)*

Sulphasalazine and its analogues (e.g. dihomosulphasalazine)
Carbenoxolone
Diphloretin phosphate (DPP)
Indomethacin
Flurbiprofen

Fig. 2.11. Structure of sulphasalazine.

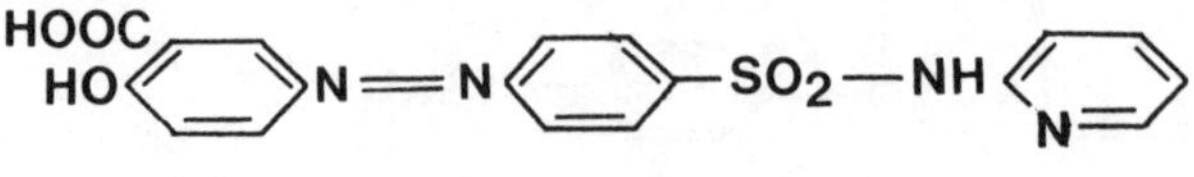

physiological pH and temperature is approximately 3 min whereas the half-life of PGI_2 incubated in human plasma under the same conditions may be in excess of 20 min. Similarly, by binding to plasma proteins, the half-life of TxA_2 in plasma is of the order of several minutes compared with only 30–40 s in aqueous buffer.

Effect of endogenous substances on eicosanoid synthesis and catabolism

Considering that prostanoids and leukotrienes exert such powerful biological actions on almost all physiological functions it is perhaps not surprising that the body should have evolved its own chemical control over arachidonic acid metabolism. Plasma from mammalian species has been found to contain uncharacterised protein factors which inhibit cyclooxygenase enzyme activity and increase the activity of the prostaglandin catabolising enzyme, 15-PGDH (Saeed et al., 1977; Moore & Hoult, 1980). In addition, human plasma and serum contain separate proteins which activate prostacyclin synthetase (called prostacyclin stimulating factor) and inhibit lipoxygenase enzyme activity (called endogenous inhibitor of leukotriene synthesis or EILS). These endogenous control mechanisms are still largely ignored by the majority of researchers even though the 'endogenous aspirin' in plasma could well have less side effects than its chemical cousin. Thus, either this substance or one of its analogues could, at some future date, provide an alternative to the currently available NSAID.

3

The cardiovascular system

Introduction

Prostanoids and leukotrienes have important actions on the cardiovascular system. Prostaglandins of the E and I series are usually vasodilator in most vascular beds while prostaglandin endoperoxides, F series prostaglandins, TxA_2 and leukotrienes are usually vasoconstrictor. The local formation of prostaglandins may be necessary for the regulation of blood flow within particular organs of the body. The effect of both cyclooxygenase and lipoxygenase metabolites of arachidonic acid on blood flow through specialised vascular beds can be found elsewhere in this volume e.g. ovary, uterus and neonate (Chapter 5), gastrointestinal tract (Chapter 6), lungs (Chapter 7), kidney (Chapter 8) and the microcirculation (Chapter 10). In this chapter, I shall concentrate on the action of prostanoids on the heart, coronary blood vessels and on blood pressure. The evidence that prostanoids are involved in disease states affecting the cardiovascular system will also be discussed. Further information may be found in the reviews of Wennmalm (1979), Nasjletti & Malik (1982) and Hirsch et al. (1981).

Biosynthesis and catabolism of prostanoids by vascular tissue

Vascular tissue displays a very low conversion (less than 1–2%) of exogenous arachidonic acid to prostanoids but avidly converts PGG_2 to PGI_2 (over 60% yield). This result suggests that blood vessels contain little cyclooxygenase enzyme activity but do utilise prostaglandin endoperoxides (possibly provided by platelets) to form PGI_2. Other prostaglandins, particularly PGE_2 and small amounts of $PGF_{2\alpha}$ and TxA_2 are also formed by some blood vessels (Forstermann, Herrting & Neufang, 1984) but it seems generally agreed that prostacyclin is the predominant cyclooxygenase product formed both by fresh vascular tissue and blood vessel microsomes. Factors which affect vascular PGI_2 formation include the animal's sex and age. Synthesis of PGI_2 by arteries of several species is greater in females than in age-matched male controls (Sinzinger et al. 1980). In addition, rat aortic PGI_2 biosynthesis is low in foetal animals and only begins to increase 4 weeks after birth to reach adult levels when the animal is about 10 weeks old (Deckmyn et al., 1983). A list of vascular tissues in which PGI_2 biosynthesis has been demonstrated is shown in Table 3.1.

There have been several attempts to determine which part of the blood vessel wall is responsible for PGI_2 biosynthesis. By carefully teasing away the layers of the blood vessel wall it has been possible to localise prostacyclin formation to the intimal layer of the wall (i.e. that part normally in direct contact with the blood). Much smaller amounts of PGI_2 are formed in the adventitial layer of the vascular wall (Moncada et al. 1977). Since the inner surface of all large blood vessels is lined with endothelial cells, this would suggest that these cells are an important site of prostacyclin biosynthesis. This conclusion has received support from several lines of evidence. Rabbit aorta which has been scraped on its inner (intimal) surface to remove the endothelial cells (de-endothelialisation) releases little or no PGI_2 on incubation. Secondly, vascular endothelial cells maintained in cell culture are an excellent source of PGI_2 (Weksler, Marcus & Jaffe, 1977). Taken together, these results point to the endothelial cell as the major site of PGI_2 formation in the blood vessel wall.

There have been fewer attempts to determine the profile of prostanoid biosynthesis formed by small blood vessels like capillaries, arterioles and venules which together make up the microcirculation. Presumably this is because of the technical difficulties involved in working with such tiny vessels. However, in one study using isolated rat brain microvessels (mainly

Table 3.1. *Examples of vascular tissue which form prostacyclin in vitro*

Tissue	Species
Aorta	Rabbit, pig, hamster, guinea pig, sheep, cow, dog, rat
Vena cava	Rabbit, pig, hamster, cow, rat
Mesenteric artery	Rabbit, pig, guinea pig, cow, dog, human, rat
Mesenteric vein	Rabbit, pig, guinea pig, cow, dog, sheep, human, rat
Femoral artery	Human
Femoral vein	Human
Coeliac artery	Rabbit, pig
Coronary artery	Rabbit, pig, cow, sheep, human
Renal artery	Rabbit, dog, human
Ductus arteriosus	Cow

capillaries) the major product formed from added arachidonic acid was again PGI_2 (Goehlert, Ng Ying Kin & Wolfe, 1981).

The heart also has feeble cyclooxygenase activity, as evidenced by the fact that cardiac microsomes convert only a fraction of added arachidonic acid to prostanoids. However, the intact beating heart does release substantial amounts of prostanoids into the cardiac effluent. Early studies in which the prostaglandins released were measured biologically suggested that the beating rabbit heart perfused with Krebs' solution by means of a cannula inserted retrogradely into the aorta (i.e. the Langendorff preparation) spontaneously released PGE_2, $PGF_{2\alpha}$ and a small amount of PGD_2. However, following the discovery of PGI_2 it was quickly realised that this prostaglandin, which had little effect on the bioassay preparations used in previous work, was the principal cyclooxygenase product of arachidonic acid released by the Langendorff-perfused guinea pig, rat and rabbit hearts (DeDeckere, Nugteren & Ten Hoor, 1977). In man also, infusion of radiolabelled arachidonic acid in to the aorta results in the appearance of radiolabelled PGI_2 and lesser amounts of radiolabelled PGD_2, PGE_2 and $PGF_{2\alpha}$ in coronary sinus blood. Whether PGI_2 biosynthesis is localised to any one part of the heart (e.g. atria or ventricles) or occurs at a similar rate in all cardiac tissue is not known. Certainly, the coronary arteries are one important site of cardiac PGI_2 formation.

Catabolism of vascular prostanoids has also been studied. Homogenised mesenteric arteries and veins contain 15-PGDH activity capable of biologically inactivating classical prostaglandins of the E and F series as well as PGI_2. Compared with organs like the lung, 15-PGDH activity in blood vessels and heart is negligible and probably plays no part in terminating the effect of these prostaglandins in the cardiovascular system. Vascular endothelial cells which synthesise and release PGI_2 are devoid of 15-PGDH activity. Interestingly, cytosolic supernates prepared from homogenised bovine mesenteric vein contain large amounts of a PG-9KR enzyme which converts PGE_2 to $PGF_{2\alpha}$ (Wong et al., 1977). It has been suggested that bradykinin, which dilates arteries but normally constricts isolated vein preparations, does so by stimulating venous PG-9KR – thus increasing the formation of $PGF_{2\alpha}$ which is a potent venoconstrictor agent (McGiff, 1981).

Vascular effects of prostanoids

PGE_2 and PGI_2 administered either intravenously or intraarterially to a number of different species, including man, produce a fall in peripheral resistance and blood pressure. Depending upon the species studied, PGI_2 may have up to 10 times the hypotensive potency of PGE_2. Since it is not inactivated by the lungs, PGI$_2$ is equipotent following both intraarterial and intravenous injection. This contrasts with E series prostaglandins which are broken down on passage across the pulmonary circulation and thus may be anything up to 100 times less potent (depending upon species and dose) when injected intravenously. E and I prostaglandins affect the heart rate of anaesthetised animals differently. PGE$_1$ and PGE$_2$, for example, always cause quickening of the heart (tachycardia) while PGI$_2$ causes the heart to slow (bradycardia). Since both elicit a fall in blood pressure, the expected effect on the heart would be a baroreceptor-mediated tachycardia as seen with PGEs. Possibly the anomalous bradycardia after PGI$_2$ injection may be caused by stimulation of cardiac sensory receptors,

Table 3.2. *Effect of prostanoids on the blood pressure and heart rate of anaesthetised animals*

Prostanoid	Species	Blood pressure[a]	Heart rate[a]	Comments
PGD$_2$		↑	↓	Not much studied (usually pressor)
PGE$_1$, PGE$_2$	All (to date)	↓	↑	Reflex tachycardia (baroreceptor)
PGF$_{1\alpha}$, PGF$_{2\alpha}$	Rabbit, cat,	↓	↑	
	rat, dog	↑	↓	Possibly due to venoconstriction
PGI$_2$	All (to date)	↓	↓	Bradycardia due to stimulation of cardiac reflexes to increase vagal tone
TxA$_2$, stable PGG$_2$ analogues	All	↑	↓	TxA$_2$ not studied directly (unstable)
PGG$_2$/PGH$_2$	All	↓	↑	Not a direct action (converted to PGI$_2$)
15-oxo-PGEs 15-oxo-PGFs 6-oxo-PGF$_{1\alpha}$ TxB$_2$	All	0	0	Negligible activity unless dose is increased 100–1000 × that of parent PG
6-oxo-PGE$_1$	All (to date)	↓	0↑	Not extensively studied – less potent than precursor PGI$_2$
PGA$_1$, PGA$_2$	All	↓	↑	Not naturally occurring but potent hypotensive agents

[a] ↑ Indicates increase; ↓ indicates decrease; 0 indicates no effect.

reflexly increasing vagal tone to lower heart rate. This would explain why atropine, a muscarinic cholinoceptor antagonist, prevents PGI_2-induced bradycardia in some species (Chapple et al., 1978).

Other prostanoids and their metabolites have also been examined for their effect on the cardiovascular system of intact animals. The results obtained are included in Table 3.2.

With one or two exceptions, the results obtained after prostanoid injection or infusion into whole animals correspond to those obtained in isolated vascular preparations. Thus, E and I prostaglandins relax vascular strips in vitro while PGFs and TxA_2 contract vascular preparations. The prostaglandin endoperoxides frequently have biphasic effects on vascular tissues in vitro. For example, the isolated bovine coronary artery is first contracted and then relaxed by PG endoperoxides. The initial transient contraction is probably a direct effect of the endoperoxide while the latter sustained relaxation has been attributed to PGI_2 formed from the PG endoperoxide by vascular PGI_2 synthetase activity (Dusting, Moncada & Vane, 1977). The action of PGE_2, PGI_2, TxA_2 and PGG_2 on several vascular strips set up in vitro is summarised in Table 3.3.

It is clear that prostanoids have marked pharmacological activity on the vasculature. There is evidence also that some prostaglandins, particularly PGI_2, may have a physiological role to play in the normal functioning of the cardiovascular system. The simplest way to estimate the physiological significance of cardiovascular prostaglandins is to study the way in which aspirin and other NSAID affect both systemic blood pressure and blood flow through individual organs. Experiments of this type have revealed that, in anaesthetised animals as well as healthy and hypertensive humans, inhibition of body prostanoid biosynthesis causes a substantial rise (more than 20% of resting blood pressure, depending upon species and the particular NSAID used) in blood pressure. In man the major areas of

Table 3.3. *Actions of prostanoids on vascular smooth muscle in vitro*[a]

Tissue	PGE_2	PGI_2	TxA_2	PGG_2
Dog renal vein	↑	↓↓	—	↑↑
Rabbit aorta	—	—	↑↑↑	↑
Rabbit mesenteric/coeliac artery	↓	↓↓↓	↑↑↑	↑↓
Dog, human or bovine coronary artery	↑	↓↓↓	↑↑↑	↑↓

[a] ↑ Indicates contraction, ↓ indicates relaxation, — no effect or not tested. The number of arrows gives a crude indication of the potency of the prostanoid.

vasoconstriction are the gut and kidney (Nowak & Wennmalm, 1978). It may be concluded that the continuous formation of PGI_2 exerts a vasodilator 'tone' in these vascular beds which, when removed (after aspirin treatment), causes a rebound vasoconstriction and rise in blood pressure. The existence of a naturally occurring anti-hypertensive prostanoid has been postulated for several years. At one point it was believed that PGA_2 formed by the kidney entered the bloodstream and circulated around the body as a hypertensive hormone. It is now clear that PGA_2 does not exist naturally in the body other than as an artifact of PGE_2 formed following extraction. Furthermore, PGI_2 is unlikely to be a circulating hormone; rather it is synthesised by vascular endothelial cells, causes vasodilation and is then broken down either chemically or enzymatically at the same site.

There is convincing evidence that prostanoids regulate blood flow through individual vascular beds (see review by Messina, Weiner & Kaley, 1976 for further information). Kidney formation of PGI_2 (and possibly PGE_2) helps to control intrarenal distribution of blood flow between the medulla and the cortex.

Two vascular beds which may also be regulated by endogenous prostanoids are the coronary and skeletal muscle circulations. In both cases, hypoxia causes a hyperaemia (i.e. the coronary/skeletal muscle vessels dilate, thereby increasing blood flow to the heart/skeletal muscle to prevent ischaemia and tissue damage). Aspirin-like drugs reduce hypoxia-induced hyperaemia in both the coronary arteries of the anaesthetised, open-chest dog and in the skeletal muscles of the human forearm (Afonso, Barchow & Rouse, 1974). Since PGI_2 relaxes isolated coronary artery preparations *in vitro* and induces coronary vasodilatation in dogs (Dusting *et al.*, 1978) as well as dilating skeletal muscle blood vessels (Messina, Weiner & Kaley, 1977) it seems likely that this prostaglandin may help to autoregulate coronary and skeletal muscle blood flow *in vivo*.

Effect of prostanoids on the heart

When administered to the whole animal the effect of prostanoids on heart rate is largely mediated through the vagus nerve. However, some prostaglandins do affect the contractility of the heart directly. In high doses PGE_1, PGE_2, $PGF_{1\alpha}$ and $PGF_{2\alpha}$ injected retrogradely into the aorta of Langendorff-perfused hearts of the frog, rat, dog, guinea pig and chicken exert a powerful positive inotropic effect (increase in force of contraction). Prostaglandins are very potent. On a weight-for-weight basis, PGE_1 is roughly equipotent with adrenaline in increasing the force of contraction of the spontaneously beating guinea pig heart. In contrast, prostaglandins have little or no chronotropic effect i.e. they do not change the beating rate of the heart *in vitro* in most species. Of the more recently discovered chemically unstable prostanoids, prostaglandin endoperoxide analogues also increase myocardial contractility *in vitro* (Rose et al., 1976) whilst PGI_2, the major prostanoid formed by the heart, has only weak and variable actions on the heart *in vitro*.

Prostanoids and pathological conditions which affect the cardiovascular system

Hypertension

Since PGE_2, PGI_2 and 6-oxo-PGE_1 are vasodilator, cause diuresis and renin release and modify noradrenaline output from sympathetic nerves, a deficiency of one or more of these substances has been proposed to contribute to the aetiology of essential hypertension in man. There have been numerous attempts to estimate prostanoid biosynthesis in humans with this condition. Most experiments of this sort have relied heavily upon measurements of urinary excretion of prostanoids which probably reflects kidney prostanoid turnover rather than whole-body prostanoid turnover. Patients with essential hypertension have reduced urinary output of PGE_2 and 6-oxo-$PGF_{1\alpha}$ with essentially normal TxB_2 excretion (Tan, Sweet & Mulrow, 1977). Thus, an imbalance in the cardiovascular formation of PGI_2 and TxA_2 in favour of vasoconstrictor TxA_2 may occur in patients with essential hypertension. Whether such an imbalance causes the hypertension or is a response of the body to the hypertension is not known. Pharmacological manipulation of the $PGI_2:TxA_2$ balance can be achieved either by administration of PGI_2 (or a stable analogue) or by selectively preventing TxA_2 biosynthesis, using a drug like dazoxiben. As yet there have been no reports of the use of dazoxiben or any other inhibitor of thromboxane synthetase in humans with essential hypertension. In contrast, PGI_2 has been infused intravenously into healthy human volunteers and brings about a graded fall in systemic and pulmonary arterial pressure (Szczeklik, Szczeklik & Nizankowski, 1980). PGI_2 has been used with some degree of success in a few patients with pulmonary hypertension (Watkins et al., 1980).

Because of the many problems which dog the accurate estimation of prostanoids in man, many researchers have preferred to study arachidonic acid metabolism in organs removed from hypertensive animals. In some cases, so-called spontaneously or genetically hypertensive rats have been used whereas in other cases hypertension has been induced surgically by removing one kidney and clipping (thereby partially occluding) the renal artery supplying the remaining kidney. The result is called a Goldblatt hypertensive rat, after the originator of the technique. It should be emphasised that animal models of hypertension are unlikely to match satisfactorily the vastly more complicated clinical picture seen in humans with essential hypertension. Results from animal experiments must therefore be interpreted cautiously. In general, vascular tissues removed from hypertensive animals synthesise more PGI_2 than age- and sex-matched controls with normal blood pressure (Pace-Asciak *et al.*, 1978). Clearly the results from hypertensive animals and hypertensive humans do not agree. In animal models PGI_2 biosynthesis is increased whilst in humans the formation of this prostanoid is depressed. It is possible to reconcile these findings. A deficient formation of PGI_2, particularly in the kidney – where it causes vasodilatation and natriuresis and blunts the vasoconstrictor, hypertensive effect of noradrenaline and angiotensin II – may accelerate the early stages of the development of hypertension. However, once the hypertensive state has become fully established (as in animal models of the disease) then vascular PGI_2 production is elevated, presumably as an adaptive response.

Although the precise role of prostanoids in human essential hypertension is not yet understood it does seem that some common, clinically used anti-hypertensive drugs may owe part of their ability to lower blood pressure to stimulation of PGI_2 biosynthesis in the body. This conclusion is based upon experiments in which pre-treatment of animals or human volunteers with NSAID reduces the anti-hypertensive effect of such pharmacologically diverse drugs as propranolol (β-blocker), verapimil (Ca^{2+} channel antagonist), frusemide (diuretic), captopril (converting enzyme inhibitor which blocks formation of angiotensin II) and clonidine (α-agonist). It is assumed that all of these drugs share the common ability to promote vascular PGI_2 formation. It is not known at which stage of the arachidonic acid cascade each of these drugs acts.

Angina pectoris

This condition is one of crippling pain in the chest caused by a lack of oxygen reaching the cardiac muscle. The prime cause of cardiac anoxia is a fall in coronary blood flow due to a combination of atherosclerosis closing the lumen of the coronary artery and coronary vasoconstriction. Prostanoids may play a part in both processes. When platelets aggregate they release TxA_2 which promotes further platelet aggregation resulting in the formation of a platelet plug which may by itself occlude the coronary artery and cut off blood to the heart. Aggregating platelets also release substances which exacerbate atherosclerosis (see Chapter 4). Additionally, the released TxA_2 may constrict the coronary artery thereby further reducing blood flow to the heart. There is considerable evidence that platelets removed from patients with angina pectoris are hyperaggregable and release larger amounts of TxA_2 when they aggregate than platelets from healthy control subjects. Whether TxA_2 is an important mediator of angina pectoris is disputed. Although peripheral venous TxB_2 concentrations do increase dramatically in patients a few minutes after an anginal attack, no change in TxB_2 level occurs just before the onset of the attack (Lewy *et al.*, 1979). Consequently it has been proposed that TxA_2 is formed as a result of cardiac ischaemia and angina but does not by itself precipitate angina. This is supported by clinical observations of angina patients pretreated with aspirin or indomethacin (Robertson *et al.*, 1981). In these studies, inhibition of cyclooxygenase activity to prevent the biosynthesis of TxA_2 in the heart did not affect the severity or discomfort of anginal pain in volunteers.

Prostacyclin has been assessed clinically for an effect in patients with angina pectoris. To date there have been few blind clinical trials comparing PGI_2 with other standard anti-anginal drugs like nitroglycerin. However, clinical studies with PGI_2 and PGE_2, albeit using relatively few patients, have produced some promising results. Both prostaglandins decrease the incidence and severity of spontaneous anginal attacks when administered as an intravenous infusion. Side effects included flushing of the face, and sometimes, headache. Further randomised trials are required to determine whether prostaglandins can be utilised routinely to treat this condition.

Myocardial infarction and cardiac arrhythmias

Numerous studies in experimental animals have revealed that large amounts of PGE_2, $PGF_{2\alpha}$, TxA_2 and PGI_2 are released from the heart during myocardial infarction. Exactly what function these prostanoids serve once released is not clear. However, indomethacin treatment causes a fall in coronary blood flow, suggesting that at least one action of the released PGI_2 (and possibly PGE_2) is to dilate the coronary arteries and thus increase blood flow to the myocardium. In addition, released prostanoids may limit myocardial ischaemia by other mechanisms e.g. stimulating myocardial adenylate cyclase and/or reducing noradrenaline release from sympathetic nerves to the heart. The effect of prostanoids on cardiac arrhythmias has also been extensively studied. In animals, arrhythmias can be induced either chemically – by injection of agents like calcium chloride, aconitine or ouabain – or surgically, by occluding the coronary artery. Both techniques have been used in a number of species including the cat, rat, rabbit and dog. The coronary-artery-ligation model probably bears a closer resemblance to cardiac arrhythmias and myocardial infarction in human beings.

Prostanoids have very complicated effects on cardiac arrhythmias. In most species, PGE_2, $PGF_{2\alpha}$ and PGI_2 diminish the number of chemically induced arrhythmias but the effect on surgically induced arrhythmias is more variable, both from one species to the next and from one laboratory to the next. For example, PGI_2 exacerbates arrhythmias in cats but in dogs this prostaglandin is an effective anti-arrhythmic agent (Au et al., 1979). In the rat, PGI_2 has been reported to increase (Coker & Parratt, 1981) and decrease the incidence of cardiac arrhythmias (Au et al., 1979). One reason for this variability is likely to reside in the different procedures used to quantify the arrhythmias by different research groups. However, since most prostanoids are anti-arrhythmic in most species, it has been suggested that these substances act as endogenous anti-arrhythmic agents. If this is the case then it may explain why massive amounts of prostanoids are released from the heart after myocardial infarction.

Peripheral vascular disease

This clinical condition is characterised by ischaemic pain in the extremities. It is usually caused by arteriosclerotic damage to those arteries which provide blood to the affected parts. A lack of blood flow frequently leads to the formation of ischaemic ulcers in the arms, legs and/or toes and, in severe cases, can lead to gangrene and mummification of the tissues. In open trials both PGE_1 and PGI_2, infused either intravenously or directly into the artery supplying the affected region, have been shown to provide relief of ischaemic pain and improved ulcer healing. The most dramatic demonstration of the beneficial effect of PGI_2 so far reported has come from the experiments of Gryglewski and his colleagues in Poland. These researchers infused PGI_2 into patients with advanced arteriosclerosis obliterans (gangrene) of the legs and who otherwise would have had the affected limb amputated (Szczeklik et al., 1979). A significant number of patients so treated were clinically improved to such an extent that amputation was avoided. Similarly encouraging results have been reported by other groups (e.g. Olsson, 1980). However, both of these studies were open-ended, uncontrolled trials in which PGI_2 was not compared with placebo treatment. It is therefore possible that some of the beneficial effects reported following PGI_2 therapy in these patients may be due to the psychological boost to the patients morale, especially since this treatment was probably offered to the patients as the last resort before surgery. Proper placebo-controlled blind trials with a sufficiently large number of patients are desperately needed. Preliminary results from one such trial, albeit using only 28 patients with peripheral vascular disease, have proved encouraging. In this study 15 patients were infused intravenously over a 4-day period with PGI_2 (about 7 ng/kg/min) and the remainder with placebo. In the PGI_2 group, 9 patients reported a reduction in pain which persisted for at least 1 month whereas only 2 patients receiving placebo were clinically improved at this time (Belch *et al.*, 1983). Obviously more trials of this sort are required. PGI_2 also causes clinical improvement in patients with Raynaud's phenomenon. Intravenous injection of PGI_2 for 3 days resulted in increased skeletal muscle blood flow, improved exercise tolerance and pain relief.

The mechanism by which PGI_2 and PGE_2 are beneficial in peripheral vascular disease is unknown. Since both of these prostaglandins inhibit platelet aggregation, they may disaggregate platelets lodged in and blocking capillaries in the damaged regions thereby improving blood supply. Alternatively, PGI_2 and PGE_1 may simply dilate blood vessels to increase blood flow and prevent ischaemia. Whatever the precise mechanism of action, both of these prostaglandins may have an important future in the treatment of this condition which until now has been largely resistant to drug therapy.

4
Prostanoids and platelets

Introduction

Platelets are specialised blood elements which circulate in the blood stream of mammals. Disc-shaped, roughly 2–3 μm in diameter and 0.7 μm thick, human blood contains about 150000–400000 platelets per microliter. Platelets have a very important part to play in cardiovascular haemostasis. By binding to damaged areas of the blood vessel wall and aggregating together they form a 'plug' and so prevent the loss of large volumes of blood from the circulation. Platelets are very much 'Jeckyll and Hyde' cells since, when activated, they release substances which cause thickening of the blood vessel wall or atherosclerosis; this is the major underlying cause of thrombotic diseases like stroke and heart attacks. Much research in the last few years has focussed upon the haemodynamic and chemical factors which regulate platelet reactivity *in vivo* since, if these are understood, there may be a chance of developing drugs capable of preventing platelet aggregation; these could be used to treat thrombotic disease. Haemodynamic factors like blood flow rate, haematocrit and viscosity are important determinants of platelet reactivity physiologically but can not readily be changed by drugs without affecting the rest of the cardiovascular system. In contrast, chemical substances which aggregate platelets in the body can be blocked by drugs. Important aggregating agents include adenosine diphosphate (ADP), collagen, thrombin and platelet activating factor (PAF). The finding that arachidonic acid also produced platelet aggregation both *in vivo* and *in vitro* and that aspirin-like drugs could reduce platelet aggregation to collagen suggested that one or more prostanoids might have a role to play in regulating platelet function.

The involvement of arachidonic acid metabolites in the control of platelet reactivity will be discussed in this chapter. This subject has been extensively reviewed (Malmsten, 1979; Moncada & Vane, 1979; Moncada,

1980; Mustard & Kinlough-Rathbone, 1980; Silver *et al.*, 1980; de Gaetano, 1981).

Platelet structure

Before embarking on a discussion of platelet adhesion and aggregation and the part that prostanoids play in these processes, it is worthwhile to consider a few aspects of the anatomy of the platelet. A cross-section through a human platelet, as seen by electron microscopy and magnified many thousands of times, is shown in Fig. 4.1.

Platelets are enclosed by a bilayer membrane to which is absorbed numerous plasma constituents including blood coagulation factors. Within the platelet are intracellular organelles such as mitochondria, lysosomes, stores of glycogen as well as alpha (α) granules (about 80 per human platelet) and darkly stained osmiophilic 'dense' granules (about 20 per human platelet). Platelets have no nucleus, RNA or DNA and thus cannot synthesise proteins. For this reason, platelets are not properly described as cells; they will survive in the bloodstream for only a short time before being destroyed in the liver. The half-life of platelets in the human circulation is about 8 days.

From a functional viewpoint, the most interesting intracellular components of the platelet are the α-granules and dense granules which are packed with biologically active substances. These granules are released

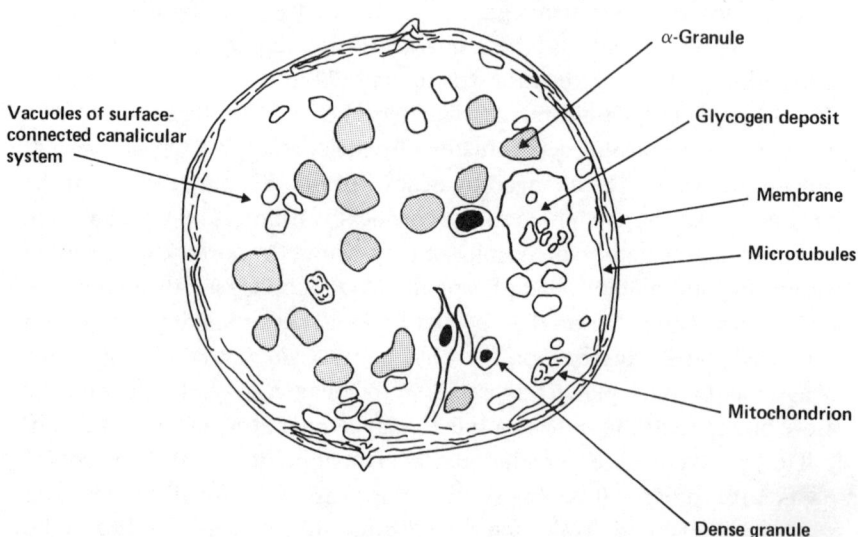

Fig. 4.1. Representation of a human platelet as seen using the technique of electron microscopy (magnified about 100000 times).

during platelet aggregation. Dense granules contain such chemically dissimilar substances as ions (K^+, Ca^{2+}), amines (serotonin, catecholamines) and macromolecules such as platelet-derived growth factor (PDGF). α-granules contain ADP, ATP, pyrophosphate, anti-plasmin, platelet factor IV and other factors which promote blood coagulation. Once released when platelets aggregate, these substances perform a range of biological functions; these include acceleration of further platelet aggregation, initiation of blood coagulation, prevention of fibrinolysis, vasoconstriction of local surrounding blood vessels and finally repair of the damaged blood vessel wall by stimulating the growth of new smooth muscle cells. All of these effects combine together to ensure that the blood loss from a torn vessel is minimised.

Platelet aggregation and adhesion
Platelets do not normally adhere to healthy vascular endothelium but will do so if the intimal layer of the blood vessel is damaged to expose subendothelial collagen fibres. Once adhesion has taken place, platelets release the contents of their intracellular granules (the platelet-release reaction) and form a platelet plug.

Most attention has been directed towards understanding the mechanism of platelet aggregation but, despite much research, this process remains very much a mystery. Most experimenters have studied platelet aggregation *in vitro* using a device called a platelet aggregometer which measures platelet aggregation in a sample of centrifuged anticoagulated blood. In principle the platelet aggregometer measures aggregation by determining the increase in light transmittance which occurs when platelets clump together. The sequence of events during platelet aggregation as studied in an aggregometer is shown in Fig. 4.2. When an aggregating agent is added, platelets become spherical and covered in both short and long protrusions resembling pseudopodia. This is referred to as the platelet-shape change (stage b) and is seen as a transitory increase in light transmittance. Thereafter primary platelet aggregation occurs (stage c) which may spontaneously reverse with individual platelets disentangling themselves (stage d) or alternatively move on to a secondary platelet aggregation phase where transmittance of light is increased even further and which is irreversible (stage e). Secondary aggregation is due to the release of ADP and other pro-aggregatory substances from the platelet and also to the synthesis by the platelet of aggregatory prostanoids like PGG_2, PGH_2 and TxA_2.

Platelets contain an actinomyosin-type protein called thrombasthenin which, when activated by Ca^{2+} ions, contracts in much the same way that smooth muscle contracts in the presence of high intracellular Ca^{2+}. The 'contraction' of the platelet causes the extrusion of the contents of the platelet granules and thereby leads to further platelet aggregation. The availability of Ca^{2+} ions inside the platelet is thus critically important for determining whether platelet aggregation occurs. In fact making this cation available in the cytoplasm for intracellular contractile proteins appears to be the end result of all agents which cause platelet aggregation. The Ca^{2+} ions necessary for platelet aggregation are already contained in the platelets' own dense granules, and thus (unlike smooth or skeletal muscle fibres) platelets do not require external Ca^{2+} ions in order to aggregate. The ability of calcium to activate the enzyme myosin kinase, which in turn causes contraction of thrombasthenin, is under the control of the adenylate cyclase/cAMP system. In the presence of a high concentration of cAMP within the platelet, myosin kinase is phosphorylated and in this state cannot bind Ca^{2+} ions and thus platelet thrombasthenin contraction and aggregation is blocked. When the concentration of platelet cAMP is

Fig. 4.2. Aggregation of human platelets induced by ADP showing the typical changes in light transmittance as determined using a platelet aggregometer (left) correlated with changes in platelet structure shown on the right hand side of the figure. (*a*) Normal, non-aggregating platelets, cigar-shaped with vesicles scattered randomly throughout the cytoplasm, (*b*) 'shape change', platelets develop pseudopodia, (*c*) primary aggregation, platelets contact each other through 'loose' fibrinogen/Ca^{2+} bridges, and (*d*) platelets drift apart or (*e*) undergo secondary aggregation following release of pro-aggregatory agents from platelet granules.

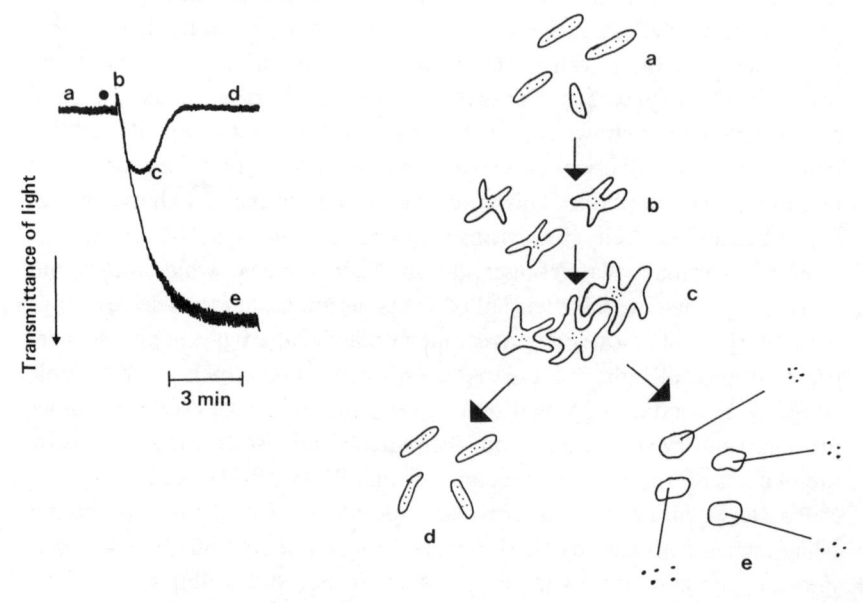

reduced myosin kinase remains unphosphorylated, binds Ca^{2+} ions avidly and platelet thrombasthenin contraction and aggregation ensues. Thus, high platelet cAMP levels lead to an inhibition of platelet aggregation while low platelet cAMP levels result in platelet aggregation.

Biosynthesis and catabolism of prostanoids by platelets

Platelets from humans and other species contain large amounts of both cyclooxygenase and lipoxygenase enzymes. The major cyclooxygenase product is TxA_2, although small amounts of PGE_2, $PGF_{2\alpha}$ and PGD_2 are also formed. Although not so well investigated, platelet lipoxygenase metabolites include 5-HETE and 12-HETE and their unstable hydroperoxy precursors (5-HPETE, 12-HPETE). Platelet cyclooxygenase has yet to be purified to homogeneity but seems to be similar if not identical in form and function to bovine seminal vesicle cyclooxygenase. The platelet enzyme occurs in the microsomal fraction (probably in the dense intracellular canalicular system), contains subunits of about 70 000 daltons and requires oxygen and haem for activity. In contrast, thromboxane synthetase (which is also membrane-bound) has been partially purified and characterised from bovine platelet microsomes (Yoshimoto & Yamamoto, 1982).

In addition to enzymes which synthesise prostanoids, platelets also contain catabolising enzymes such as 15-PGDH, PG-9KR and PG-9HDH. All of these enzymes are cytoplasmic. Their physiological function in the platelet, if any, is not known. Since most prostaglandins do not readily cross the platelet membrane it is unlikely that these enzymes inactivate circulating prostaglandins. Furthermore, the biological activity of PGE_2 does not decline when incubated for several hours in human blood or plasma. Presumably the major role of these prostaglandin-catabolising enzymes is to 'mop up' any excess prostanoids formed in the platelet. Platelet PG-9HDH also converts PGI_2 to 6-oxo-PGE_1.

Prostaglandin endoperoxides, TxA_2 and platelet aggregation

The possibility that prostanoids could cause platelets to aggregate together was suggested by the experiments of Silver *et al.* (1973) who found that arachidonic acid caused rapid and irreversible platelet aggregation when added to human platelet-rich plasma (PRP). No other unsaturated fatty acid produced a similar effect. Addition of thrombin to human PRP also caused platelet aggregation and resulted in the appearance in PRP of PGE_2 and $PGF_{2\alpha}$, presumably formed from aggregating platelets. Aspirin pre-treatment inhibited thrombin-induced platelet aggregation and abolished the formation of both prostaglandins which led to the proposal that the two events were causally related i.e. a prostaglandin was released by platelets which caused platelet aggregation. Since neither PGE_2 nor $PGF_{2\alpha}$ directly affected platelet function (Kloeze, 1967), another cyclooxygenase product had to be responsible. Willis and his colleagues (1974) provided evidence for the existence of a chemically unstable product(s) of arachidonic acid intermediate in the formation of PGE_2 and $PGF_{2\alpha}$ which had powerful platelet-aggregatory activity. This metabolite(s) was named labile aggregation stimulatory substance(s) or LASS for short. It is now clear that LASS is a mixture of PGD_2, PGH_2 and TxA_2 and it is these three prostanoids (or a combination thereof) which are responsible for arachidonic-acid-induced platelet aggregation.

Addition of PGG_2 or PGH_2 (10–300 ng/ml) to stirred human PRP induced platelet aggregation and the platelet-release reaction. PGG_2 was about 3 times more potent than PGH_2 and aggregation was not altered by aspirin or indomethacin pretreatment (Hamberg *et al.*, 1974). In a later study, Hamberg, Svensson & Samuelsson (1975) demonstrated that aliquots of PRP incubated with either arachidonic acid or PGG_2, if transferred to a fresh sample of PRP, caused platelets in that sample to aggregate together. However, this 'transferability' was lost if there was a long delay before aggregating platelets were added to fresh PRP. In fact the aggregating ability of both arachidonic acid and PGG_2 declined with the same half-life as TxA_2 (about 30 s). This result was interpreted to mean that TxA_2 was mainly responsible for platelet aggregation induced by arachidonic acid.

Whether PG endoperoxides aggregate platelets in their own right or must first be enzymatically converted to TxA_2 has been a controversial issue. The rapidity with which exogenous PGG_2 causes platelet aggregation and the release reaction when added to human PRP (Claesson & Malmsten, 1977), coupled with the powerful aggregatory activity of synthetic PGH_2 analogues like U46619 which do not get converted to TxA_2, suggests that PG endoperoxides do have intrinsic platelet-aggregatory activity. Theoretically, an excellent way to determine whether PG endoperoxides cause platelet aggregation would be to study PGG_2-induced platelet aggregation in the presence of a selective thromboxane synthetase inhibitor. If it could be demonstrated that PGG_2 still aggregated platelets in the presence of such a drug then this would suggest that it did not need to be converted to TxA_2 to cause aggregation. Several such experiments have been performed but, unfortunately, conflicting results have been reported. For example when imidazole was used to inhibit thromboxane synthetase activity, TxA_2 biosynthesis was found to be unnecessary for platelet aggregation (Needleman *et al.*, 1977). In contrast, when thromboxane synthetase activity was prevented with 9,11-azoprost-5,13-dienoic acid, it was concluded that TxA_2 formation was a prerequisite for PGG_2-induced platelet aggregation (Gorman et al., 1977a).

Even more confusing is the finding that dazoxiben, at micromolar concentrations, virtually abolished TxA_2 formation (measured by RIA of TxB_2) in PRP from human volunteers but inhibited platelet aggregation in only a few of the subjects tested. These subjects were labelled 'responders' (Heptinstall et al., 1980). In other individuals (the 'non-responders') platelet aggregation was normal. A possible explanation for these results may be afforded by considering how thromboxane (Tx) synthetase inhibitors affect the formation of other metabolites of arachidonic acid in the platelet. In the presence of dazoxiben, formation of TxB_2 by platelets is virtually abolished in all subjects. This is accompanied by a redirection in PG endoperoxide catabolism towards PGE_2 and PGD_2, the latter having anti-aggregatory activity. Thus when platelet Tx synthetase activity is prevented in the subject who is a 'non-responder' then build-up of PG endoperoxides in the platelet is sufficient to maintain a perfectly normal platelet-aggregation response. However, in 'responder' subjects, PG endoperoxides which accumulate in the platelet are converted to PGD_2. The formation of anti-aggregatory PGD_2, coupled with the removal of pro-aggregatory PGG_2 and PGH_2, combine to reduce or prevent the platelet-aggregation response in that individual.

Other approaches have been used to assess the relative importance of PG endoperoxides and TxA_2 for platelet aggregation. One of these has been to block TxA_2 receptors on the platelet with a prostaglandin analogue such as 13-azoprostanoic acid. This substance prevents arachidonic-acid-induced human platelet aggregation (Bertele et al., 1981), suggesting that TxA_2 is the mediator of platelet aggregation. However, this conclusion must be tempered somewhat by the finding that 13-azoprostanoic acid also reduces platelet aggregation due to U46619, suggesting that it blocks not only receptors for TxA_2 but also those for the PG endoperoxides. Since PGG_2, PGH_2 and TxA_2 probably act on the same receptor this is not surprising. Consequently, such experiments cannot be used to distinguish PGG_2 from TxA_2.

A further approach has been to identify and characterise individuals with a congenital deficiency of platelet thromboxane synthetase. If TxA_2 is vitally important for platelet aggregation then such individuals should exhibit much reduced platelet reactivity both in vitro and in vivo. Machin et al. (1981) studied one such family. Compared with normal control subjects each member of this family showed only a feeble platelet aggregation response to arachidonic acid and their platelets formed much smaller amounts of TxB_2 during aggregation. Clinically each had a moderate bleeding tendency and a prolonged bleeding time. Thus, even in patients devoid of TxA_2 synthesis, platelet reactivity is reduced but still occurs; this suggests that, at least in these individuals, PG endoperoxides are capable of causing platelet aggregation in the absence of TxA_2.

In summary, it appears that PGG_2 and PGH_2 are normally converted to TxA_2, which is responsible for platelet aggregation. If this conversion does not take place, because a thromboxane synthetase inhibitor or TxA_2 receptor blocker is present or because of a genetic deficiency of the enzyme which forms TxA_2, then platelet aggregation will still procede brought about by PG endoperoxides.

Interaction of prostanoids with other important aggregating agents

As we have seen, platelets are aggregated by exposure to arachidonic acid *in vitro*. This occurs when platelets enzymatically convert arachidonic acid to PGG_2 and TxA_2 which are both potent aggregating agents. Virtually all of the endogenous arachidonic acid in the body is bound in membrane phospholipids and thus is unlikely to be an important trigger for platelet aggregation in the circulation. Physiologically important platelet-aggregating agents *in vivo* include collagen, thrombin, ADP and possibly adrenaline and platelet-activating factor (PAF). An important question is, therefore, do any of these substances cause platelets to

Fig. 4.3. The three tiers of platelet aggregation.

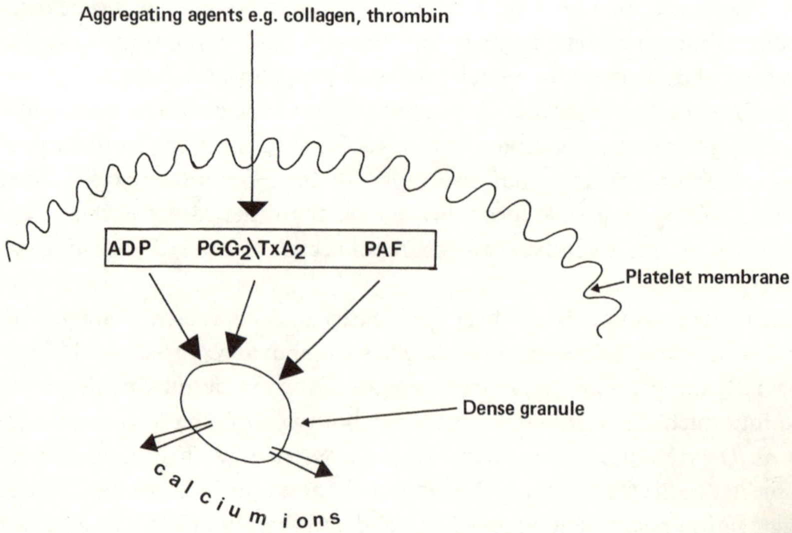

aggregate by stimulating the formation of PGG_2 and/or TxA_2 by the platelet itself?

There is good evidence that platelet aggregation and the platelet-release reaction induced by collagen and thrombin can be inhibited by pre-treating the platelets with aspirin and other NSAID (O'Brien, 1968), thromboxane synthetase inhibitors (Kam et al., 1979) and TxA_2 receptor antagonists (Fitzpatrick et al., 1978). Thus the platelet response to these two agonists at least is dependent upon their ability to stimulate platelet TxA_2 biosynthesis. This stimulation is brought about by activating platelet phospholipase enzymes, thereby releasing free arachidonic acid for conversion. In this way, platelet aggregation may be considered a three-tier mechanism (Fig. 4.3). According to this scheme collagen, thrombin and adrenaline are 'first messengers', producing platelet aggregation which is boosted by one or more of the 'second messenger' systems i.e. ADP released from α-granules, or PGG_2/TxA_2, or PAF synthesised in the process of platelet aggregation. It is likely that all three 'second messenger' systems ultimately cause mobilisation of Ca^{2+} from intracellular dense granules, a phenomenon initiated by decreasing intracellular levels of cAMP (for review, see Vargaftig, Chignard & Benveniste, 1981).

Prostaglandins as inhibitors of platelet aggregation and adhesion

PGI_2, PGE_1, PGD_2 and 6-oxo-PGE_1 are potent inhibitors of human platelet aggregation, as shown in Fig. 4.4. The concentration required to produce 50% inhibition of ADP-induced human platelet aggregation are 0.3 ng/ml, 46 ng/ml, 20 ng/ml and 4.0 ng/ml respectively. Thus PGI_2 is some 15 times more potent that 6-oxo-PGE_1 and over 60 times more potent than PGD_2. These potency ratios depend, to a large extent, upon the species studied and the choice of aggregating agent. For example, although PGD_2 prevents human platelet aggregation it has little or no anti-aggregatory activity if rat or rabbit platelets are used (Smith, Ingerman & Silver, 1976). Similarly PGI_2 is 6 times more potent than PGD_2 in inhibiting adrenaline-induced human platelet aggregation but only 1.3 times more potent when arachidonic acid is used as the aggregating agent (DiMinno, Silver & De Gaetano, 1979).

In addition to the results of such studies *in vitro*, PGI_2 has been shown to produce a dose-dependent inhibition of platelet aggregation when infused intravenously into human volunteers. The threshold inhibitory effect occurs at between 2 and 4 ng/kg/min and disappears within 20–30 min of stopping the infusion (Fitzgerald et al., 1979). Inhibition of platelet aggregation in these studies was well correlated with an increase in platelet cAMP concentration, as shown in Fig. 4.5. In this particular

Fig. 4.4. Inhibition of platelet aggregation *in vitro* by prostaglandins. PGI_2 (circles), 6-oxo-PGE_1 (filled squares), PGD_2 (triangles) and PGE_1 (open squares). Prostacyclin is the most potent naturally occurring anti-aggregatory substance discovered to date.

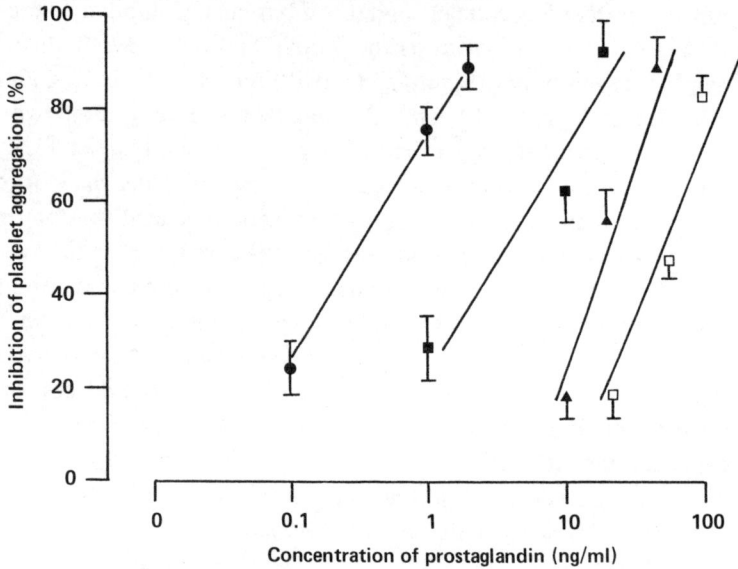

Fig. 4.5. Effect of PGI_2 infusion (10 ng/kg/min) on human platelet cAMP levels. Histograms represent cAMP measurements at timed intervals before (0), during (20, 40 min) and after (70, 90 and 120 min) PGI_2 administration. Notice that platelet cAMP levels are still elevated 30 min after stopping the PGI_2 infusion (data from Gorman, 1982).

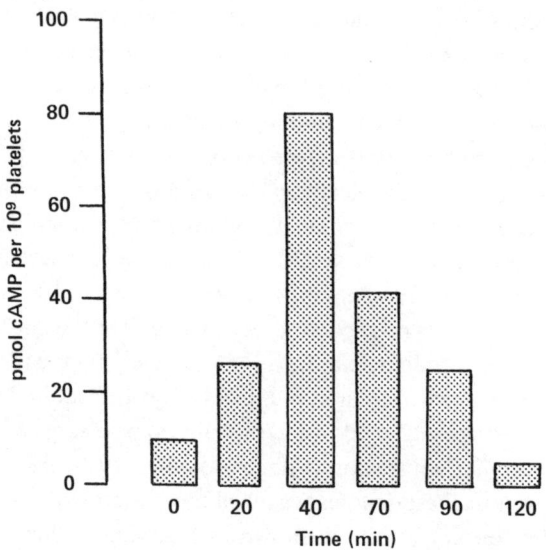

subject, infusion of 10 ng/kg/min PGI_2 in glycine buffer (pH 10) produces an immediate 8-fold increase in platelet cAMP concentration during the period of infusion between 20 and 40 min. When the infusion is stopped, both platelet reactivity and cAMP level return rapidly to normal.

Prostanoids also influence other platelet functions. PGE_1 and PGI_2 inhibit platelet adhesion to foreign surfaces and damaged blood vessel endothelium, although at concentrations higher than those needed to inhibit platelet aggregation (Cazenave et al., 1979). Thus PGI_2 may permit platelets to adhere to a damaged blood vessel wall and initiate the repair process but then prevents or limits the formation of a thrombus. Both 6-oxo-PGE_1 and PGI_2 have the interesting and potentially clinically important ability to disaggregate platelet clumps once formed either in plasma or in extracorporeal circuits where aggregates have formed on tendon collagen strips. Additionally, PGI_2 enhances the process of fibrinolysis in the dog lung and in human skin fibroblasts and also prevents white blood cells from adhering to the blood vessel wall (Higgs, Moncada & Vane, 1978b).

Inhibition of platelet aggregation by prostaglandins *in vivo* and *in vitro* is correlated with stimulation of membrane adenylate cyclase and an increase in platelet intracellular cAMP concentration (Gorman, Bunting & Moncada, 1977b). The potency ratio for activating adenylate cyclase in human platelets matches the anti-aggregatory potency for PGI_2, PGD_2 and PGE_1. In addition to promoting cAMP formation, prostaglandins may also elevate cAMP by inhibiting the enzyme cAMP phosphodiesterase which normally breaks down cAMP. The maximum increase in cAMP within the platelet occurs about 1 min after adding PGI_2 or 6-oxo-PGE_1 to human PRP (Miller et al., 1981).

The PGI_2:TxA_2 balance in health and disease

Moncada & Vane (1979) proposed that the balanced synthesis and action of pro-aggregatory, vasoconstrictor TxA_2 by platelets and anti-aggregatory, vasodilator PGI_2 by vascular tissue was of physiological importance in maintaining the fluidity of circulating blood. According to this theory, platelets may approach and collide with the healthy blood vessel wall but do not adhere since normal healthy endothelium is 'protected' by the continuous synthesis and release of anti-aggregatory PGI_2. However, this does not occur if the blood vessel wall is damaged. In this case the endothelial cells, which are largely responsible for vascular PGI_2 biosynthesis, are either disrupted or completely removed and subendothelial layers of the wall containing pro-aggregatory collagen fibres and smooth muscle cells are exposed (Fig. 4.6.). Platelets which approach and collide with the damaged part of a blood vessel wall will, in the absence of PGI_2, rapidly adhere to exposed collagen fibres, aggregate and form a platelet plug to prevent further loss of blood through the torn vessel. The loose platelet plug will later be reinforced by strands of fibrin to form a thrombus.

The shape of the thrombus is also of interest, covering only the area of damage and spreading neither up nor down the vessel wall, nor growing across to occlude the vessel entirely. Clearly, if the thrombus was to enlarge significantly, then the blood flow through that vessel would be reduced and this could result in ischaemic damage to the tissues serviced. The growth or spread of the thrombus may well be limited by PGI_2 released from healthy vascular endothelial cells on either side of the damaged area and also from those of the opposite blood vessel wall.

Since blood vessel endothelial cells avidly convert PG endoperoxides but only feebly convert arachidonic acid to PGI_2 (Bunting *et al.*, 1976), it has been proposed that platelets 'feed' the vascular wall with precursor PG endoperoxides which endothelial cells then convert into PGI_2. According to this proposal, platelet cyclooxygenase enzyme synthesises PG endoperoxides from endogenous arachidonic acid for conversion to TxA_2 which causes platelet aggregation. However, if the platelet is in close contact or

Fig. 4.6. Prostanoids and the platelet/blood vessel wall interaction. (*a*) Platelets are repulsed from healthy blood vessel wall due to the release of PGI_2 from vascular endothelial cells, (*b*) Platelets adhere to damaged blood vessel walls which are no longer protected by anti-aggregatory PGI_2. Aggregating platelets release TxA_2 which, in turn, recruits more platelets to form a homeostatic plug (the function of which is to seal off the torn blood vessel and prevent the loss of large amounts of blood).

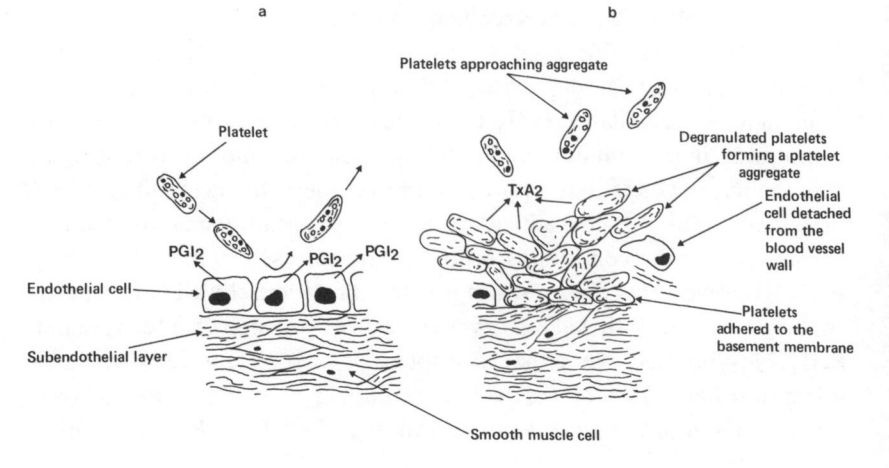

approaching the blood vessel wall then some of the PG endoperoxides are 'stolen' by endothelial cells and converted to PGI_2 which repulses the approaching platelet, preventing it from sticking to the blood vessel wall (Fig. 4.7). In this way, PGI_2 formation by the vascular endothelium would be pulsatile in nature, occurring only when required to prevent platelet adhesion. Clearly, if the vascular endothelial cells do 'steal' PG endoperoxides in this way, then it represents a very intimate biochemical cooperation between the different cell types.

Whether or not the $PGI_2:TxA_2$ balance is necessary for the regulation of platelet reactivity *in vivo* and whether blood vessels do actually utilise platelet PG endoperoxides remains to be seen. Certainly tipping the $PGI_2:TxA_2$ balance in favour of PGI_2 by treatment with drugs which prevent TxA_2 biosynthesis or by administering PGI_2 or its analogues reduces platelet reactivity *in vivo* and prolongs the bleeding time. Furthermore, several disease states in which altered platelet reactivity occurs have been related to an imbalance in the $PGI_2:TxA_2$ system. Some of these disease states are discussed below.

Diabetes mellitus

Patients with insulin-dependent diabetes mellitus exhibit enhanced platelet aggregability associated with increased susceptibility to both macrovascular (myocardial infarction, stroke) and microvascular (retinopathy, nephropathy) thrombotic disease. One of the underlying causes of platelet hyperreactivity in this condition may be a disturbance in the

Fig. 4.7. A scheme for the proposed biochemical cooperation between platelets and the blood vessel wall. As the platelet approaches the vessel wall it releases PGG_2 for conversion to PGI_2 by the lining endothelial cells. PGI_2, in turn, prevents adhesion of the platelet to the vessel wall. PGG_2 released from platelets not close to the blood vessel wall is converted by platelets to TxA_2.

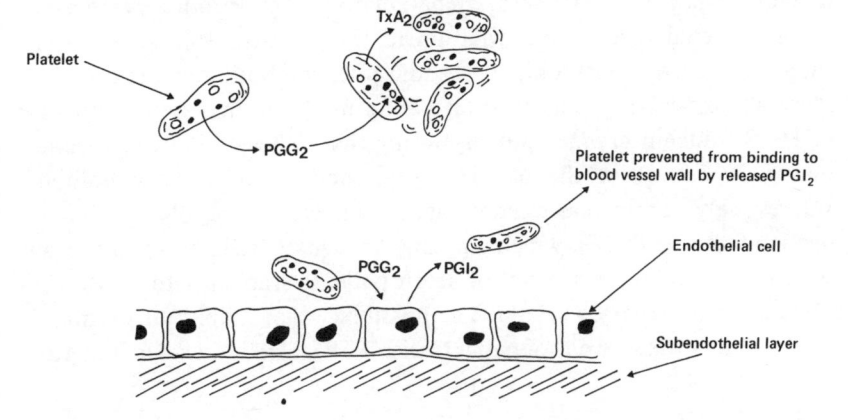

$PGI_2:TxA_2$ balance in favour of pro-aggregatory TxA_2. Increased TxA_2 biosynthesis by platelets challenged with arachidonic acid has been demonstrated in insulin-dependent diabetics (Ziboh et al., 1979) as well as in streptozotocin-induced diabetes in rats and in rats which are genetically diabetic. Synthesis of PGI_2 by biopsy specimens of human forearm vein and arteries from diabetic patients was much reduced compared with vascular tissue from healthy, matched non-diabetic controls. Measuring plasma 6-oxo-$PGF_{1\alpha}$ levels as an index of PGI_2 turnover in diabetics has been attempted with extremely variable results. The concentration of this prostaglandin is reportedly normal, increased or decreased in plasma from diabetic subjects.

Coronary artery disease

Patients who have survived myocardial infarction release more TxA_2 from platelets challenged with arachidonic acid *in vitro* and their platelets are more sensitive to aggregating agents than platelets from matched control human volunteers (Szczklik et al., 1978b). Additionally, plasma TxB_2 and urinary dinor TxB_2 are increased in patients with variant angina (Robertson et al., 1981). Thus, in patients with coronary artery disease the $PGI_2:TxA_2$ balance seems to be shifted in favour of TxA_2. Although there have been no reports of deficient vascular PGI_2 biosynthesis in these patients, platelets removed from patients during an anginal attack are not only more aggregable than normal but are also less sensitive to the inhibitory effect of PGI_2. Thus the increased formation of pro-aggregatory TxA_2 in these patients may be compounded by a reduction in sensitivity of platelets to the anti-aggregatory effect of PGI_2.

Thrombotic microangiopathy

Thrombotic thrombocytopaenic purpura (TTP) and haemolytic uraemic syndrome (HUS) are characterised by thrombocytopaenia, anaemia, renal disease and widespread thrombosis. Both diseases are usually classified under the blanket heading of thrombotic microangiopathy. Plasma 6-oxo-$PGF_{1\alpha}$ concentration is abnormally low in patients with TTP or HUS (Machin et al., 1980) suggesting that deficient PGI_2 production may play a part in the platelet hyperreactivity of these conditions. Interestingly, normal plasma contains a substance, possibly platelet-derived growth factor (PDGF), which stimulates vascular PGI_2 biosynthesis but which was lacking in five out of seven patients studied with TTP/HUS (Jorgenson & Pedersen, 1981). The possibility that inability to synthesis PGI_2 in blood vessel walls of patients with TTP/HUS is due to an inherited deficiency of a plasma PGI_2-stimulating factor is strengthened by the finding that patients with these syndromes are clinically improved by infusion of normal plasma which contains normal amounts of this plasma factor.

From this brief discussion of prostanoid biosynthesis in thrombotic diseases it seems clear that thrombosis may be caused at least in part by enhanced platelet TxA_2 and/or reduced vascular PGI_2 formation. However, a disadvantage of all the studies described here is that it is not possible to determine whether the aberrant prostanoid biosynthesis causes or is caused by the change in platelet reactivity. Furthermore, recent refinements in radioimmunoassay and gas-chromatography–mass-spectrometry techniques have suggested that the actual plasma concentration of both 6-oxo-$PGF_{1\alpha}$ and TxB_2 in healthy human beings are either very low (less than 2–3 pg/ml) or not detectable at all. This must shed serious doubt on many of the experimental results discussed above in which normal plasma levels of these prostanoid metabolites are frequently as high as 200–300 pg/ml. It is now realised that measurement of plasma prostanoids is technically much more difficult than was originally envisaged. Great care must be taken when body fluids are obtained by an invasive technique like venepuncture. Also assays must be carefully standardised to avoid erroneous results due to contaminants. Furthermore, it may well be that 6-oxo-$PGF_{1\alpha}$ is not the major breakdown product of endogenous PGI_2 formed naturally in the body. This honour may belong to an enzymatic metabolite of PGI_2 such as 6,15-dioxo-$PGF_{1\alpha}$ which could therefore turn out to be a more reliable index of endogenous PGI_2 turnover.

The effect of diet on the PGI_2:TxA_2 balance

The fatty acid composition of the diet may also play a part in regulating the PGI_2:TxA_2 balance *in vivo*. Eicosapentaenoic acid (20:5) is a substrate for cyclooxygenase and gives rise to prostaglandins containing 3 double bonds, e.g. PGE_3, PGI_3 and TxA_3. Vascular cyclooxygenase converts eicosapentaenoic acid in exactly the same way that it converts arachidonic acid to a substance chemically identical to PGI_3 (Gryglewski *et al.*, 1979). PGI_3 inhibits platelet aggregation and is virtually equipotent with PGI_2. Although platelet cyclooxygenase also converts eicosapentaenoic acid to PGH_3 and then TxA_3, both of these prostanoids have only weak aggregatory activity compared with PGH_2 and TxA_2 (Raz, Minkes & Needleman, 1977). Thus any diet which is rich in eicosapentaenoic acid will cause this fatty acid to be incorporated into plasma membranes at the expense of arachidonic acid. The result is a shift in the PGI_2:TxA_2 balance towards anti-aggregatory prostanoids like PGI_2 and PGI_3 and away from aggregatory prostanoids like TxA_2 and TxA_3. Foods rich in eicosapentaenoic acid are fish and fish products like cod liver oil.

Several attempts have been made to link intake of this fatty acid in the diet with a fall in platelet reactivity. Human volunteers eating a diet consisting mainly of mackerel (which is an excellent source of eicosapentaenoic acid) exhibit increased bleeding time, reduced platelet aggregability with decreased formation of TxB_2. With these changes in platelet function, biochemical analysis has also revealed an increase in the amount of eicosapentaenoic acid in platelet membranes and a corresponding decrease in arachidonic acid. That diet can change platelet reactivity by affecting the $PGI_2:TxA_2$ balance has also been suggested by experiments on platelets from Eskimoes. This group survive mainly on a fish diet rich in eicosapentaenoic acid and their platelet membranes have very high levels of this fatty acid (about 8% of total fatty acids). In contrast, a matched group of Danes living on a Westernised diet containing much less fish had platelets which contained relatively small amounts of this fatty acid in their membranes (about 0.5% of total fatty acids). These biochemical changes are correlated with differences in platelet reactivity between Eskimoes and Danes. For example, Eskimoes have prolonged bleeding times, reduced platelet reactivity and are much less likely to suffer from thrombotic disease such as myocardial infarction and diabetes mellitus than are the group of Danes (Dyerberg et al., 1979).

Perhaps a very simple way of altering the $PGI_2:TxA_2$ balance in a favourable direction and thus reducing the risk of thrombosis is to alter the diet to eat more fish and less red meat. Changing the diet of an individual may also affect other aspects of platelet function and blood coagulability which are not related to the $PGI_2:TxA_2$ balance. For example, a fish diet may ensure that Eskimoes have low plasma cholesterol, triglyceride and low density lipoprotein (LPL) concentrations, all of which may contribute to their reduced susceptibility to atherosclerosis compared with other ethnic groups.

Effect of drugs which displace the $PGI_2:TxA_2$ balance on platelet aggregation

Several drugs are available which modify platelet and/or vascular arachidonic acid metabolism in such a way as to tip the $PGI_2:TxA_2$ balance in favour of PGI_2. This may be done by administration of PGI_2 analogues, TxA_2 receptor blockers or cyclooxygenase or thromboxane synthetase inhibitors. The expected result of treatment with such drugs is inhibition of platelet reactivity. Many of these drugs (e.g. aspirin, sulphinpyrazone, dipyridamole) have been clinically tried and tested. Some, like dazoxiben, are undergoing clinical trials in humans and yet other drugs (e.g. naphazotrom, nitroglycerin) have been tested only in animals. Clearly a number of drugs affecting the arachidonic acid cascade are available to the clinician and this number is likely to increase dramatically in the next few years. The pharmacology and clinical applications of these drugs will be discussed.

Aspirin. The most widely studied anti-thrombotic drug is aspirin which acetylates and irreversibly inhibits cyclooxygenase, thus preventing both platelet TxA_2 and vascular PGI_2 formation. In theory therefore aspirin is not the ideal anti-platelet drug since although it prevents TxA_2 formation it also blocks the biosynthesis of protective, anti-aggregatory PGI_2. Any procedure which confers selective inhibition of TxA_2 formation on aspirin would be an advantage. Since vascular endothelial cells which contain a nucleus, RNA and DNA (and can thus make proteins) can synthesise new cyclooxygenase enzyme after aspirin treatment the inhibitory effect on PGI_2 biosynthesis would be expected to be short-lived. In contrast, platelets have no nucleus and thus cannot regenerate fresh cyclooxygenase enzyme after administration of aspirin. Consequently, a single dose of aspirin blocks platelet TxA_2 formation for as long as the platelet is in the circulation.

Since platelets have an average half-life in the human bloodstream of about 8 d it should be feasible to work out a dose regimen of aspirin which is sufficient to inhibit TxA_2 production but which has little or no effect on PGI_2 biosynthesis. In practice this 'ideal' aspirin dose regimen has proved elusive. 150–300 mg aspirin daily given orally to healthy, human volunteers produced an almost parallel reduction in biopsy blood vessel PGI_2 formation and platelet TxA_2 production. Lowering the dose to 81 mg daily inhibited PGI_2 formation to the same extent as the higher dose whereas a single dose of 40 mg/day had no effect on vascular PGI_2 biosynthesis (Hanley et al., 1981). Unfortunately, this low dose of aspirin did not consistently inhibit platelet cyclooxygenase activity, and thus TxA_2 formation, in all subjects. Further clinical trials to determine a dose of aspirin which blocks TxA_2 biosynthesis but leaves PGI_2 formation unchanged are sorely needed.

Most of the clinical trials with aspirin have employed daily doses of between 1 g and 1.5 g. We now know that such amounts of aspirin are sufficient to prevent not only TxA_2 and PGI_2 biosynthesis but also lower the plasma levels of vitamin-K-dependent clotting factors. Thus, at this sort of dose, aspirin will almost certainly affect both platelet aggregation and blood coagulation. However, despite this dual activity, clinical experience with aspirin in treating thromboembolic disorders in man has been disappointing. Of the six long-term preventative studies in patients who had survived one myocardial infarction, aspirin reduced the incidence of a second heart attack in five trials and actually increased mortality in the sixth. However, in none of these trials did the changes observed reach a statistically significant level. Details of these trials and of the doses of aspirin used are shown in Table 4.1.

Aspirin has also been used to treat other thrombotic diseases like stroke, transient ischaemic attacks (TIAs) and venous thromboembolism. Although most strokes are caused by thrombus formation in cerebral blood vessels, a significant number may be due to rupture of small blood vessels in the brain producing an intracranial haemorrhage. Treatment with anti-platelet drugs could be useful to treat strokes due to cerebrovascular thrombosis but equally could be disastrous if the stroke was caused by intracranial bleeding. Since it is sometimes difficult to distinguish clinically between these two causes of stroke in the individual patient, there have been few studies of the effect of anti-platelet drugs in the treatment or secondary prevention of stroke. However, aspirin has been tested for ability to reduce the incidence and severity of another form of cerebrovascular thrombosis, the transient ischaemic attack (TIA). Characterised by brief periods of loss of motor function and sometimes blindness, a TIA results from arterial thromboembolism. Aspirin (1300 mg/d) reduced the incidence of TIAs in male patients by 48% but, remarkably, was without effect in female patients (Canadian Cooperative Study Group, 1978). Even though this was one of the largest of such trials (585 patients studied) many

Table 4.1. *Clinical trials of aspirin in the prevention of myocardial infarction*

Author	Dose (mg/day)	Number of patients	Reduction in mortality[a] (%)
Elwood *et al.* (1974)	300	1239	25
Coronary Drug Project Research Group (1976)	975	1259	30
Breddin *et al.* (1979)	1500	626	18
Elwood & Sweetnam (1979)	900	1682	17
AMIS (1980)	1000	4324	−11
PARIS (1980)	975	1216	18

[a] In no trial was the reduction in mortality after aspirin therapy statistically significant. Notice that aspirin actually increased mortality in the AMIS trial.

more patients, possibly 2000 or more, would have to be included before analysis of the results could indicate a statistically significant trend.

A similar problem has plagued most studies of aspirin in venous thrombotic diseases. Over 20 clinical trials have been carried out, usually with insufficient numbers of patients and with widely different dose regimens. The concensus from all of these trials seems to be that none of the anti-platelet drugs affecting arachidonic acid metabolism affect post-operative venous thrombosis. However, several studies have suggested that there are sex differences in the way in which individuals respond to aspirin therapy. For example, male patients receiving aspirin showed a 67% decrease in venous thrombosis after total hip replacement operations while the corresponding beneficial effect in women was only 4% (Harris et al., 1977). The explanation for this apparent advantage which men hold over women with respect to aspirin treatment in both TIAs and venous thromboembolism is not known.

Sulphinpyrazone. Sulphinpyrazone and its active thioether metabolite are weak cyclooxygenase inhibitors which apparently produce little or no inhibition of vascular PGI_2 formation at doses which prevent TxA_2 biosynthesis (Livio, Villa & De Gaetano, 1980). Both ticlopidine and itanoxone given orally to rats also appear to discriminate between TxA_2 and vascular PGI_2 biosynthesis in a similar way (Ashida & Abiko, 1978; Livio et al., 1980). Theoretically, drugs with this profile of activity on the arachidonic acid cascade should be better anti-platelet agents than aspirin. In practice this is not the case. In a large, multicentred, double blind randomised trial (Anturane Reinfarction Trial Research Group, 1978), sulphinpyrazone (800 mg/day) was compared with placebo in over 1600 patients with a myocardial infarction 22–35 d previously. Initial results suggested that death from a second myocardial infarction was significantly less likely to occur in sulphinpyrazone-treated patients than in patients taking the placebo. Unfortunately, despite much early enthusiasm, analysis of the way in which the trial was designed, conducted and reported revealed very substantial flaws which have rendered the results almost impossible to interpret properly. Whether sulphinpyrazone has a beneficial effect in the post-myocardial-infarction period will have to be decided by another properly designed clinical trial. In other thrombotic disorders sulphinpyrazone has proved to be a disappointment, producing a small (less than 10%) but statistically non-significant reduction in TIAs (Canadian Cooperative Study Group, 1978) and no effect on the development of deep-vein thrombosis after surgical operation (Gruber et al., 1977).

Dipyridamole. Dipyridamole may prevent platelet aggregation by inhibiting platelet cAMP phosphodiesterase activity and thus potentiating the anti-aggregatory effect of PGI_2 (Moncada & Korbut, 1978). Additionally, dipyridamole may owe some of its anti-aggregatory activity to a variety of other actions. These include: prevention of adenosine uptake into red blood cells, thus making more anti-aggregatory adenosine available at the platelet; stimulation of vascular PGI_2 production and inhibition of platelet thromboxane synthesis. Which of these mechanisms of action is responsible for the anti-aggregatory activity of this drug remains to be seen. Clinical experience with dipyridamole (trade name: Persantine) has proved a disappointment. The Persantine–Aspirin Reinfarction Study (PARIS) which reported in 1980 suggested that the combination of aspirin with dipyridamole was especially beneficial in patients if administered after a first heart attack. Without aspirin, dipyridamole had no effect on the prognosis for myocardial infarction and did not influence the incidence of TIAs or post-operative venous thrombosis.

Prostacylin. Prostacylin is commercially available as a stable, freeze-dried powder (epoprostenol) which may be reconstituted in glycine buffer (pH 10) for administration to humans. Several clinical or potential clinical applications of PGI_2 have been suggested including the treatment of peripheral arteriosclerotic vascular disease (see Chapter 3) as well as maintaining the patency of the ductus arteriosus in the foetus (see Chapter 5). In this chapter only the clinical potential of the anti-platelet activity of PGI_2 will be discussed.

The major interest in PGI_2 for the clinician is the prevention of platelet loss from the body during extracorporeal circulation of the blood in processes such as renal dialysis, charcoal haemoperfusion and cardiopulmonary bypass. When blood is brought into contact with artificial surfaces (such as dialysis membrane) platelets avidly adhere to it and form platelet aggregates. This results in a reduced blood platelet count (thrombocytopaenia) in the patient and a dialysis machine clogged with platelet aggregates which may embolise and flush back into the body with potentially disastrous consequences. More often than not the effect on platelets of extracorporeal circulation is more subtle. This may involve accelerated consumption of platelets and some loss of their ability to adhere and aggregate. Heparin is normally used to prevent platelet aggregation and blood clotting in dialysis machines but, because heparin is only slowly degraded in the body, suffers from the disadvantage of remaining in the bloodstream for some time after the dialysis has been stopped. During this period, the patient is very susceptible to bleeding.

Furthermore, heparin may cause bone wasting. Prostacyclin infusion prevents thrombocytopaenia and loss of platelet function in animal models of extracorporeal circulations, in humans with liver failure undergoing liver charcoal haemoperfusion (Gimson *et al.*, 1980) and in humans undergoing renal dialysis and cardiopulmonary bypass during open-heart surgery (Longmore *et al.*, 1981). Since PGI_2 is degraded in the liver and other sites it does not have a prolonged effect on platelet function and blood clotting, does not affect bone-calcium metabolism and thus may prove to be a viable alternative to heparin.

Others. Selective inhibitors of thromboxane synthetase may be the future drugs of choice for treatment of thromboembolic diseases especially with the introduction of dazoxiben, a potent and selective inhibitor of this enzyme which is active when administered orally in man (Tyler *et al.*, 1981). Clinical trials of this compound in thrombotic diseases in man are awaited eagerly. Several drugs have been reported to enhance vascular PGI_2 synthesis *in vivo* and/or *in vitro* by a direct action on the vascular endothelial cells. These drugs include naphazotrom (BAY g 6575, 1-(2-naphthyloxy)ethyl-3-methyl-2-pyrazolin-5-one) (Vermylen, Chamone & Verstraete, 1979), verapimil, bendrofluazide and nitroglycerin. Of these, nitroglycerin is probably of some interest since it stimulates cultured human vascular endothelial PGI_2 formation at doses which can be achieved clinically during treatment of anginal attacks (0.1–10 ng/ml). The anti-thrombotic drug, trapidil, has dual action inhibiting thromboxane synthetase and potentiating the anti-aggregatory effect of PGI_2, possibly by inhibiting platelet cAMP phosphodiesterase activity. No doubt the next decade will determine whether any or all of these drugs prove to be useful in the clinic for the treatment of thrombotic disease.

5

The reproductive system

Introduction

Since their discovery in human semen over 50 years ago the possibility that prostaglandins may affect reproductive function has been exhaustively researched. A mass of information on the way in which prostaglandins affect the male and female reproductive tracts has been gathered and summarised in numerous reviews, monographs and books (Karim, 1975, 1979; Horton & Poyser, 1976; Poyser, 1981). The aim of the present chapter is to collate this wealth of experimental evidence and thereby provide an overview of the effect of prostanoids on reproduction.

Occurrence of prostaglandins in the reproductive organs

In the female, prostanoid biosynthesis and catabolism has been most studied in the uterus, ovary and placenta. Prostanoid metabolism by these three organs depends not only upon age and species but more importantly upon the local concentration of oestrogens and progestogens. The amount and type of prostanoid formed will therefore change throughout the oestrus cycle in vertebrates and, likewise, throughout the menstrual cycle in primates. With these factors in mind, PGI_2 appears to be the major product of arachidonic acid metabolism in the myometrium (the muscular part) of the uterus, the ovarian corpus luteum and Graafian follicle and the placenta. PGI_2, measured as 6-oxo-$PGF_{1\alpha}$, is also formed by foetal amniotic and chorionic membranes which no doubt explains the occurrence of this prostaglandin in such large amounts in amniotic fluid. In contrast, the endometrium (secretory lining) of the uterus synthesises predominantly $PGF_{2\alpha}$ in most species, with lesser amounts of PGI_2 and PGE_2. Interestingly PGD_2 is found in large concentrations in the human placenta which in addition is also an excellent source of the prostaglandin-catabolising enzyme, 15-PGDH. The function of placental 15-PGDH may be to protect the foetus from prostaglandins circulating in the mother's bloodstream.

Human seminal plasma is the richest single source of classical prostaglandins of any body fluid. The following prostaglandins occur in human semen: PGE_1, PGE_2, $PGF_{1\alpha}$, $PGF_{2\alpha}$ and their corresponding 19-hydroxylated and 8-iso-derivatives, as well as PGE_3. The mean total prostaglandin concentration of human semen may be as high as 400 µg/ml. A few years ago A and B series prostaglandins were also believed to occur in semen but these are now known to be artefacts formed from PGEs following extraction. As far as is known, human semen does not contain PGI_2, leukotrienes or TxA_2. Prostaglandins also occur in semen from primates (gorilla, orang-utan and chimpanzee) and in smaller amounts in non-primate species such as sheep, goat, rabbit and rat.

In the same way that uterine and ovarian prostanoid biosynthesis is controlled by female sex hormones, the content of prostaglandins in semen is regulated by the male sex steroid (testosterone). Hypogonadal males, in whom testicular development is retarded, have only about 20% of the semen prostaglandin levels of normally developed males. Testosterone-replacement treatment in these patients increases the concentration of prostaglandins in semen to normal or near-normal levels.

The prostaglandins which are found in semen are synthesised in the seminal vesicle. This organ is a very rich source of cyclooxygenase enzyme. Incubation of homogenates or resuspended microsomes of ram seminal vesicle with high concentrations (more than 10 µg) of arachidonic acid yields mainly PGE_2 and $PGF_{2\alpha}$. At lower substrate concentrations, arachidonic acid is converted chiefly to PGI_2. Other parts of the male reproductive tract also synthesise prostaglandins under the control of testosterone. The order of prostaglandin biosynthesis in the testis and vas deferens of several species, for example, is $PGI_2 > PGE_2 > PGF_{2\alpha}$.

Prostanoids in the female reproductive system

Cyclooxygenase metabolites of arachidonic acid influence virtually every aspect of female reproductive function including:
(1) the release of gonadotrophic hormones from the pituitary gland;
(2) maturation of the egg (ovum) and its release from the ovary;
(3) destruction of the corpus luteum (luteolysis);
(4) menstruation;
(5) fertilisation of the egg with sperm and implantation of the fertilised egg into the uterine endometrium;
(6) placental function;
(7) parturition;
(8) lactation.

The role of prostanoids in each of these functions will be discussed in turn.

Effect of prostaglandins on LH/FSH release from the pituitary

Before embarking on a discussion of the role of prostaglandins in these processes, a brief outline of the factors governing the release and actions of LH and FSH will be provided.

Luteinising hormone (LH) and follicle stimulating hormone (FSH) are gonadotrophic hormones liberated from the anterior pituitary to influence the function of the ovary and testis. In the female they cause ovum maturation, ovulation and the subsequent formation of the corpus luteum. These hormones are liberated from the pituitary in a cyclic manner under the control of the median eminence region of the hypothalamus which elaborates a decapeptide called gonadotrophin-releasing hormone (GRH). This peptide is transported to the pituitary gland via the hypophysial portal blood vessels where it causes the release of LH and FSH. A simplified diagram to represent the neuroendocrine control of gonadotrophic hormone release is shown in Fig. 5.1.

Oestradiol formed by the ovary and released into the bloodstream affects the formation and release of GRH from the hypothalamus in at least two ways:
(1) by acting on the median eminence itself to inhibit GRH output (i.e. a negative-feedback process);
(2) at higher concentrations oestradiol stimulates the pre-optic area of the brain which is connected to the median eminence by a nerve tract causing GRH output (i.e. a positive-feedback process).

Prostaglandins stimulate FSH and LH release from the pituitary gland and may be responsible for the oestradiol-induced positive-feedback mechanism mentioned above. In the rat, for example, injection of PGE_1 or PGE_2 either systemically, or into the third ventricle or directly into the median eminence region of the hypothalamus all elicit a surge in plasma LH concentration (Spies & Norman, 1973) whilst injection of either of these prostaglandins directly into the pituitary had no such effect (Eskay *et al.*, 1975). Thus prostaglandins of the E series cause LH release by an action directly on the hypothalamus to stimulate GRH release into the hypophysial blood vessels and thus to the pituitary. Direct evidence for this conclusion has come from studies in which GRH has been measured by radio-immunossay in hypophysial portal blood draining the hypothalamus (Fig. 5.2).

Further evidence that prostaglandins stimulate GRH release has come from studies using aspirin-like drugs to block prostaglandin formation in the median eminence. If large doses of indomethacin are given systemically to rats, plasma LH levels are decreased. Much smaller amounts of indomethacin injected directly into the median eminence also consistently decreases plasma LH (Sato *et al.*, 1975). These results strongly suggest that endogenous prostaglandins, perhaps PGE_2 (which appears to be the most abundant cyclooxygenase product of arachidonic acid formed by the hypothalamus), play a part in regulating oestradiol-induced release of LH into the bloodstream. They do this by causing GRH release from the hypothalamus. Whether prostaglandins have a similar effect on FSH release from the pituitary gland is not so clear since indomethacin injected

Fig. 5.1. Feedback control of gonadotrophin (GRH) release from the hypothalamus. Notice that oestradiol may exert a positive and negative influence over GRH output.

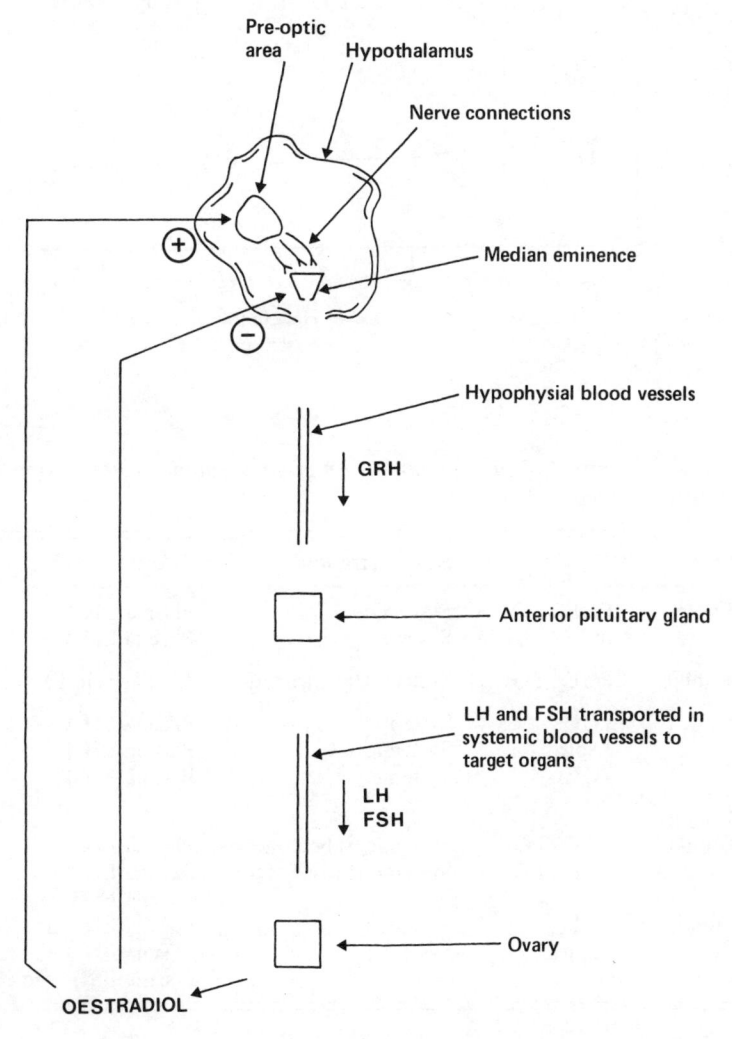

Fig. 5.2. Effect of intraventricular injection of 20 μg PGE_2 in anaesthetised rats on plasma GRH concentration measured in hypophysial portal blood draining the hypothalamus (triangles) and in arterial blood (squares). Data from Eskay et al., 1975.

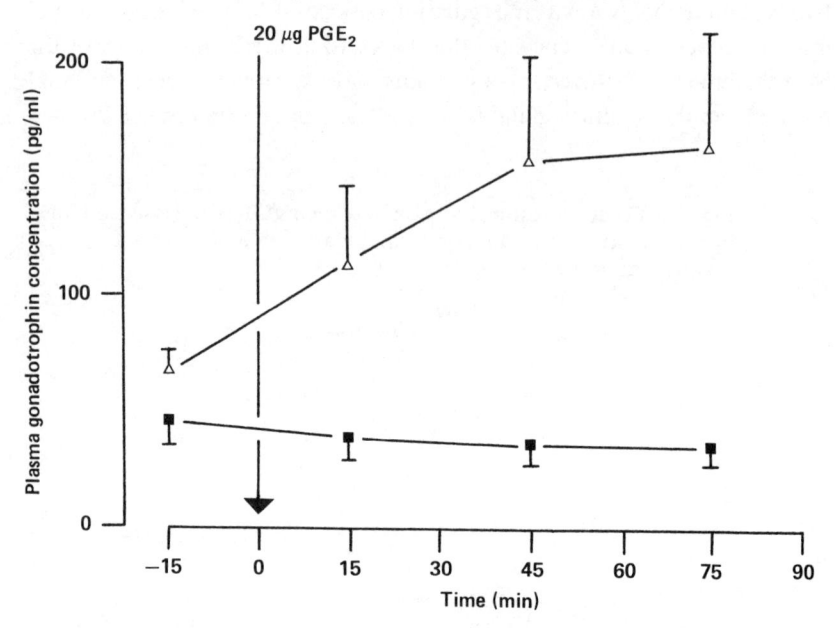

Table 5.1. *Effect of prostaglandins on gonadotrophin release from the pituitary gland*

Species	Drug	Administration	Effect
Sheep	$PGF_{2\alpha}$	Systemic	Plasma LH ↑
	Indomethacin	Systemic	Plasma LH ↓
Rabbit	PGE_2, $PGF_{2\alpha}$	Systemic/intracarotid	Jugular vein LH ↑
Cow	$PGF_{2\alpha}$	Systemic	Plasma LH ↑
	Aspirin	Systemic	Plasma LH ↓
	PGE_2	Systemic	Plasma LH ↓
Primates			
Monkey	PGE_2, $PGF_{2\alpha}$	Systemic/subcutaneous	Plasma LH ↑
	PGE_1, PGE_2	Systemic/intracarotid	Plasma LH ↑
			Plasma FSH – no change
Human	$PGF_{2\alpha}$	Systemic	Plasma LH – no change
	Aspirin	Systemic	Plasma LH – no change
			Plasma FSH – no change

directly into the median eminence did not lower plasma FSH concentration even though plasma LH levels were markedly suppressed. Although the majority of experiments of this kind have been performed using rats, broadly similar observations have been made in other species (Table 5.1).

In some species (e.g. the cow) it is not absolutely clear whether prostaglandins exert their effort on LH release by acting on the hypothalamus (as in the rat) or directly on the pituitary. Furthermore, the effect of other prostaglandins like PGI_2, PGD_2 and the PG endoperoxides and TxA_2 have not yet been determined. One feature of Table 5.1 which should be pointed out is that neither treatment with aspirin nor with $PGF_{2\alpha}$ changed plasma LH levels in human females. On the basis of these results, it is not yet clear whether prostaglandins have a role to play in the physiological regulation of LH release in the human.

Ovulation – a role for prostaglandins?

In women, ovulation occurs in the middle of the menstrual cycle on one of days 14–16 and is brought about by the action of LH released from the pituitary. As we have seen, this LH surge at mid-cycle may be modulated in some species by hypothalamic prostaglandins. However, this is not the only way in which prostaglandins may affect ovulation. This is shown by the finding that indomethacin treatment of rats at the time of the expected LH surge from the pituitary blocks ovulation and that this block cannot be reversed by injecting LH (Armstrong & Grimwich, 1972). In this case, indomethacin clearly does not prevent LH release but does prevent the ability of the ovary to respond to LH. Thus, prostaglandins made in the ovary itself may also modulate ovulation in the rat. Exactly how ovarian prostaglandins modulate ovulation in response to LH is a matter for some debate. It has been pointed out that rat ovarian follicle cells contain fibres of smooth muscle and this has prompted the suggestion that prostaglandins 'contract' these muscle fibres thereby increasing the pressure within the follicle making it burst open and release the ovum into the Fallopian tube. Further experiments to clarify the precise mechanism of action of prostanoids and to identify exactly which prostanoid is responsible are clearly required.

Luteolysis – a role for prostaglandins?

As described previously, ovulation in humans occurs in midmenstrual cycle under the influence of LH released from the pituitary gland. Once the ovum is released into the Fallopian tube the vacated ovarian follicle, under the continuing influence of LH, develops rapidly into a large yellowish structure called the corpus luteum. If the ovum encounters a sperm and is fertilised the now-fertilised egg is transported to the uterus and becomes embedded in its secretory lining (endometrium). Following fertilisation the corpus luteum develops fully and secretes large amounts of progesterone into the blood stream. This hormone is required to maintain the uterine endometrium in a suitable state to safeguard and nourish the developing embryo. In humans the corpus luteum serves this function for about 6–7 weeks after which progesterone production is taken over by the placenta. The process whereby the corpus luteum stops secreting progesterone is called functional luteolysis; this is followed closely by the destruction and resorption of the corpus luteum, which is known as structural luteolysis. If fertilisation does not occur, then the corpus luteum persists only for a day or two before luteolysis occurs and the whole procedure is repeated in the next cycle.

A great deal has been written about luteolysis and probably more is known about this process than any other aspect of female reproductive function. One reason for the interest in luteolysis is probably that if a way could be found to block corpus luteal activity after fertilisation has taken place then this may provide a novel and effective means of post-coital contraception. A major advantage of this form of contraceptive treatment would be that it need only be taken once a month. By the same reasoning, a luteolytic drug could, theoretically, be used to induce abortion in early pregnancy. That some prostaglandins produce luteolysis has been known for 15 years and a vast amount of information on this subject has been accumulated (see Labhsetwar, 1975; Horton & Poyser, 1976, for reviews). However, several questions still remain to be answered about the precise role of prostanoids in luteolysis. These questions include:

(1) Which prostanoids are luteolytic and how do they work?
(2) Are prostanoids luteolytic in all species?
(3) Are endogenous prostanoids responsible for physiological luteolysis which occurs spontaneously if pregnancy does not occur?
(4) Can the luteolytic effect of prostanoids be exploited clinically either as a contraceptive or as an abortifacient?

Some of the experiments designed to answer these important questions are described below.

Luteolytic activity of different prostaglandins. Of the naturally occurring prostaglandins which have been studied, $PGF_{2\alpha}$ has the most potent luteolytic effect. PGE_1 and PGE_2 are also luteolytic in some species but are generally about 5–10 times less potent. PGI_2 has very weak luteolytic activity in the rat which is the only species in which this particular prostanoid has been studied to date. Other prostaglandins and leukotrienes have received only cursory study or have not been investigated at all. A single subcutaneous injection of 25 µg $PGF_{1\alpha}$ in hamsters can produce an almost immediate luteolysis with a significant drop in plasma progesterone concentration in as little as 15 min (Gutknecht, Duncan & Wyngarden, 1972). In other species a similar decrease in the plasma concentration of this steroid may take several hours. The sensitivity of the corpus luteum to the luteolytic effect of $PGF_{2\alpha}$ depends also upon its age. During the early part of its lifespan the corpus luteum is not susceptible to $PGF_{2\alpha}$. For this reason $PGF_{2\alpha}$ does not cause luteolysis until 4 d after fertilisation in the cow and sheep and 24 d after fertilisation in the bitch. Furthermore, the corpus luteum seems to be more resistant to $PGF_{2\alpha}$ when the animal is pregnant, possibly because the placenta in such animals may secrete hormones like chorionic gonadotrophin which prolongs the lifespan of the corpus luteum (i.e. a luteotrophic as opposed to a luteolytic action).

There is little doubt that, in some species, $PGF_{2\alpha}$ is a potent luteolytic agent. Exactly how $PGF_{2\alpha}$ produces luteolysis remains controversial. Several theories have been put forward although none satisfactorily explains all the experimental data. Phariss, Cornette & Gutknecht (1970) suggested that $PGF_{2\alpha}$ constricted the uteroovarian vein thereby reducing blood flow through the ovarian artery and effectively starving the corpus luteum of oxygen and nutrients. This proposed mechanism is now considered to be unlikely for several reasons, not least that PGE_1 and PGE_2 (which are luteolytic in rats, hamsters and sheep) do not constrict the uteroovarian vein in these species (McCracken, Baird & Goding, 1972 b). Furthermore, several authors using a range of different techniques have frequently failed to demonstrate that $PGF_{2\alpha}$ decreases ovarian blood flow. The discovery of specific, high-affinity binding sites, which may represent receptors for $PGF_{2\alpha}$ actually on the luteal cell membranes of several species including women (Powell et al., 1974), has shifted attention from an indirect mechanism of action on the vasculature to a direct effect of $PGF_{2\alpha}$ on the corpus luteum itself. The nature of this direct effect has so far proved elusive but may involve cAMP. Unlike prostaglandins of the E series which activate adenylate cyclase in rat luteal membranes in vitro and thus mimic LH in raising intracellular cAMP levels, $PGF_{2\alpha}$ blocks LH- (but not PGE-) induced stimulation of adenylate cyclase, thereby reducing the concentration of cAMP inside luteal cells (Lahav, Freud & Lindner, 1976). It has been proposed that a fall in luteal cell cAMP concentration causes the enzyme cholesterol esterase to be converted from its active phosphorylated form to its inactive dephosphorylated form (Henderson & McNatty, 1977). Inhibition of this enzyme prevents progesterone biosynthesis. Other authors have found that $PGF_{2\alpha}$ treatment produces a long-term fall in the number of LH receptors on luteal cell membranes (Berhman, Grimwich, Hichens & MacDonald, 1978). In conclusion, the luteolytic activity of $PGF_{2\alpha}$ may have two components. Firstly, and in the short term, $PGF_{2\alpha}$ binds to receptors on the luteal membrane in such a way as to prevent LH from activating a membrane-bound adenylate cyclase enzyme. Secondly, and in the longer term, $PGF_{2\alpha}$ decreases the number of LH receptors on luteal membranes. These mechanisms are depicted diagrammatically in Fig. 5.3.

Luteolytic effect of prostaglandins in different species. The luteolytic potency of $PGF_{2\alpha}$ is very species-dependent. To date, $PGF_{2\alpha}$ has been shown to cause luteolysis in almost all commonly used laboratory animals (e.g. rat, hamster, gerbil, cat, dog, guinea pig, mouse) and farm animals (e.g. sheep, cow, mare). The amount of $PGF_{2\alpha}$ required, the optimum route of administration, the change in sensitivity of the corpus luteum with age and pregnancy, all vary from one species to the next. In some of these species (e.g. rat, hamster) PGE_1 and PGE_2 are also luteolytic. The situation in primates is completely different. In some species of monkey $PGF_{2\alpha}$ will cause luteolysis resulting in premature menstruation if administered in

Fig. 5.3. Interaction of $PGF_{2\alpha}$ (PGF) and luteinising hormone (LH) in the control of ovarian luteolysis. Luteal cells contain separate receptors for LH and $PGF_{2\alpha}$. (*a*) Stimulation by LH activates luteal cell adenylate cyclase (AC) activity to increase cAMP levels intracellularly. Increased cAMP within the cell promotes progesterone biosynthesis. (*b*) Occupation of its receptor by $PGF_{2\alpha}$ prevents LH activation of luteal AC and so progesterone biosynthesis declines followed by luteolysis.

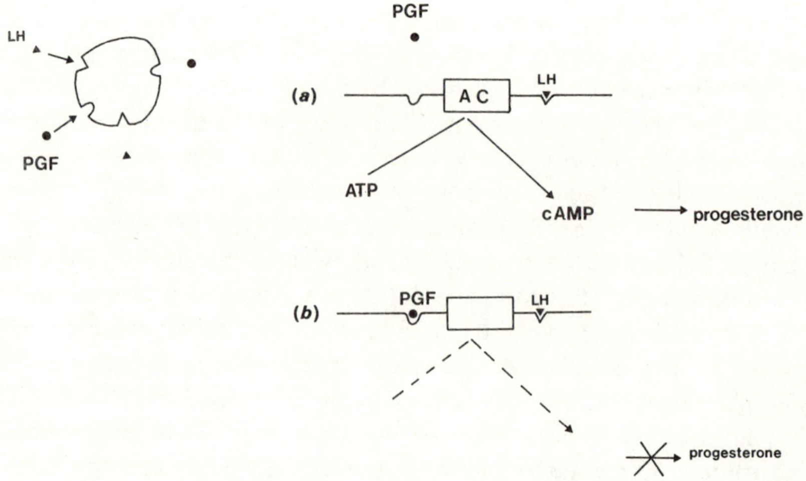

combination with oestradiol (Saikh & Klaiber, 1974). Furthermore, $PGF_{2\alpha}$ produced abortion in early pregnant monkeys by destroying the corpus luteum and thereby removing the progesterone support for the developing foetus. However, several other studies have failed to show $PGF_{2\alpha}$-induced luteolysis in monkeys. In humans also $PGF_{2\alpha}$ generally has only a very weak or no luteolytic action at all following administration to non-pregnant women. $PGF_{2\alpha}$ does cause abortion in women in early pregnancy but this is attributed to mechanical stimulation of the uterus to dislodge the conceptus rather than a luteolytic effect. To summarise, as a general rule, primates are resistant to the luteolytic action of $PGF_{2\alpha}$ while sub-primate species readily undergo luteolysis when treated with this prostaglandin.

Is $PGF_{2\alpha}$ a natural luteolytic agent? As we have seen, administration of $PGF_{2\alpha}$ can cause luteolysis in sub-primates. In certain cases $PGF_{2\alpha}$ synthesised and released by the uterus may be important physiologically for terminating the corpus luteum. That the uterus is capable of forming a potent luteolytic agent at the appropriate time of the menstrual or oestrus cycle has been known since the 1920s. For many years this substance was called luteolysin. There is now considerable evidence that luteolysin and $PGF_{2\alpha}$ are one and the same thing (see review by Horton & Poyser, 1976). This evidence briefly consists of identification of $PGF_{2\alpha}$ in the uterine vein during luteolysis and prevention of luteolysis by treatment with anti-$PGF_{2\alpha}$ antibodies. It should be stressed that the uterus plays a part in determining the lifespan of the corpus luteum probably only in sub-primate species. While removing the uterus from species like the sheep or cow retards luteolysis, a similar surgical operation has no effect on the development and destruction of the corpus luteum in women and other primates. Thus only the sub-primate uterus cyclically releases luteolysin.

A major difficulty with the idea that the uterus releases an endogenous luteolytic agent has been an inability to explain how this substance passes from the uterus to the ovary since there is no direct portal blood system linking these two organs. The answer to this question may lie in the unique arrangement of the vasculature in this part of the body (Fig. 5.4).

In the sheep, the uterus and ovary share a single common vein, the uteroovarian vein which drains blood from both organs. The ovarian artery runs over the surface of the uteroovarian vein. Clearly any substance released from the uterus into the uterine vein could, theoretically, be transported from the uteroovarian vein into the ovarian artery and thus to the ovary itself. Indeed if the ovarian artery is carefully teased away from the uteroovarian vein in an anaesthetised sheep and the animal allowed to recover it thereafter cannot undergo luteolysis (Barrett et al., 1971). A countercurrent mechanism of transfer of $PGF_{2\alpha}$ between the uteroovarian vein and the ovarian artery has been proposed (McCracken et al., 1972a). Although the efficiency of transfer is only 2–10%, taking into account the concentration of $PGF_{2\alpha}$ found in the uterine vein this is still sufficient to bring about a complete luteolysis. Such a countercurrent transfer mechanism appears to exist in sheep, cow, rat, hamster and pig. However, in species such as the guinea pig and rabbit the vascular arrangement of the uterine and ovarian blood vessels make countercurrent transfer of $PGF_{2\alpha}$ virtually impossible. The fact that in these species also the uterus releases an endogenous luteolytic substance which is, again, almost certainly $PGF_{2\alpha}$ suggests that there must be more than one mechanism for transferring this prostaglandin from the uterus to the ovary.

Clinical uses of luteolytic prostaglandins. $PGF_{2\alpha}$ brings about abortion in pregnant women during the first 7 weeks of pregnancy. This process is usually not referred to as abortion but by the term menstrual regulation since the results are similar to a normal menstrual period. Numerous

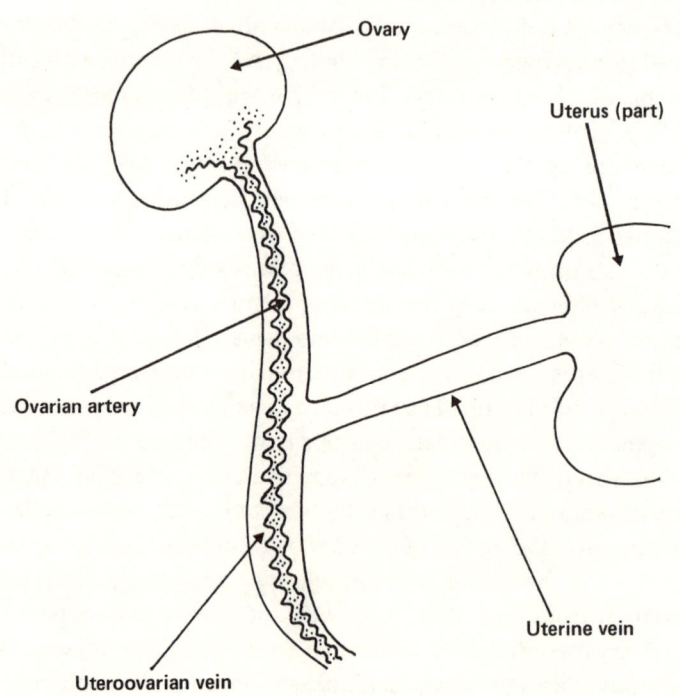

Fig. 5.4. Vascular arrangement of the uterus and ovary in the sheep.

clinical trials have reported successful menstrual regulation with PGE_2, $PGF_{2\alpha}$, 15-methyl-$PGF_{2\alpha}$, 16,16-dimethyl-PGE_2 and 16,16-dimethyl-$PGF_{2\alpha}$ (see Bygdeman, 1979). Intrauterine and vaginal administration of prostaglandins are equally effective, which raises the possibility that prostaglandins could be self-administered by a woman who suspects she may be pregnant. Since prostaglandins are only weakly luteolytic in women it is likely that menstrual regulation in this way is the result of increased uterine contractions, resulting in the expulsion of the foetus.

Luteolytic prostaglandins do have a veterinary use in animal husbandry to synchronise oestrus. Intrauterine implantation of $PGF_{2\alpha}$ or PGE_2 (or an appropriate analogue) to normally cycling cows or sheep causes luteal regression which is followed 2–3 d later by a normal ovulation and the animal coming once more into heat. By treating animals in this way the veterinarian can ensure that the whole herd is on heat more or less at the same time. Such an arrangement is more convenient for the farmer who can then, for example, make sure that the animals are artificially inseminated when they are on heat and thus when the chances of successful fertilisation are highest. The use of prostaglandins and their analogues in this way are discussed in detail by Cooper, Hammond & Schulz (1979).

Menstruation – a role for prostaglandins?

Menstruation occurs in primates and represents the process of blood loss through the vagina which marks the end of the menstrual cycle. The function of menstruation is to remove the unwanted uterine endometrium which, at this stage, is thickened and highly vascularised in preparation for the receipt of a fertilised egg. If fertilisation does not occur then the corpus luteum is broken down, plasma progesterone levels fall, the endometrium is shed and the menstrual cycle starts again.

Prostaglandins may be involved in physiological menstruation and may be responsible for some of the unpleasant symptoms of menstrual disorders such as dysmenorrhoea and menorrhagia. $PGF_{2\alpha}$ and PGE_2 occur in the endometrium and in menstrual blood (Pickles, 1957). Human endometrial levels of these prostaglandins change with the different stages of the menstrual cycle, as shown in Fig. 5.5. Small amounts of both prostaglandins are found in the proliferative phase, increasing in concentration during the luteal phase, and reaching maximum levels just before menstruation. Interestingly, the rise in endometrial $PGF_{2\alpha}$ during the latter part of the menstrual cycle is considerably more dramatic than the modest increase in endometrial PGE_2 concentration. Whether or not this difference in the rate of biosynthesis of the two prostaglandins has physiological relevance for menstruation remains to be seen.

Menstrual blood loss is significantly reduced by pre-treatment with NSAID like indomethacin (Anderson et al., 1976). Since menstruation in women is not prevented or delayed by aspirin-like drugs it is clear that prostaglandins formed by the uterus do not by themselves cause menstruation. Rather prostaglandins probably act as secondary modulators or 'fine tuners' of menstruation. Exactly how prostaglandins facilitate menstruation is not known, but at least two mechanisms may be involved. Firstly, $PGF_{2\alpha}$ constricts spiral arteries in the endometrium, thereby depriving large parts of the endometrium of oxygen. These arteries are known to constrict just before the onset of menstruation in women. $PGF_{2\alpha}$ may promote blood loss and endometrial shedding simply by contracting the uterus and so physically rupturing the delicate endometrial blood vessels. In contrast to $PGF_{2\alpha}$, PGE_2 and PGI_2 (the major cyclooxygenase metabolites of arachidonic acid in the uterus) are vasodilator in most vascular beds and thus are most unlikely to constrict spiral arteries in the endometrium.

Up until this point I have concentrated upon those events in the menstrual cycle which become important only if fertilisation does not occur. However, if the ovum is successfully fertilised by a sperm in the

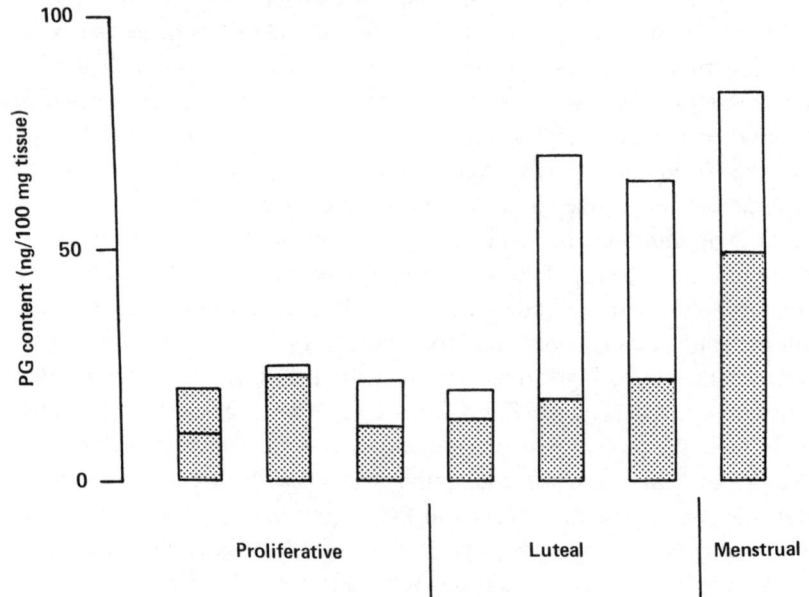

Fig. 5.5. PGE_2 (stippled) and $PGF_{2\alpha}$ (blank) levels in uterine tissue from women during different stages of the menstrual cycle (data from Downie, Poyser & Wunderlich, 1974).

ampulla region of the Fallopian tube then a different course of events follows, leading eventually to pregnancy and childbirth. Prostaglandins are almost certainly involved in the processes which take place in the body both during and after the pregnancy.

Fertilisation and implantation

After fertilisation the developing embryo or blastocyst is wafted along the Fallopian tube to the uterus, aided by the rhythmic contractions of the tube. These contractions are initiated by activity in noradrenergic autonomic nerves. In women, the transit time for the egg along the Fallopian tube is about 5 d and in other species about 3–4 d. Administered prostaglandins affect egg transport by contracting or relaxing the smooth muscle of the Fallopian tube. $PGF_{2\alpha}$, for example, hastens the arrival of the fertilised egg in the uterus when administered 10 h before mating in rabbits; in this (and other) species, $PGF_{2\alpha}$ contracts the oviduct *in vitro* (Chang, Saksena & Hunt, 1974). It is not known whether endogenous prostaglandins formed by the cells lining the Fallopian tube also cause it

Fig. 5.6. The role of prostaglandins in fertilisation and implantation. (1) ovulation, (2) fertilisation in the ampulla region of the Fallopian tube, (3) transport of the fertilised egg aided by prostaglandins from the wall of the Fallopian tube, (4) prostaglandins osmotically lyse blastocyst, (5) implantation.

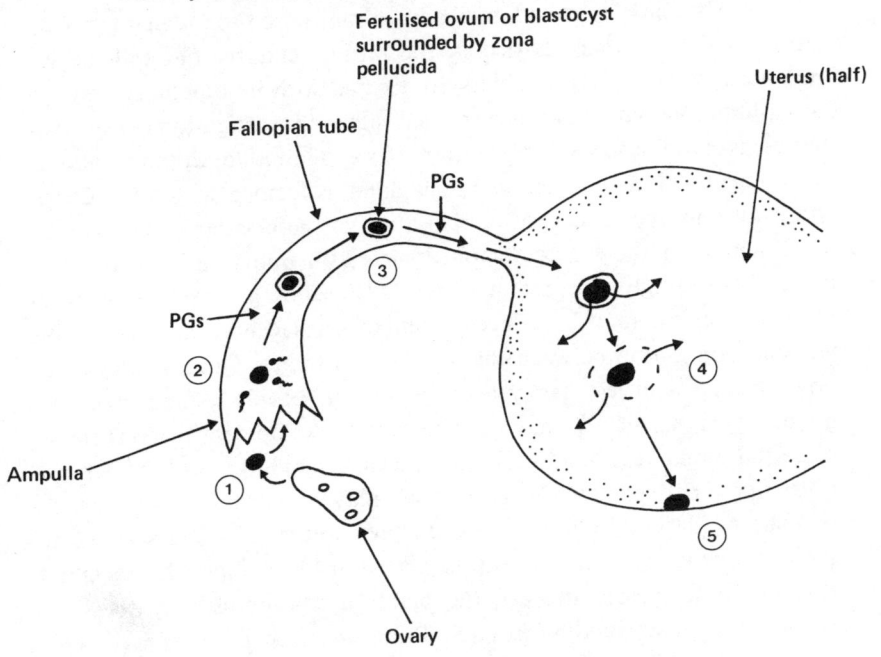

to contract but the finding that rabbits treated with indomethacin show normal egg-transit times along the Fallopian tube to the uterus argues against a physiological role for prostaglandins.

Once the blastocyst arrives in the uterus it becomes embedded in the endometrium. In species which give birth to several young at one time, the blastocysts attach to the endometrium at neatly spaced intervals. Before a blastocyst can embed itself in the endometrium it must first lose its outer protective glycolipoprotein coating called the zona pellucida. Enzymes secreted by the endometrium assist in this process, as do prostaglandins synthesised by the blastocyst (Dickmann & Spilman, 1975). Blastocyst prostaglandins may produce a massive influx of water into the blastocyst from the uterine liquor, causing it to burst thereby removing the zona pellucida. Certainly aspirin-like drugs prevent mouse blastocyst hatching *in vitro*, suggesting a physiological role for prostaglandins (presumably of the E series, since other prostaglandins are not known to promote fluid transport across epithelial cells) in the actual implantation process. This conclusion would certainly agree with the frequently reported observation that indomethacin treatment delays implantation in species such as the hamster and rabbit. A summary of the way in which prostaglandins may influence fertilisation and implantation can be found in Fig. 5.6.

Development and function of the placenta

As we have seen, indomethacin administered to pregnant rabbits before the egg becomes buried in the uterine endometrial wall delays implantation. However, if rabbits are treated with indomethacin several days after implantation has occurred subsequent investigation reveals that the uterus contains fewer implantation sites and fewer developing foetuses. This indicates that endogenous prostaglandins formed within the uterus affect not only the process of implantation of the blastocyst but also the development of the blastocyst once successful implantation has taken place. Interestingly, the major effect of F series prostaglandins post-implantation is to retard the development of the placenta, which is the vital vascular connection between the mother and foetus. Once again exactly how this is achieved has yet to be unequivocably established. One possibility is that $PGF_{2\alpha}$ may slow placental development by constricting placental blood vessels and thereby starving the placenta of oxygen and nutrients. Endogenous prostanoids do play a part in controlling the patency of placental blood vessels. In pregnant rabbits and sheep indomethacin produces a sharp drop in placental blood flow which suggests that, in these species at least, the placenta continuously synthesises a vasodilator prostaglandin (probably PGI_2) which maintains the placental blood vessels in a dilated state. When PGI_2 formation is prevented the placental blood vessels constrict, leading to slower development of the blastocyst and fewer foetuses.

Parturition

Parturition is the process of childbirth. There is good evidence derived from studies in several species that prostaglandins are involved in both parturition and the softening and dilatation of the cervix which immediately precedes parturition. This evidence has been derived from three types of experiment.

Prostaglandin biosynthesis by the uterus during pregnancy. Large amounts of prostaglandins (PGE_2, $PGF_{2\alpha}$, PGI_2) occur in amniotic fluid of women in late pregnancy and during labour (Salmon & Amy, 1973). As labour proceeds and the cervix becomes progressively dilated, in readiness for childbirth, amniotic fluid concentrations of both PGE_2 and $PGF_{2\alpha}$ increase steadily (Fig. 5.7).

Increased synthesis and release of $PGF_{2\alpha}$ and, to a lesser extent, PGE_2 and PGI_2 by the uterus and/or placenta just prior to parturition has now

Fig. 5.7. Amniotic fluid prostaglandin concentrations in women during labour. $PGF_{2\alpha}$ (open triangles) and PGE_2 (filled triangles) levels are increased towards the end of labour as the cervix becomes maximally dilated. Data from Karim (1975).

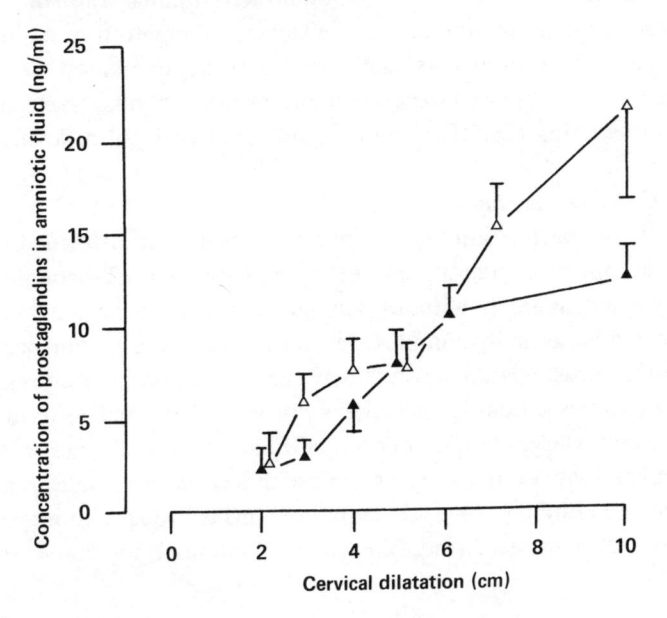

been demonstrated in several species including sheep, rabbit, cow, monkey, guinea pig and of course in humans.

Effect of prostaglandins on uterine contractility. The administration of PGE_2 or $PGF_{2\alpha}$ to women at any stage of their pregnancy causes powerful contractions of the uterus and expulsion of the foetus. Use is made of this action in the clinic to produce abortion and induce labour at term. These aspects are dealt with in detail later in this chapter. The majority of pharmacological studies have been performed using the isolated uterus from pregnant rats sacrificed shortly before the expected onset of parturition. On this preparation, $PGF_{2\alpha}$, PGE_2, PGD_2 and PGI_2 are all spasmogenic although PGD_2 and PGI_2 are considerably less potent than $PGF_{2\alpha}$ or PGE_2 (Williams, El Tahir & Marienkiewicz, 1979). In addition to contracting the uterus directly, PGI_2 (which is the major arachidonic acid product detected in uterine myometrium) also potentiates the ability of oxytocin to contract the rat uterus. Since this peptide is released from the posterior pituitary during labour and is known to stimulate uterine PGI_2 biosynthesis (Williams & El Tahir, 1980) it may well be that an interaction between oxytocin and PGI_2 on the myometrium coupled with the direct stimulant effect of $PGF_{2\alpha}$ and PGE_2, combine to ensure the rapid delivery of the foetus during labour.

Effect of NSAID on uterine contractility. If endogenous prostaglandins do promote uterine contractility during labour then administration of NSAID to pregnant animals should have two effects: prolongation of labour and delay of parturition to increase the duration of the pregnancy. Both such effects have been observed experimentally in rats, rabbits, hamsters and in women receiving NSAID chronically for the treatment of arthritis.

Cervical ripening

Before parturition begins the cervix must be transformed from a narrow, fibrous tube to a dilated, soft, compliant channel through which the baby can pass safely without damage. This process is called cervical ripening and is normally complete several hours before labour begins. It is likely that prostaglandins formed by the cervix during late pregnancy contribute to this softening and opening process since the late-term sheep cervix generates large amounts of both PGE_2 and PGI_2 and treating women intravaginally with PGE_2 causes ripening of the cervix during late pregnancy (Mackenzie & Embrey, 1978). Exactly how this is brought about remains something of a mystery since this organ contains only sparse smooth muscle, contraction or relaxation of which has little or no effect on the width of the cervix. One possibility is that prostaglandins change the structure of collagen material in the cervix which gives it its fibrous properties (Liggins, 1978). The clinical applications of prostaglandins for pre-cervical ripening in women in whom, for one reason or another, this process does not occur naturally is discussed later in this chapter.

Lactation

Pharmacological doses of PGE_2 and $PGF_{2\alpha}$ administered systemically to lactating women provoke copious milk ejection. This is a central effect caused by stimulation of oxytocin release from the posterior pituitary gland. Because of the magnitude of the doses of prostaglandin required and the finding that aspirin-like drugs do not influence milk release in lactating women it is unlikely that prostaglandins have a physiological function in lactation.

Clinical uses of prostaglandins and aspirin-like drugs in the female reproductive tract

Prostaglandins

Abortion. Of the naturally occurring prostaglandins, PGE_2 and $PGF_{2\alpha}$ are used clinically to induce therapeutic abortion in the second trimester of pregnancy. Sometimes this is necessary because the foetus has died in the uterus or, alternatively, because screening of the aminiotic fluid has revealed that the foetus has a crippling disease (like spina bifida). Prostaglandin analogues have also been used as abortifacients. These analogues include 16-phenoxy-ω-17,18,19,20-tetranor-PGE_2-methylsulphonylamide (sulprostone), 15(S)-15-methyl-PGE_2 and the same derivative of $PGF_{2\alpha}$ and 16,16-dimethyl-PGE_2-p-benzaldehyde semicarbazone among many others. The time interval between administration of a prostaglandin and the induction of abortion, as well as the incidence and severity of side effects, varies with the route of administration.

A great deal of effort has gone into identifying the optimum route of administration. Intravenous infusion of PGE_2, $PGF_{2\alpha}$ and sulprostone causes abortion after 8–16 h but the incidence of side effects (vomiting, pyrexia, diarrhoea) is high, sometimes unacceptably so. Other routes of administration which result in absorption of large amounts of prostaglandins into the systemic circulation (oral, intramuscular, intravaginal) also produce abortion after a time lag of 10–20 h with a higher incidence of side effects. The preferred route of administration of prostaglandins for abortion is probably intraamniotic. Although this procedure requires the skills of a trained doctor the incidence of side effects is comparatively low as the prostaglandins are localised within the uterus and the time to abortion is also somewhat shorter (about 8–10 h) than with other routes. Whichever route is chosen, prostaglandin-induced abortion is a relatively safe procedure. In the USA, between 1972 and 1975, the incidence of death of the mother following $PGF_{2\alpha}$ abortion of the foetus was only 10.5 per 100 000 abortions. This compares favourably with an alternative method of mid-trimester abortion, injection of hypertonic saline, in which the maternal death rate is 17.7 per 100 000 successful abortions (Gates et al., 1977).

Although, at first sight, the way in which prostaglandins induce abortion would appear obvious (i.e. by increasing uterine contractions which physically dislodge and expel the foetus) this is probably not the entire story. Since the time interval between the administration of prostaglandin and abortion can be anything from 7 to 30 h, it is most unlikely that any prostaglandin will still be in the uterus when the abortion takes place. Possibly prostaglandins constrict the uteroplacental blood vessels, thereby starving the placenta and corpus luteum (in some species) of blood and preventing the formation of progesterone. In the absence of this hormone the uterus is rapidly changed in character from a quiescent organ to a powerfully contracting one and abortion follows.

Induction of labour. Occasionally if natural childbirth is delayed it is necessary to induce labour. This may be achieved using oxytocin which is the natural stimulant to parturition in women. PGE_2, $PGF_{2\alpha}$ and their chemical analogues are also used clinically to induce labour on their own and administered with oxytocin. Since prostaglandins sensitise the near-term human uterus to oxytocin, the administration of oxytocin and prostaglandins at the same time allows the dose of each to be reduced thus cutting down the incidence of side effects such as vomiting and diarrhoea.

Pre-cervical ripening. Prostaglandins probably play a part in the normal, physiological opening and softening of the cervix which immediately precedes childbirth. In some near-term pregnant women the cervix does not ripen properly and, if parturition is imminent, then some other means of opening up the cervix must be found. This process can be done manually with metal instruments inserted into the cervix which is then physically forced open. In recent years, PGE_2 either in tablet form or inserted directly into the vagina as a gel has provided an effective alternative for cervical ripening in the majority of women so treated.

Ductus arteriosus. The ductus arteriosus is a blood vessel found in the foetus which links the pulmonary artery with the ascending aorta (see Fig. 5.8.).

The foetal lungs are unused *in utero* and the pulmonary vasculature is almost maximally vasoconstricted. Thus, the pressure of blood in the pulmonary artery is very high. In contrast, since over 40% of blood flowing through the foetal aorta is passed to the placenta, an organ of low resistance, the arterial blood pressure and thus the pressure in the aorta is low. The effect of a high pressure pulmonary artery linked to a low pressure aorta means that blood will flow from the pulmonary artery through the ductus arteriosus into the aorta i.e. in the direction of the arrows in Fig. 5.8. This vascular arrangement is useful in the foetus since it by-passes the lungs and ensures a good flow of blood through the placenta, where oxygen is picked up from the maternal blood. After birth the ductus arteriosus constricts and shuts down completely, the pulmonary blood vessels dilate and blood flows through the lungs in the normal way. Exactly how these changes are brought about is uncertain. However, the very rapid increase in blood oxygen which occurs with the infant's first breath is known to be important. Two clinical conditions affecting the ductus arteriosus have been described:

Fig. 5.8. Vascular anatomy of the heart and lungs in the foetus. Notice the position of the ductus arteriosus which links the pulmonary artery to the ascending aorta thereby diverting blood away from the non-functioning lungs towards the placenta where gas and nutrient exchange occurs.

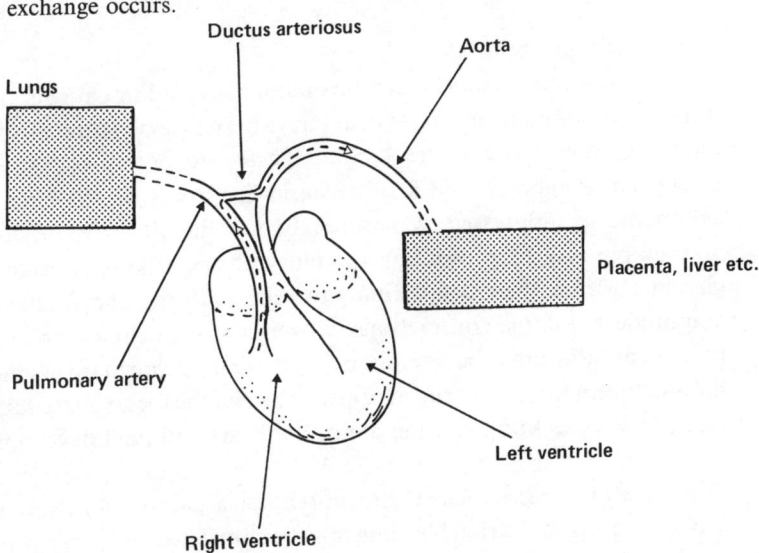

(1) in infants born prematurely, the ductus arteriosus may not close entirely leading to underperfusion of the lungs via the pulmonary artery, reduced uptake of oxygen into the bloodstream and respiratory distress;
(2) some infants may be born with a heart defect in which the right side of the heart does not pump properly, or alternatively, the pulmonary artery is badly damaged (in this case not enough blood passes through the lungs and the baby rapidly becomes ischaemic).

In the first case a treatment is required which closes the ductus arteriosus. Many drugs have been tested for this purpose but none has proved completely successful. However, the treatment of these premature infants with indomethacin causes a rapid closure of the ductus arteriosus with only minimal side effects (Heymann, Rudolph & Silverman, 1976). The treatment is simple, effective and avoids the necessity of surgery.

In the second case the opposite strategy is required. In such infants the ductus arteriosus must be kept open to allow some blood from the aorta to pass 'the wrong way' along the ductus arteriosus and force blood through the lungs for oxygenation. If the ductus arteriosus cannot be made patent, then surgery must be performed rapidly to correct the right side heart defect. In these circumstances such surgery is hazardous. Administration of vasodilator prostaglandins such as PGE_2 by intravenous infusion has been used successfully in infants with obstructive right heart malfunction prior to surgical correction of the defect (Olley, Coceani, Bodach, 1976).

Aspirin-like drugs

Premature labour. Premature labour is a leading cause of perinatal mortality. Four main groups of drugs have been assessed clinically for their ability to inhibit premature labour. These are progestogens, ethanol, adrenoceptor agonists and finally aspirin-like drugs. In the latter group, indomethacin suppresses premature labour in the majority of women. It presumably acts by preventing the biosynthesis of spasmogenic prostaglandins within the myometrium, thereby reducing the frequency and amplitude of uterine contractions. However, recent clinical trials suggest that great care must be exercised when treating pregnant women with indomethacin since this drug may also constrict the foetal ductus arteriosus (see previous section), causing abnormal heart and lung development.

Disorders of menstruation. Dysmenorrhoea, a painful affliction affecting women during menstrual bleeding may be connected with over-production of prostaglandins by the uterus. In women experiencing dysmenorrhoea plasma levels of 13,14-dihydro-15-oxo-$PGF_{2\alpha}$ are raised as are $PGF_{2\alpha}$ levels in the uterine endometrium. Treatment with aspirin, indomethacin, naproxen or mefanamic acid alleviates the pain and increased uterine contractility associated with this condition.

Prostanoids in the male

Although seminal plasma contains far and away the highest concentration of prostaglandins of any mammalian fluid or tissue, their physiological function or functions remain a mystery. Numerous roles for semen prostaglandins have been proposed although none, as yet, has been backed by concrete evidence. For this reason the role of prostaglandins in male reproductive function will be dealt with only briefly in this chapter. Further information and speculation is available in Karim & Rao (1975). Erection and ejaculation are processes in the male which some authors have suggested require either the presence of prostaglandins or, alternatively, prostaglandin biosynthesis by the seminal vesicle and vas deferens. Although an attractive hypothesis, several lines of evidence imply that prostaglandins are not involved in these processes. Firstly, several species (notably the rabbit, which seems to have no difficulty in reproducing), have little or no prostaglandins in their semen. Secondly, if male volunteers take aspirin in high dose for several days then, although semen prostaglandins levels fall by over 80%, no deleterious effect on erection/ejaculation are observed. Prostanoids have also been proposed to affect sperm migration, the inherent movements of the spermatozoa (which allow these cells to wriggle through the mucus layers in the cervix and the point at which the Fallopian tube joins the uterus). Again, no direct evidence for this claim has been put forward. Indeed, there is no correlation between sperm motility in a particular sample of seminal fluid and the concentration of prostaglandins in that sample. Furthermore, adding exogenous prostaglandins (other than at non-physiological concentrations) to seminal fluid does not affect sperm motility.

It is possible that, by affecting the smooth muscle of the female reproductive tract, seminal fluid prostaglandins can speed up the rate at which sperms are passed from the vagina to the ampulla region of the Fallopian tube where fertilisation occurs. This process is usually called sperm transport. Most prostaglandins which occur in seminal fluid (PGE_1, PGE_2, $PGF_{2\alpha}$, 19-hydroxy-$PGF_{2\alpha}$) contract the uterus. Only 19-hydroxy-PGE_1 causes the human uterus to relax *in vivo*. Perhaps seminal prostaglandins contract the uterus and this contraction acts to 'suck up' sperm deposited in the vagina following copulation, thus accelerating sperm transit. Once in the uterus, the sperm must still negotiate the uterotubule junction and penetrate to the Fallopian tube before fertilisation can take place. Prostaglandins have complex effects on Fallopian tube mobility. They either contract (PGFs) or relax (PGEs) tubal smooth muscle *in vivo*. Thus, whether seminal fluid prostaglandins promote sperm transport along the Fallopian tube will presumably depend upon the relative proportions of contractor and relaxant prostaglandins in the semen mixture.

Whatever functions prostaglandins have in seminal fluid, it is possible that they are necessary for normal male fertility. Several years ago it was observed that semen samples from men in infertile liaisons contained less E series prostaglandins than did samples from men of proven fertility. Later studies showed that seminal 19-hydroxy-PGE_1 and PGE_1 were indeed lower in infertile men. In the many subsequent studies, some researchers have confirmed deficient prostaglandins in the semen of infertile males whereas others have found no correlation between male fertility and seminal PG levels.

6
Prostanoids and the gut

Introduction
Ever since the discovery of prostanoids some 50 odd years ago, use has been made of isolated gut smooth muscle as bioassay preparations for prostanoids. Despite the introduction of highly selective and sensitive chemical and immunological assays a vast amount of prostanoid research is still carried out using the humble rat stomach strip or guinea pig ileum preparations set up in an organ bath.

Over and above these practical applications many other aspects of gastrointestinal function may to some extent depend upon the generation of endogenous prostanoids or are influenced by the administration of exogenous prostanoids. The list of gut functions so affected includes secretion of acid by the stomach lining, movement of water and ions in the colon and intestine and the regulation of secretions from organs like the gall bladder and pancreas. Evidence has also accumulated which suggests a pathophysiological role for prostanoids in several diseases of the gastrointestinal tract.

Synthesis and catabolism of prostanoids in the gut
Prostaglandin-like material has been found in the alimentary canal of all species studied to date. The amount and type of prostanoid depends upon the species, region of gut and experimental methods used. For example, whole (i.e. uncentrifuged) homogenates of dog gastric corpus mucosa incubated with radiolabelled arachidonic acid form large amounts of PGE_2 with much smaller amounts of PGD_2, TxB_2 and 6-oxo-$PGF_{1\alpha}$. However, if these homogenates are first centrifuged and the microsomal pellet obtained resuspended in buffer before incubation with arachidonic acid then a different array of prostanoids are formed with PGD_2 the major product (Ahlqist *et al.*, 1982). Similarly, rat stomach homogenates convert

arachidonic acid chiefly to PGE_2 (Pace-Asciak & Wolfe, 1970) but intact pieces of rat stomach agitated gently in physiological solution for a few minutes release large amounts of PGI_2 (Whittle, Moncada & Vane, 1978a).

There are probably several reasons for these contradictory results. One of the most obvious arises from the natural desire of the researcher to increase the yield of prostanoids in his/her experiments. To do this many researchers include in the incubation cofactors like reduced glutathione (GSH) or hydroquinone. Unfortunately these cofactors appear to redirect the breakdown of PG endoperoxides towards classical prostaglandins of the E, F and D series and away from PGI_2 and thromboxanes. Consequently the prostanoids detected in these experiments may not be the major products formed naturally in the body. A second complicating factor may be the presence in gut mucosa of as-yet uncharacterised factors which inhibit various enzymes involved in the arachidonic acid pathway (LeDuc & Needleman, 1979). These factors are present in the cell cytoplasm and thus are found in organ homogenates but not therefore in microsomes. Clearly great care must be taken when attempting to interpret the results of experiments in which prostanoid formation in the gut has been estimated *in vitro*.

Since prostanoids may have a physiological role to play in regulating contractility of the stomach and controlling gastric acid secretion it is not surprising that most studies of gut prostanoid biochemistry have concentrated on the stomach. Conversion of arachidonic acid to prostanoids occurs in intact tissue, homogenates and microsomes of stomachs removed from several species, as shown in Table 6.1. Prostanoids containing 1 or 3 double bonds are either altogether absent or are found in only trace amounts in the gut. Although only the major prostanoid(s) are shown in Table 6.1, the different parts of the gastrointestinal tract are capable of

Table 6.1. *Formation of prostaglandins in the stomach*

Species	Major PGs detected
Human	$PGI_2 = PGE_2$
Rabbit	$PGI_2 \gg PGE_2 > PGF_{2\alpha}$
Guinea pig	$PGI_2 \gg PGE_2$
Cat	PGE_2
Rat	$PGI_2 > PGE_2$
Dog	$PGI_2 = PGE_2$
Cow	$PGI_2 > TxB_2$
Mouse	PGI_2

synthesising small amounts of virtually all prostanoids e.g. PGE_2, $PGF_{2\alpha}$, PGI_2, TxA_2, PGD_2 as well as lipoxygenase products such as 5-HETE. PGE_2 has also been detected by radioimmunoassay in extracted human gastric juice. The principal site of prostanoid formation in the stomach is almost certainly the corpus region, both the mucosal and serosal muscular layers of which are able to synthesise prostanoids. Whether the mucosa or the muscle is the major site of synthesis has not been unequivocably established although, in humans, intact gastric mucosa does form vastly more PG-like material (determined biologically) than the underlying muscle layer (Bennett, Stamford & Unger, 1973).

As well as synthesising prostanoids, the gastrointestinal tract also has the capacity to catabolise prostanoids once formed. Human, rat and rabbit stomach mucosa exhibit high 15-PGDH and PG-Δ13R activity *in vitro*. Surprisingly perhaps rabbit stomach antrum homogenates have approximately 6 times the 15-PGDH activity of rabbit lung homogenates, an organ always associated with very high PG-catabolising activity.

Other parts of the gastrointestinal tract have also received attention from the prostanoid researcher. Intact rat jejunum converts arachidonic acid mainly to PGD_2 with smaller amounts of PGE_2 and 6-oxo-$PGF_{1\alpha}$. Human small intestine also converts arachidonic acid to these three prostaglandins although the rate of formation is only about 10% that in the rat (Peskar *et al.*, 1981). Human and rat colon homogenates contain mainly PGD_2 and 6-oxo-$PGF_{1\alpha}$, as determined by gas-chromatography–mass-spectrometry; homogenates of human rectal biopsy material incubated with arachidonic acid yield both PGE_2 and $PGF_{2\alpha}$ in a ratio of 3:1 (Hawkey, 1982). Similarly, PGE_2 has also been found in extracts of human rectal biopsies and in human stools. Like the stomach, the lower parts of the gut are also excellent sources of prostanoid-catabolising enzymes like 15-PGDH. Prostanoid synthesis also occurs in organs associated with the gut. Human gall bladder strips form PGE_2 when incubated with arachidonic acid; isolated rat pancreatic acinar cells convert arachidonic acid to 5-HETE, 6-oxo-$PGF_{1\alpha}$, PGE_2 and $PGF_{1\alpha}$, indicating both cyclooxygenase and lipoxygenase activity (Rubin *et al.*, 1979).

The synthesis and release of prostanoids can also be demonstrated from intact gut segments mounted in warmed, oxygenated Krebs' solution in organ baths. Some isolated pharmacological preparations (e.g. rabbit jejunum and rat stomach) exhibit spontaneous, rhythmic contractions and relaxations. This so-called myogenic tone is believed to reflect spontaneous nervous activity in the myenteric plexus. However, the contractions of these isolated preparations are also associated with release of prostanoids (mainly PGE_2 and $PGF_{2\alpha}$). These may contract gastrointestinal smooth muscle directly or influence gut motility indirectly by affecting the firing of neurones in the myenteric plexus (Eckenfels & Vane, 1972). Thus part of the 'myogenic tone' may be due to prostaglandins released from the tissue. The finding that aspirin-like drugs reduce or abolish the spontaneous movements of rabbit jejunum and rat stomach strip preparations in organ bath supports the idea that prostanoids contribute to myogenic contractility of these tissues.

Biological actions of prostanoids in the gut

Although the gut is an excellent source of a number of different prostanoids, their precise role in gastrointestinal function has been debated for many years. Three major physiological functions have been proposed:

(1) control of gut motility;
(2) control of gastrointestinal secretions;
(3) protection of the gut against ulcers.

Effect of prostanoids on gastrointestinal motility

There are two major approaches to the study of the effect of drugs on the motility of the gut. The classical pharmacological approach involves removal of a section of the gut from an experimental animal to an organ bath or a cascade superfusion bioassay set up, and studying its response to agonist and antagonist drugs. The main advantage of these experiments *in vitro* are that (*a*) the longitudinal and circular muscle layers of the gut may be studied separately, (b) the results obtained are uncomplicated by factors like metabolism of the drug and effects of the drug on blood flow to the gut and, finally, (*c*) that gut smooth muscle preparations are usually sensitive to very small amounts of agonists and are relatively inexpensive and easy to work with. The second approach, estimation of the effect of drugs on gut motility in the whole animal *in vivo*, involves more complicated and time-consuming techniques but does give a valuable indication of what happens in the whole animal.

Most prostanoids contract longitudinal smooth muscle of the gastro-intestinal tract both *in vitro* and *in vivo*. Exceptions to this general rule are the rat and frog intestine which are relaxed by PGE_1 and human stomach and colon which are relaxed by PGE_2. Prostaglandins of the E and F series are more potent spasmogens on gut longitudinal smooth muscle than PGD_2 or PGI_2. The prostaglandin endoperoxides and TxA_2 have only a weak spasmogenic effect on most gut smooth muscle preparations (e.g. guinea pig ileum, rat stomach strip) although both PGG_2 and PGH_2 potently contract the isolated chick rectum preparation. TxB_2 and 6-oxo-$PGF_{1\alpha}$ are virtually inactive.

Leukotrienes also contract longitudinal muscle of the gut. LtC_4, LtD_4 and LtE_4 are thousands of times more potent than histamine on the guinea pig ileum with an order of potency $LtD_4 > LtC_4 > LtE_4$ (Lewis *et al.*, 1980). Leukotrienes also contract the rat stomach preparation. All of these actions on the gut may be blocked with the leukotriene receptor antagonist FPL 55712.

Prostanoids also affect circular smooth muscle of the gut. In order to study circular smooth muscle, spiral strips of ileum or colon are cut and preparations mounted in organ baths in the normal way. Alternatively, rat or rabbit stomach may be cut transversely across the longitudinal muscle layer in order to study contractions and relaxations of the circular smooth muscle. Whittle, Mugridge & Moncada (1979) have examined the effect of several prostanoids on the superfused transverse stomach strip of the rabbit. This preparation has been found to relax in the presence of PGE_2 and PGD_2 (threshold concentration about 1 ng) but contracts to PGI_2 and $PGF_{2\alpha}$ (threshold concentration about 10 ng). PG endoperoxides and TxA_2 have little effect even at doses as high as 40 ng. In general similar results are obtained in other circular smooth muscle preparations in that PGEs are nearly always relaxant and PGFs nearly always cause a contraction. The effect of prostanoids on the motility of gut smooth muscle *in vitro* is summarised in Fig. 6.1.

When injected into the intact animal, prostaglandins increase gastro-intestinal motility and can cause diarrhoea. In the anaesthetised rat, for example, intravenous injection of PGE_2 (1–2 μg/kg) contracted longitudinal smooth muscle and increased the intraluminal pressure. $PGF_{2\alpha}$ had a similar action but much higher doses (10 μg/kg) were required to produce the same rise in intraluminal pressure. To complicate matters, definite species variations do occur. As in the rat, $PGF_{2\alpha}$ injected intraarterially in anaesthetised dogs increased jejunal motility whilst PGE_1 administered by the same route decreased jejunal motility. Since prostaglandins of the E series contract longitudinal smooth muscle of the gut and relax the circular smooth muscle *in vitro*, the effect of PGEs in the intact animal may be a balance between these opposing effects. Interpreted in this way, in the intact rat the contractile effect of PGE_1 on intestinal longitudinal smooth muscle is predominant whereas in the dog the effect on circular muscle is greatest. The idea that the effect of prostanoids on the gut *in vivo* is simply the sum of their actions on the longitudinal and circular muscle layers is a convenient one but is almost certainly an oversimplification. In the intact animal, prostanoids may affect gut contractility not only directly but also indirectly by influencing the activity of sympathetic and parasympathetic nerves within the gut, by affecting gut blood flow, by releasing other local hormones or even possibly acting in the brain. All of these possibilities should be, but rarely are, considered when attempting to understand the complicated effects of individual prostanoids on gastrointestinal contractility *in vivo*.

Unlike many other body systems, the effect of prostanoids on the gut can be quite easily investigated in man. Orally administered PGE_1 (0.8–3.2 mg) to human volunteers increased gut motility, caused colic (a painful spasm of the gut) and the passage of loose watery stools (Horton *et al.*, 1968). Interestingly the gastric juice of these volunteers frequently contained bile, after treatment with PGE_1; this suggested that this prostaglandin also contracts the gall bladder and/or relaxes the cardiac sphincter, thus allowing reflux of the contents of the intestine into the

Fig. 6.1. Effects of prostanoids on gastrointestinal motility in mammals. ↑, contraction; ↓, relaxation; 0, not determined.

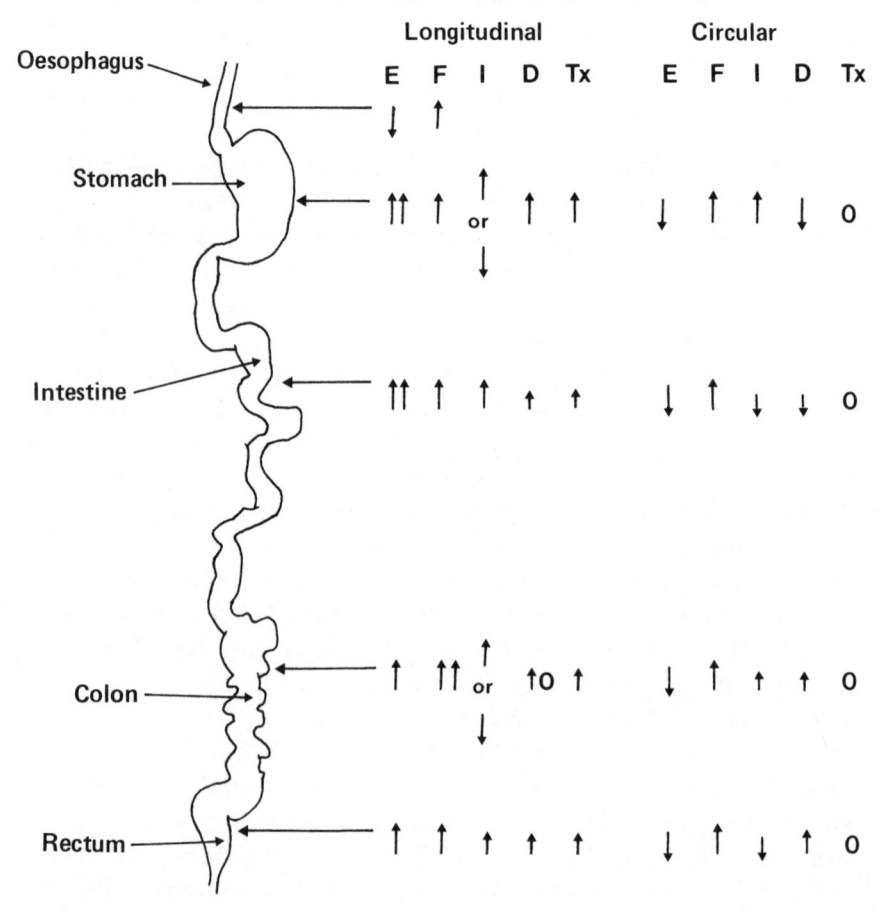

stomach. In a later study Misiewicz et al. (1969), using more objective techniques like radiotelemetry (swallowing a capsule which contains a radioactive source which may be monitored to locate its position in the gut) and balloons inserted into the colon and rectum connected to pressure transducers, demonstrated that PGE_1 (2 mg orally) increased propulsive movements of the intestine and colon. Patients receiving PGE_1 complained of colic and not surprisingly, experienced the desire to defaecate.

Synthetic analogues of prostaglandins have also been assessed for their effect on gut motility. A 5 min infusion of the PGE_2 analogue, 16,16-dimethyl-PGE_2 (7–140 ng/kg) increased gastric emptying and reduced (not increased) small intestinal transit. In man, oral 16,16-dimethyl-PGE_2 delayed intestinal transit, measured along a 70 cm jejunal segment. These latter two reports suggest that orally administered 16,16-dimethyl-PGE_2 and PGE_2 have opposite effects on intestinal contractility. Naturally occurring PGE_2 enhances contractility while its dimethylated derivative inhibits contractility. However, both PGE_2 and 16,16-dimethyl-PGE_2 enhanced gastric motility in man; similar results have been obtained using 15-methyl-PGE_2 and 15-methyl-$PGF_{2\alpha}$ in conscious Rhesus monkeys (Nompleggi et al., 1980).

There have been few reports of the effect of the more chemically labile prostanoids on gastrointestinal motility in vivo. Prostacyclin injected either intravenously or intraarterially into monkeys or rats inhibits gastric emptying and either has little effect or slightly increases intestinal transit (Shea-Donohue et al., 1980). There have as yet been no reports of the effect of PGG_2, PGH_2 or TxA_2 on gastrointestinal smooth muscle in vivo although intravenous infusion of the PGH_2 analogue, U46619 (0.2 µg/kg/min) increased gastric emptying in Rhesus monkeys, as deter-

Table 6.2. *Effect of a range of prostaglandins and their analogues on gastrointestinal contractility in vivo[a]*

Prostanoid	Gastric emptying	Intestinal transit	Colonic transit
PGE_1, PGE_2	↑	↑↓	↑
$PGF_{2\alpha}$	↑	↑	—
PGI_2	↓	↑	—
U46619	↑	—	—
15-methyl-PGE_2	↑	↑	—
16,16-dimethyl-PGE_2	↑	↓	—

[a] ↑ Indicates contraction, ↓ indicates relaxation, — no effect.

mined by a dye-dilution method. The effect of prostanoids on gastrointestinal contractility *in vivo* is summarised in Table 6.2.

Mechanism of action of prostanoids on gut smooth muscle. Prostaglandins act directly on smooth muscle although part of their effect on isolated gastrointestinal preparations may be mediated by nervous mechanisms or by the release of other biologically active substances. Contractions of the guinea pig ileum to E series prostaglandins (but not to F series prostaglandins) are partially reduced but not abolished by atropine, tetrodotoxin or procaine but are unaffected by the ganglion-blocking drug, hexamethonium. From this it would appear that the nervous component of the action of PGE_2 on this preparation involves stimulation of postganglionic cholinergic nerves in the myenteric plexus. In some other tissues (human ileum) prostaglandins also have a direct effect on gut smooth muscle, combined with an indirect effect on the enteric nerves. In contrast, the relaxant effect of PGE_1 on the isolated rat intestine is almost certainly an indirect action which may be blocked by a combination of α and β adrenoceptor antagonists. Thus, in this preparation, PGE_1 may be assumed to cause relaxation by releasing endogenous catecholamines from sympathetic nerves within the ileum.

The first step in the contraction of gut smooth muscle is the combination of the prostanoid with a selective receptor on the smooth muscle cell membranes. Gastrointestinal tissue of both man and experimental animals contains several prostanoid receptors on smooth muscle and nerves, the type and density of which varies between species, different parts of the gut and between longitudinal and circular muscle. As yet no satisfactory means of distinguishing these receptors is available despite the fact that several prostanoid receptor antagonists (e.g. SC19220, dibenzoxazepine, diphloretin phosphate, 7-oxo-prostaglandins, AH19437) have been developed and tested for their effect on the response of various gut smooth muscle preparations. In addition little is known of the events which take place within smooth muscle cells following combination of a prostanoid with its receptor. Some preliminary information about these post-receptor events has been gleaned from experiments using electrophysiological methods. Relaxation of the circular smooth muscle of the guinea pig stomach by PGE_2 was associated with membrane hyperpolarisation, whilst contraction of the same muscle by $PGF_{2\alpha}$ followed a depolarisation of the membrane (Mishima & Kuriyama, 1976). It is likely that these changes in membrane potential arise from changes in the fluxes of cations like Ca^{2+} and Na^+ across the membranes of smooth muscle cells. Direct evidence for this comes from experiments using gut smooth muscle and frog skin epithelium in which prostaglandins cause release of Ca^{2+} ions (Cociani et al., 1969).

Another way in which prostanoids may affect gastrointestinal motility is by an action on membrane cyclase enzymes to change the intracellular concentrations of cyclic nucleotides like cAMP and cGMP. Unfortunately, experiments to study this possibility have produced conflicting results. In tissues like the rat intestine, contractions due to PGE_1 are associated with a fall in cAMP in smooth muscle cells whereas contractions due to PGI_2 are not. In other tissues e.g. rabbit intestine, both PGE_1 and PGI_2 cause contraction without affecting intracellular levels of cAMP. The role of cyclic nucleotides in smooth muscle contractility is outside the scope of this chapter but has been reviewed by Bolton (1979).

Effect of prostanoids on secretions from the gastrointestinal tract

Gastric acid secretion. Prostaglandins of the E and I series and their analogues inhibit gastric acid secretion in man and in experimental animals. The naturally occurring prostaglandins are inactive when administered by mouth because of their very rapid catabolism in the gut. PGI_2, PGE_1 and PGE_2 administered intravenously inhibit basal, pentagastrin-, histamine- or tetragastrin-induced gastric acid secretion in man, dogs and rats. The order of potency is $PGI_2 > PGE_2 > PGE_1$. In general, the ED_{50} concentration (i.e. the amount of prostaglandin required to bring about 50% inhibition of gastric acid secretion) is about 0.5–1 $\mu g/kg/min$. These prostaglandins also reduce gastric acid secretion induced by a variety of other stimuli including food, insulin and 2-deoxyglucose. $PGF_{2\alpha}$ has no significant anti-secretory effect in man or experimental animals. Subcutaneous administration of the prostaglandin endoperoxide analogue U46619 did not affect gastric acid secretion in monkeys (Shea-Donohue et al., 1980).

The lack of effect of naturally occurring prostaglandins given orally and their short duration of action has encouraged the development of metabolically stable analogues which are better suited to chronic administration in humans. The structure of some of these analogues is shown in Fig. 6.2. Analogues with a methyl group at C-15 or C-16 are poor substrates for 15-PGDH and have been extensively studied for their anti-secretory effect. $15(R)$-15-methyl-PGE_2 and its methyl ester administered orally reduced basal and pentagastrin-induced gastric acid secretion in man (Karim et al., 1973). An inhibition of basal acid output of 70% was observed with a 150 μg dose of this analogue. The stereoisomer, $15(S)$-15-methyl-PGE_2 also inhibited gastric acid secretion in man and was slightly more potent than the 15(R)-isomer (Nylander, Robert & Andersson, 1974) but had the disadvantage of causing some nausea and vomiting. Other researchers have studied the effect of a prostaglandin containing two methyl groups namely 16,16-dimethyl-PGE$_2$ and its methyl ester. This analogue inhibited basal acid secretion in man as well as acid secretion

Fig. 6.2. Chemical structures of two gastric anti-secretory prostaglandin analogues.

15-methyl-PGE$_2$

16, 16-dimethyl-PGE$_2$

Fig. 6.3. Effect of 15-methyl-PGE$_2$ on gastric acid secretion induced by infusion of histamine in unanaesthetised Heidenheim pouch dogs. 15 methyl PGE$_2$ (0.25 µg/kg, triangles), 0.5 µg/kg (circles) or vehicle (squares) were injected intravenously. Data from Robert et al., 1976b.

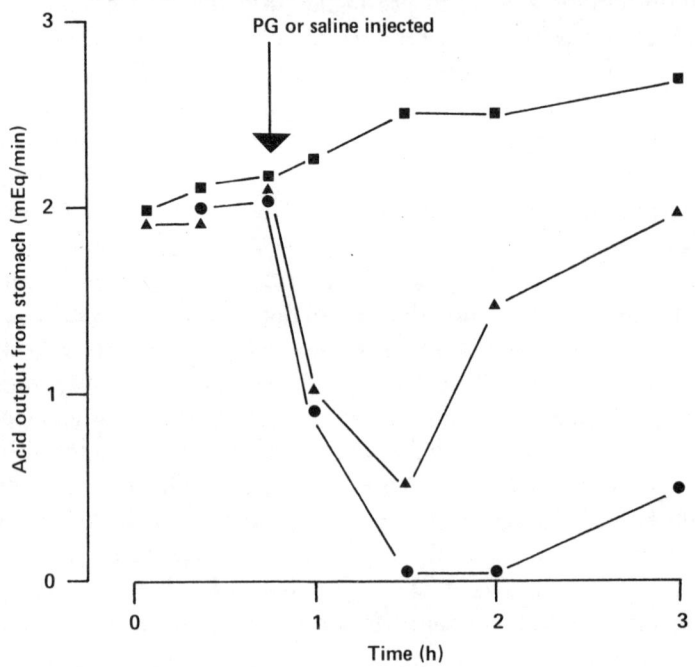

stimulated by histamine, food and pentagastrin all at doses as low as 1 μg/kg orally and without causing vomiting and diarrhoea. The effect of 16,16-dimethyl-PGE_2 on gastric acid secretion in intact dogs is illustrated in Fig. 6.3. Another analogue, 11-methyl-PGE_2 is also an effective anti-secretory drug given orally. Other anti-secretory prostaglandin analogues include 11,15-dihydroxy-16-methyl-16-methoxy-PGE_1 (DL 646), 15-deoxy-16-hydroxy-PGE_1 and 5-cyano-16-methyl-PGI_2 (nileprost).

Mechanism of gastric anti-secretory effect. An early attempt to explain how prostanoids inhibit gastric acid secretion centred upon their effect on gastric mucosal blood flow. It was proposed that one or more of the prostaglandins formed within the stomach mucosa constricted mucosal blood vessels causing a fall in blood flow, effectively 'starving' the acid-secreting parietal cells of oxygen and nutrients. As a result these cells were presumed to stop secreting H^+ ions and thus acid secretion declines. Early work seemed to support this possibility. In dogs surgically operated to produce a denervated (Heidenheim) stomach pouch, the fall in acid secretion caused by PGE_1 was accompanied by a sharp decline in mucosal blood flow (Wilson & Levine, 1969). However, it was later revealed that whilst mucosal blood flow was reduced by PGE_1, the ratio of blood flow to acid secretion was unaltered; this suggested that the effect on blood flow to the mucosa was a consequence, and not the cause, of the reduction in acid output. Later experiments have shown that PGEs also reduce H^+-ion release from the isolated bullfrog gastric mucosa, in which blood flow cannot be a factor.

Another possible mechanism of action is by an effect on cAMP and/or cGMP in the gastric mucosa. Prostanoids are known to interfere with the formation of cAMP and cGMP in a number of target tissues. There is much evidence that parietal cell cAMP levels are very important in the control of acid secretion. Thus it is not unreasonable to suppose that cyclic nucleotides may also be involved in the response to prostanoids. Early experiments along these lines produced confusing results in that both PGE_1 (which decreases acid output) and histamine (which increases acid output) caused a stimulation of adenylate cyclase to increase cAMP in guinea pig gastric mucosa. In later experiments, homogenates of gastric fundus mucosa from the dog were treated with enzymes to separate the parietal cells from other cells in the mucosa. Using such isolated and concentrated parietal-cell suspensions, Soll (1980) and Soll & Whittle (1981) showed that PGE_2 (IC_{50}, concentration for 50% of the maximum inhibitory effect is 10 nM) reduced histamine-induced uptake of [^{14}C]aminopyrine (used as an index of parietal-cell acid secretion) and decreased histamine-stimulated cAMP production. Higher concentrations (> 1 μM) of PGE_2 increased cAMP levels, most probably by an action on contaminating non-parietal cells. Prostacyclin and two of its analogues 6-β-PGI_1 and 5(α)-5,9-epoxy-16-phenoxy-$PGF_{1\alpha}$ also reduced histamine-stimulated aminopyrine uptake and cAMP formation. There now seems little doubt that some gastric acid secretagogues (e.g. histamine, pentagastrin) act by raising cAMP levels in gastric mucosal parietal cells, while anti-secretory (E and I series) prostaglandins are inhibitory by blocking adenylate cyclase activity thereby reducing parietal cell cAMP.

Intestinal secretion. PGE_1, PGE_2, PGA_1 and $PGF_{2\alpha}$ infused directly into the jejunum of conscious dogs dose-dependently inhibited the absorption of water and electrolytes (Matuchansky & Bernier, 1971). Similar results have been obtained in humans given either natural prostaglandins or the prostaglandin analogue, 16,16-dimethyl-PGE_2. Prostanoids appear to inhibit fluid absorption in the gut in two ways. Firstly, they inhibit the active reabsorption of Na^+ and Cl^- ions from the lumen across the brush border of the intestine wall and, secondly, they stimulate the secretion of these ions into the lumen from the intestine. Both of these processes occur when prostanoids stimulate adenylate cyclase, thereby increasing cAMP levels in intestinal cells. It is believed that a rise in cAMP in the villus cells lining the intestine inhibits Na^+ coupled Cl^- uptake at the brush border, while a similar rise in cAMP in the intestinal crypt cells opens up a Cl^- ionophore in the membrane of these cells to allow Cl^- ions to escape into the lumen.

Prostaglandin-induced accumulation of fluid in the small intestine is called enteropooling (Robert *et al.*, 1976*a*). If sufficient enteropooling occurs then large amounts of enteropooling fluid are carried to the colon and expelled as diarrhoea. Interestingly, cholera toxin produces a similar enteropooling effect. Both prostaglandins and cholera toxin promote the net secretion of fluid and electrolytes into the lumen of the small intestine by a similar mechanism i.e. stimulation of mucosal adenylate cyclase to increase cAMP levels. Not all prostaglandins cause enteropooling and thus diarrhoea. PGI_2 and PGD_2, for example, have exactly the opposite effect in that they prevent enteropooling due to $PGF_{2\alpha}$ and 16,16-dimethyl-PGE_2 as well as that caused by cholera toxin (Robert, 1981).

Further down the gastrointestinal tract, prostanoids also influence absorption of water and electrolytes in the large intestine although once again the response varies with the prostanoid under study. For example, PGE_2 and 16,16-dimethyl-PGE_2 added to segments of rat colon *in vitro* reduced the reabsorption of Na^+, Cl^- and water.

Pancreatic secretion. Prostanoids interfere with both the exocrine and endocrine functions of the pancreas. PGE_1 (0.5–5 µg/kg/min) infused into dogs with chronic duodenal fistulae reduced the volume and bicarbonate concentration of exocrine pancreatic juice secreted in the resting and in the secretin-stimulated state (Rudick *et al.*, 1971). The ED_{50} for inhibition of bicarbonate output was some 3–4 times that required to inhibit gastric acid secretion by a similar amount. The endocrine function of the pancreas (i.e. the secretion of insulin) has also been studied. Prostaglandins of the E series inhibit basal secretion and glucose-stimulated insulin release from the β cells of the islets of Langerhans (Robertson & Chen, 1977).

Additionally there is some evidence that prostaglandins synthesised locally in the pancreas may help to regulate insulin secretion *in vivo*. Aspirin and indomethacin both enhance insulin secretion and potentiate the surge in insulin output which occurs in response to a rise in plasma glucose, suggesting that these drugs inhibit the production of a prostanoid which is normally made in the pancreas and which reduces insulin release. The identity of the prostanoid responsible for this 'tonic inhibition' of insulin release has not been established but both PGE_2 and PGI_2 are formed when sonicates of rat pancreatic islets are incubated with arachidonic acid and therefore must be good candidates for the role. In contrast to the inhibition of insulin release which has been demonstrated with prostanoids, Morgan & Pek (1982) have obtained evidence that lipoxygenase products of arachidonic acid (possibly leukotrienes) exert a tonic stimulatory effect on insulin release in man. Although the situation is still confused, metabolites of arachidonic acid may serve an important function in controlling insulin release from the pancreas.

Protection of the gut against ulcers – cytoprotection

Prostaglandins of the I and E series inhibit gastric acid secretion in humans and experimental animals and additionally prevent ulcer formation and promote the healing of existing ulcers. Anti-ulcer activity was at one time ascribed totally to the ability to block acid secretion in the stomach. Since some prostaglandins are anti-ulcer at concentrations which do not influence gastric acid secretion and protect against ulcers which do not respond to therapy with antagonists acting on H_2 histamine receptors or to antacids, it was concluded that not all of the anti-ulcer activity of prostaglandins could be attributed to inhibition of acid secretion. The protective action of prostaglandins towards gastric ulcers was termed their cytoprotective effect and the process called cytoprotection. Originally cytoprotection was used to describe only the anti-ulcer activity of prostaglandins. More recently, cytoprotection has been invoked to explain also the ability of prostaglandins to protect organs such as the liver and pancreas from noxious stimuli.

Gastric cytoprotection. The first indication that prostaglandins could prevent ulcer formation in animals by a mechanism unrelated to inhibition of gastric acid secretion came with the experiments of Robert (1975). Aspirin administered orally to groups of rats produced haemorrhagic ulcers in the stomach. Not surprisingly prior administration of prostaglandins of the E series prevented ulcer formation, most likely because sufficiently high doses had been given to exert significant acid anti-secretory effect. A more unexpected finding was that $PGF_{2\alpha}$ and $PGF_{2\beta}$, which do not influence acid secretion by the stomach, also reduced the formation of ulcers caused by aspirin or indomethacin.

Prostaglandins were later shown to protect against gastric and intestinal ulcers induced by a number of other surgical and chemical stimuli (see Table 6.3; Robert, 1975). Some of the most interesting and dramatic experiments have been carried out on anaesthetised rats. In these studies large volumes of noxious substances were instilled directly into the stomach by means of a tube. For example, ethanol administered in this way caused massive necrosis and haemorrhage of large portions of the stomach mucosa. However, oral or subcutaneous injections of PGE_2, 16,16-dimethyl-PGE_2 or 15(*R*)-15-methyl-$PGF_{2\alpha}$ 1 or 2 min before ethanol produced a dose-dependent inhibition of ulcer formation and necrotic

Table 6.3. *Treatments which cause gastrointestinal ulcers in experimental animals and man*[a]

Aspirin, indomethacin and other NSAID
Prednisolone and other corticosteroids
HCl (0.6 N)
NaOH (0.2 N)
Bile salts
Reserpine
Serotonin
Histamine
Carbachol
Pentagastrin
Boiling water
Taurocholate

[a] Prior administration of prostaglandins prevents or reduces the incidence or severity of ulcers in each case.

lesions. In this sort of study prostaglandins are very potent gastric cytoprotective agents. 16,16-dimethyl-PGE_2 completely prevented ethanol-induced tissue damage at a dose (0.5 µg/kg) which was only 1% of the dose required for a significant anti-secretory effect. Using the same experimental procedure, prostaglandins have been shown to protect the gastric mucosa from damage caused by 0.6 M HCl, 0.2 M NaOH, 25% NaCl, acidified aspirin and even boiling water (Robert et al., 1979).

Gastric cytoprotection is rapid in onset (80% inhibition of ethanol-induced gastric damage by prostaglandins given orally to rats 1 min before ethanol) and provides protection for about 2 h. Cytoprotection occurs following several routes of administration (oral, intravenous, subcutaneous) and has been found to occur in several species including rat, rabbit, guinea pig, dog and even in the frog. PGE_2 has also been shown to exert a cytoprotective, anti-ulcer action in man. Orally administered PGE_2 (1 mg, 3 times daily) reduced faecal blood loss (indicative of gastrointestinal bleeding) in patients taking indomethacin to treat rheumatoid arthritis, pelveospondylitis and reactive arthritis (Johansson et al., 1980). In a more extensive, randomised, double-blind trial on 27 healthy male volunteers, Cohen, Cheung & Lister (1980) studied faecal blood loss using the ^{51}Cr-labelled red blood cell technique. Aspirin (600 mg, 4 times daily) for 7 d induced faecal blood loss which in turn was significantly reduced by PGE_2. Since PGE_2 was administered orally in these studies (and thus probably had little or no anti-secretory activity) the anti-ulcer effect observed presumably represented a cytoprotective action. Similar cyto-protection was also observed in an endoscopic study in humans in which 15(R)-15-methyl-PGE_2 protected against gastric mucosal damage caused by a single dose of 1300 mg aspirin (Gilbert et al., 1981).

Intestinal cytoprotection. As well as causing haemorrhagic ulceration of the stomach, NSAID also produce multiple lesions in the intestine extending from the upper jejunum to the lower ileum. Such lesions have been documented in man and experimental animals. Perforations and peritonitis, which can be fatal, may also occur after NSAID therapy. Subcutaneous or oral administration of a number of different prostaglandins dose-dependently prevent aspirin-induced intestinal ulceration, perforation and peritonitis. Prostaglandins are also cytoprotective in the intestine against large doses of anti-inflammatory corticosteroids like prednisolone. Studies in man have shown that orally administered 16,16-dimethyl-PGE_2 (1 µg/kg) in 11 healthy volunteers prevents the generation of ethanol-induced duodenal mucosal damage (determined endoscopically).

Colonic cytoprotection. There have been few reports of cytoprotection by prostaglandins in the large intestine. However, clindamycin (an antibiotic) when administered to hamsters causes a syndrome similar to human ulcerative colitis with the exception that in the hamster this syndrome is usually fatal. Both the colitis and the death of the animal can be prevented by oral administration of 16,16-dimethyl-PGE_2 (Robert *et al.*, 1976*b*). This effect of the PGE_2 analogue may be attributed to cytoprotection. Whether or not similar cytoprotection occurs in the colon of human beings has yet to be studied in, for example, patients with ulcerative colitis or Crohn's disease.

Cytoprotection in other organs associated with the gastrointestinal tract. Carbon tetrachloride (CCl_4) or α-napthylisothiocyanate (ANIT) produce, within 24 h of administration, accumulation of fatty deposits in the liver with extensive liver necrosis. 16,16-dimethyl-PGE_2 (75 μg/kg) injected subcutaneously 24 and again 0.5 h before the hepatotoxin partially reduced the liver damage in these rats (Ruwart *et al.*, 1981). Although cytoprotection is not the only explanation for these results (16,16-dimethyl-PGE_2 may, for example, decrease the absorption or block the catabolism of CCl_4 or ANIT) it does provide a rational interpretation of the results especially since earlier studies had revealed that the same prostaglandin analogue protected rats against liver injury caused by galactosamine which is chemically dissimilar to either CCl_4 or ANIT. The liver is not the only organ in which cytoprotection by prostaglandins has been shown. Pancreatic damage produced in mice by feeding a choline-poor diet was reduced by concomitant treatment with PGE_2 (Manabe & Steer, 1980). Thus, prostaglandin-induced cytoprotection is not necessarily restricted to the gut but may occur in other organs as well. It is too early as yet to predict whether cytoprotective prostaglandins could be of clinical use in conditions affecting the liver or pancreas.

Mechanism of cytoprotection. At present there is no entirely satisfactory explanation for the cytoprotective actions of prostaglandins in the stomach, intestine, colon, liver and pancreas. It is not even certain whether a single mechanism of action exists or whether cytoprotection occurs by different mechanisms in different organs. At the cellular level the only mechanism of action, of the many that have been put forward, which could explain all the cytoprotective effects of prostaglandins was proposed by Chaudhury & Jacobsen (1978). It was suggested that NSAID, and possibly also ethanol, damaged the gastric mucosa by inhibiting the active transport of Na^+ ions from the mucosal to the serosal sides of gut mucosal cells. This would have the effect of raising intracellular Na^+-ion concentration, causing osmotic swelling of the cell and thus cell damage. By stimulating Na^+-ion transport, prostaglandins may prevent this sequence of events and thus maintain normal cellular integrity and function. Such a mechanism could apply equally to all parts of the gut where prostaglandins are known to be cytoprotective.

Considerable attention has been focussed upon one cytoprotective function of prostaglandins i.e. the prevention of stomach ulcers. It is probable that the ability of prostaglandins to reduce gastric ulceration is a combination of their anti-acid secretory action and probably one or more of the following cytoprotective mechanisms.

Mucus production

Prostaglandins increase gastric mucus formation in man, rat and dog. 16,16-dimethyl-PGE_2 increased the thickness of the mucus layer of the rat stomach by 140% (Wilson et al., 1975). The function of the mucus layer remains unclear. It has been suggested that the mucus layer delays diffusion of H^+ ions from the gastric mucosa into the mucosal cells. The possibility that mucosal cells release HCO_3^- ions, thus adding to the effectiveness of the mucus barrier by neutralising any H^+ ions present, has also been put forward.

Effects on the gastric mucosal barrier

The gastric mucosal barrier is a theoretical concept proposed by Davenport in the early 1960s and based originally on experiments in dogs with surgically prepared Heidenheim stomach pouches (Davenport, Warner & Code, 1964). By filling these pouches with solutions of differing ionic composition Davenport demonstrated that the gastric mucosa very efficiently prevented the absorption of H^+ ions and coined the term 'gastric mucosal barrier' to describe this property. Aspirin, ethanol, acetic acid and bile salts (so-called barrier breakers) in some way damage mucosal cells, thus reducing the effectiveness of the gastric mucosal barrier and allowing the absorption of large quantities of H^+ ions into the mucosa. Once inside the mucosal cells, H^+ ions release histamine which acts on H_2 histamine receptors to stimulate acid secretion and thus exacerbate ulcer formation. In addition, histamine dilates mucosal capillaries and increases their permeability to fluid. These effects on the capillaries produce oedema, inflammation and rupture of some capillaries to cause bleeding into the stomach.

The permeability of the mucosa to H^+ ions can be assessed by measuring the gastric mucosal potential difference which is normally 40 mV negative compared with the gastric serosa. Drugs which break the gastric mucosal barrier and increase the absorption of H^+ ions decrease the negativity of the potential difference. Monitoring the mucosal potential difference has thus been used by several workers to determine the mucosal barrier breaking potency of drugs and other substances. According to Davenport's theory, aspirin-like drugs produce ulceration not by depleting endogenous prostanoids in the stomach but by breaking the mucosal barrier. On the other hand prostaglandins prevent ulceration by strengthening the mucosal barrier. Evidence for this proposal is based primarily on the finding that prostaglandins of the E and I series prevent the increase in H^+-ion absorption into gastric mucosal cells caused by aspirin, bile salts or alcohol in rats and dogs (Whittle, 1977).

Alkali secretion
Garner & Heylings (1979) showed that 16,16-dimethyl-PGE_2 and $PGF_{2\alpha}$ stimulated alkali secretion from the frog antral and fundic mucosa *in vitro* and suggested that a similar mechanism to neutralise gastric acidity may underlie gastric cytoprotection in mammals. However, these results could not be repeated by other authors with 16,16-dimethyl-PGE_2, even at concentrations as high as 10^{-5} M (Schiessel *et al.*, 1980). In mammals, prostaglandins do appear to stimulate HCO_3^- secretion from gastrointestinal mucosal cells. For example, PGE_2 increased epithelial transport of alkali HCO_3^- in segments of guinea pig and cat duodenum which may contribute to its anti-ulcer activity in this part of the gut (Flemström & Garner, 1982). Stable PGI_2 analogues like 16-phenoxy-(5α)-5,9-epoxy-$PGF_{1\alpha}$ and 16,16-dimethyl-PGI_2 also stimulate gastric alkali secretion in both rat and dog stomach *in vitro* and PGE_1 stimulates HCO_3^- output from guinea pig gall bladder epithelial cells. Johansson *et al.* (1982) measured gastric HCO_3^- secretion in human volunteers before and after ingestion of 2 mg PGE_2. Basal HCO_3^- output (about 16 μmol/min) was dramatically increased (about 56 μmol/min) after oral PGE_2 and remained increased for about 3 h. Thus there is certainly evidence that some prostaglandins may reduce gastric acidity by increasing the flow of HCO_3^- ions into the stomach, thereby serving to neutralise H^+ ions which are present.

Effect on blood flow
Changes in blood flow probably play little or no part in the gastric anti-secretory action of prostaglandins, as shown previously. However, this is no reason why an action on the gastric vasculature should not contribute to the cytoprotective activity of prostaglandins. As it happens, most experimental evidence suggests that there is no link between

cytoprotection and any change in mucosal blood flow. For example, $PGF_{2\alpha}$ is a vasoconstrictor of mucosal blood vessels but is also cytoprotective. Furthermore, 16,16-dimethyl-PGE_2 is cytoprotective in sacs of isolated frog gastric mucosa in which an effect on blood flow can be excluded.

Lysosomal fragility

PGE_1 stabilises lysosomal membranes *in vitro* although high concentrations which would not be expected to occur naturally are required. Stabilisation of gastric mucosal lysosomes and prevention of the release of proteolytic enzymes stored therein have been proposed to underlie the inhibition of serotonin-induced gastric acid secretion by PGE_1 (Ferguson, Edmonds & Starling, 1973).

Physiological functions of prostanoids in the gut – a summary

Since prostanoids are synthesised in the gut and have potent effects on gastrointestinal motility it is not surprising that they should have been put forward as candidates for a physiological role in the control of gut contractility. As we have seen, some isolated pharmacological preparations (like the rat stomach strip mounted in an organ bath) exhibit spontaneous contractions and relaxations for some hours and, at the same time, release copious amounts of classical prostaglandins into the bathing medium. Addition of NSAID reduces the spontaneous contractility and prevents release of prostaglandins. It is well known that mechanical deformation of the rat stomach strip (e.g. by pinching or stretching) releases prostaglandins and thus PG-mediated tone in these isolated preparations may be an artifact of the trauma associated with cutting and removing the tissue from the body. However the possibility that deformation of cells lining the stomach and intestines as a result of the stretching effect of food causes prostaglandins to be released which, in turn, contract the stomach and intestine to maintain muscular tone and oppose the action of the food is a very attractive one (Piper & Vane, 1971).

Prostaglandins may also have a role to play in controlling the contractile response of the gut to other substances. For example, ATP-induced contraction of intestine smooth muscle may be prevented by indomethacin, suggesting that ATP causes prostaglandin biosynthesis and that it is these PGs which contract the intestine (Kamikawa, Serizawa & Shimo, 1977). In a similar way, the large rebound contraction of the intestine which frequently follows the inhibitory effect of stimulating non-cholinergic non-adrenergic nerves may also be due to prostaglandins (Burnstock *et al.*, 1975). Prostaglandins are known to have a pre-junctional action to reduce the release of neurotransmitter from both sympathetic and parasympathetic nerves (Chapter 9) and are released when enteric nerves are electrically stimulated. Thus, prostaglandins may also influence gut motility *in vivo* by regulating release of acetylcholine and noradrenaline from the intrinsic myenteric nerves. In addition to their pre-junctional effect, some prostaglandins also act on the post-junctional membrane of gastrointestinal smooth muscle cells to potentiate the contractions due to a variety of spasmogens like histamine, acetylcholine, serotonin and even other prostaglandins.

The enteropooling, anti-secretory and cytoprotective effects of exogenous prostaglandins and their analogues have already been described. Whether or not endogenous prostanoids formed locally in the gut have a physiological role to play in these processes has also been studied. Simple mechanical handling of the intestines following laparotomy in rats causes enteropooling which may be prevented by indomethacin (Robert, 1981) implying that endogenous prostaglandins formed as a result of membrane deformation during handling are responsible. Whether or not the normal peristaltic movements of the gut are sufficient to initiate prostaglandin formation and thus to exert a continuous inhibitory effect on fluid and electrolyte reabsorption in the intestine and colon has not been determined.

In the gastric mucosa lining the stomach, a continuous synthesis of prostaglandins may be important to exert a fine control over gastric acid secretion, to provide endogenous cytoprotective 'cover' and in the regulation of mucosal blood flow. Certainly, the fact that both steroid and non-steroid anti-inflammatory drugs (which share the ability to inhibit prostanoid biosynthesis) produce both gastric and intestinal ulceration suggests that endogenous prostaglandins are natural cytoprotective agents and inhibitors of gastric acid secretion. The finding that indomethacin and aspirin vasoconstrict mucosal blood vessels, by the same reasoning, suggests that locally formed prostanoids (probably PGI_2 and PGE_2) are tonically dilating these blood vessels *in vivo*.

The role of prostanoids in diseases affecting the gastrointestinal tract

The importance of prostanoids in disease of the gut has been thoroughly researched in recent years. Metabolites of arachidonic acid may well contribute to the aetiology of four major gastrointestinal diseases: ulcers, diarrhoea due to several causes, bowel cancer and pancreatitis.

Ulcers

The observation that drugs which deplete the gut mucosa of prostaglandins cause ulcer formation and that very small amounts of exogenous prostaglandins prevent this ulceration implies that a deficiency of mucosal prostaglandins predisposes to ulcers. Direct evidence for such a deficiency in humans with gastric ulcers has come from several sources. Firstly, PGE levels (measured by radioimmunossay) in gastric juice of patients with duodenal ulcer were significantly lower than the corresponding levels in age- and sex-matched ulcer-free volunteers (Hinsdale, Engel & Wilson, 1974). These same authors have also shown that venous plasma obtained from patients with duodenal ulcer contained less PGE than plasma from healthy control subjects. In addition the PGE concentration (again measured by radioimmunoassay) in stomach homogenates from patients with gastric ulcer was considerably lower than in homogenates from patients without an ulcer operated, for example, for removal of a tumour (Wright et al., 1982). Similar results have been obtained in rats. In these experiments stomach PGE and 6-oxo-$PGF_{1\alpha}$ levels were significantly lower in rats with restraint-induced gastric ulceration than in tissue from control animals without ulcers.

Diarrhoea

Prostaglandins have been implicated in the aetiology of diarrhoea associated with medullary carcinoma of the thyroid gland, ulcerative colitis, menstrual cycle disorders like dysmenorrhoea as well as that following administration of endotoxins and after irradiation with X-rays. The evidence linking prostaglandins with each of these conditions is discussed below.

Medullary carcinoma of the thyroid. Williams, Karim & Sandler (1968) were the first to report that human medullary carcinoma of the thyroid gland contained higher concentrations of both PGE_2 (18–674 ng/g) and $PGF_{2\alpha}$ (17–844 ng/g) than did normal thyroid tissue. Additionally these authors demonstrated raised peripheral blood prostaglandin concentrations in two patients with this carcinoma and very high prostaglandin levels in blood draining the tumour of one patient. Since 1968 several studies have reported blood or plasma prostaglandin levels in patients with this condition and the results obtained have varied tremendously. Raised plasma PGE levels have been confirmed by some authors but not by others. It should be remembered that in all studies plasma prostaglandin concentrations are probably over-estimated due to prostanoids released from platelets during extraction (see Chapter 1 for details). Consequently, measurement of plasma levels of prostaglandins in this condition is probably meaningless. Since medullary carcinoma is associated with a high incidence of watery diarrhoea and these tumors synthesise and apparently release large amounts of prostaglandins into the blood, it would seem logical to reduce diarrhoea in this condition with drugs which prevent prostaglandin formation. In this context, nutmeg (which has been used for many years as a folk remedy for the diarrhoea of medullary carcinoma of the thyroid) is a potent inhibitor of fatty acid cyclooxygenase (Barrowman et al., 1975).

Endotoxins. Cholera toxin and prostaglandins both stimulate secretion of water and electrolytes into the jejunum. This action is probably mediated by stimulation of mucosal adenylate cyclase to cause an increased accumulation of cAMP intracellularly. At one time it was proposed that choleratoxin-induced diarrhoea in animals was mediated by the release of prostaglandins from the gut (Bennett, 1971) and this claim was supported by experiments in which the diarrhoea following cholera-toxin administration in rats and cats was partly prevented by NSAID (Finck & Katz, 1972). Furthermore, purified cholera toxin was found to increase prostaglandin synthesis and release from human or guinea pig ileum *in vitro*, as measured by bioassay on the rat stomach strip. However, further studies in man and experimental animals other than the rat and cat have generally failed to show a significant beneficial effect of ibuprofen and indomethacin in cholera diarrhoea. On balance it seems unlikely that prostaglandins have a major part to play in the diarrhoea of cholera. Other bacterial toxins may produce diarrhoea by stimulating intestinal prostaglandin formation. For example, intravenously injected *Salmonella enteritides* endotoxin produced diarrhoea in mice, which was prevented by the prior administration of indomethacin. Furthermore, intestine removed from rabbits injected intravenously with *Escherichia coli* endotoxin released more prostaglandin-like biological activity into the bathing solution than did intestine from control, saline-treated animals (Herman & Vane, 1975). Thus a case can be made that the diarrhoegenic effect of some, but not all, bacterial endotoxins may result from the stimulation of large amounts of endogenous prostaglandins from the gut.

X-irradiation. The treatment of cancer with X-rays is always associated with a very high incidence of unpleasant side effects including diarrhoea. Mennie & Daley (1973) studied patients receiving X-irradiation therapy for cancer of the cervix and observed that the diarrhoea which many of them experienced, although resistant to conventional treatment, was reduced by aspirin. Eisen & Walker (1976) have provided a scientific explanation for this effect of aspirin. In their experiments they exposed mice to a single dose of whole-body X-irradiation, killed the animals 4 d later and measured the activity of several enzymes involved in the biosynthesis and catabolism of prostaglandins in the lung, spleen, brain, seminal vesicles and blood. A large reduction in the activity of 15-PGDH was observed in the lung and spleen; this was correlated with a significant increase in the amount of prostaglandins measured in extracts of these organs. These results suggest that X-irradiation in mice causes a reduction in 15-PGDH levels in some organs, thereby increasing prostaglandin levels in these organs. A rise in gut prostaglandin concentration may therefore contribute to the diarrhoea observed and underlie the therapeutic usefulness of aspirin in this condition.

Ulcerative colitis. Ulcerative colitis is an inflammatory disease of the large bowel characterised clinically by short periods of active disease in which patients have diarrhoea, bloody stools and ulcers in the colon punctuating longer periods of remission in which the patient feels perfectly healthy and the endoscopic appearance of the colon appears normal. The acute disease is associated with over-production of prostanoids in the gut, as evidenced by raised plasma and urine prostanoid concentrations and increased prostaglandin biosynthesis by rectal biopsy tissue. The acute disease is best treated with anti-inflammatory steroids while sulphasalazine, an inhibitor of 15-PGDH, significantly prolongs the periods of remission possibly by increasing the colonic concentration of cytoprotective prostaglandins.

Menstrual cycle disorders. The possibility that some or all of the gastro-intestinal distress (diarrhoea, nausea, dyspepsia) which occurs in 30–40% of women with dysmenorrhoea may be secondary to raised gut prostaglandin biosynthesis was suggested by Kaupilla & Ylikorkala in 1977. They observed that the diarrhoea of dysmenorrhoea was relieved by treatment with indomethacin in 83% of subjects compared with those receiving a placebo (46% relief). Indomethacin therapy also moderated the nausea and vomiting in many of these patients.

Bowel cancer

Prostaglandins have a part in the aetiology of some cancers (see Chapter 10). In the gut, human colonic and rectal cancers synthesise larger amounts of prostaglandins than does normal colonic or rectal tissue. Also, cultured human colonic cancer cells release more PGE-like material into the incubation medium than do normal mucosal cells from the colon (Jaffe, 1974) and prostaglandin biosynthesis is reportedly greater in human colonic cancer tissue than in homogenised normal mucosa or submucosa (Bennett & del Tacca, 1975). Prostaglandins are also synthesised in tumours of other organs associated with the gut. Raised plasma prostaglandin levels occur in some patients with Vernier–Morrison syndrome (characterised clinically by watery diarrhoea and hypokalaemia) due to a tumour of the pancreas. PGE_2 and $PGF_{2\alpha}$ also occur in tumours of the pancreatic alpha cells in man (Sandler, Karim & Williams, 1968). It seems likely that prostaglandin formation by the tumour cells may contribute both to the control of the growth of the tumour (Chapter 10) and to the diarrhoea seen in patients with tumours of the gastrointestinal tract.

Pancreatitis

Pancreatitis, inflammation of the pancreas, is probably caused by digestion of the pancreas by its own enzymes (i.e. autodigestion). It may therefore be treated by drugs which inhibit trypsin activity. The condition is a very painful one and can be fatal. A clinical syndrome analogous to human pancreatitis can be induced in dogs by injecting a mixture of bile salts and trypsin into the main pancreatic duct. Prostaglandin-like material released into the pancreatic vein of such animals was increased significantly following induction of experimental pancreatitis as was the release of prostaglandins into the peritoneal fluid (Glazer & Bennett, 1974). These experiments relied upon a biological assay for prostaglandins and thus provided no information about the metabolites of arachidonic acid (e.g. PGI_2, 5-HETE) which have only feeble spasmogenic activity. This may be important since the chief products of arachidonic acid in the rat pancreas *in vitro* are PGI_2 with lesser amounts of PGE_2. Whether prostaglandins have a specific role in pancreatitis or are formed as a general response to inflammation of the pancreas is not clear.

Clinical uses of prostanoids in the gastrointestinal tract

The major clinical and potential clinical applications of prostaglandins in the gut are in the treatment of ulcers which has been discussed previously in this chapter, prevention of paralytic ileus after surgery and as an aid in the visualisation of the vascular circulation of various parts of the gut (which is a process known as angiography).

Paralytic ileus

This condition is characterised by delayed gastrointestinal propulsion after any form of abdominal surgery, even occurring in those patients whose recovery from surgery is otherwise uncomplicated. The cause of post-operative ileus is not known but may be the result of stimulation of sympathetic nerves to the gut or suppression of cholinergic nerve activity or both. If not successfully treated the condition may proceed to paralytic ileus which is potentially fatal.

Since prostaglandins contract the intestine *in vivo*, reduce noradrenaline release from sympathetic nerves and augment release of acetylcholine from the guinea pig ileum, they would appear to be possible candidates for the clinical treatment of post-operative ileus. Laparotomised rats show reduced gastric emptying and slowed small intestine and colon transit for a period of up to 2 d after the initial surgery. This condition is analogous to but unlikely to be identical with post-operative ileus in humans. PGE_2, $PGF_{2\alpha}$ and 16,16-dimethyl-PGE_2 given orally, subcutaneously or intraduodenally increased gastrointestinal propulsion in the stomach, intestine and colon of laparotomised rats (Ruwart, Klepper & Rush, 1980). Remarkably, 16,16-dimethyl-PGE_2 (5 μg/kg intravenously) completely restored gastric emptying in these animals for up to 4 h. $PGF_{2\alpha}$ has also been tested in the treatment of post-operative ileus in man. In these studies, 50 patients with post-operative ileus resistant to conventional therapy were infused intravenously with $PGF_{2\alpha}$ at 10–20 μg/min over a 4 h period. Within 12 h 75% of patients had started defaecation. This result was accomplished with no change in the patients' blood pressure, heart rate or blood or urine biochemistry. Despite these promising early results many more controlled, double-blind clinical trials of this sort are required before prostaglandins can be employed routinely in post-operative ileus.

Angiography

Opacification of the venous anatomy of an organ can be routinely achieved by close arterial injection of contrast medium into the organ being studied. Better results can be obtained by prior injection of a vasodilating drug such as tolazoline and/or bradykinin but these may produce unwanted haemodynamic side effects and reduce the quality of the angiograph, possibly by altering vascular permeability to fluid. $PGF_{2\alpha}$ has been used successfully for superior mesenteric, pancreatic and hepatic angiography, thus allowing detailed examination of the small arteries supplying the bowel, pancreas, liver and gall bladder and thus improved chances of identifying the existence of tumours of these organs. $PGF_{2\alpha}$ produced little or no hypotension or bradycardia and was apparently well tolerated by all patients. The amount of $PGF_{2\alpha}$ used (less than 100 μg) was insufficient to produce diarrhoea. The applications of $PGF_{2\alpha}$ and other prostaglandins in angiography has been reviewed by Legge (1979).

7
The respiratory system

Introduction

The lungs have two important physiological functions:
(1) the interchange of oxygen and carbon dioxide between the body and the environment;
(2) as an endocrine organ liberating biologically active substances into the circulation and acting as a 'chemical filter' for substances present in venous blood (by taking up and inactivating agents like 5-hydroxytryptamine and bradykinin the lungs serve to protect the arterial tree from their potent vasoactive effects).

Since the early 1960s, when human lung was shown to contain large amounts of prostanoids, interest has centred upon the action of these local hormones on bronchopulmonary smooth muscle and also upon the release and breakdown of prostanoids and leukotrienes in this organ. These aspects of prostanoid pharmacology are reviewed and summarised in this chapter. The reader is advised to refer to some of the many excellent and exhaustive reviews on this subject for additional information (e.g. Vane, 1969; Mathe, 1977; Cuthbert & Gardiner, 1981; Hyman et al., 1981; Hedqvist, Dahlen & Björk, 1982).

Synthesis and release of prostanoids and leukotrienes from the lung

Lung homogenates and subcellular fractions

$PGF_{2\alpha}$ was identified in extracts of sheep and swine lung by Bergström and his colleagues in 1962 (Bergström et al., 1962). Later work revealed that homogenates of chicken, rabbit, guinea pig, rat, cat and dog lung also contained relatively large amounts of $PGF_{2\alpha}$ and less PGE_2. These experiments were based upon a simple bioassay determination after thin-layer chromatography to separate PGE_2 from $PGF_{2\alpha}$ and probably reflect not prostaglandin content of the lung as was first envisaged but the capacity of the lungs to synthesise prostaglandins from endogenous arachidonic acid during homogenisation. More recently it has become apparent that such classical prostaglandins of the E and F series are only minor products of arachidonic acid metabolism in the lungs of most species. Table 7.1 shows the percentage conversion of radiolabelled PGH_2 to five prostanoids (6-oxo-$PGF_{1\alpha}$ and TxB_2 – representing PGI_2 and TxA_2 respectively – PGE_2, $PGF_{2\alpha}$ and PGD_2) following incubation with 100 000 g microsomal preparations from the lungs of several species. The major prostanoid formed by the lungs under these conditions is PGI_2, with smaller amounts of TxA_2 and classical prostaglandins of the E, F and D series. The exception to this general rule is the guinea pig lung which avidly converts PGH_2 to TxA_2 at the expense of other prostanoids. Interestingly, the ratio $PGI_2:TxA_2$ is approximately 1 for the other species studied. A different profile of lung prostanoid formation occurs in newly born animals. PGE_2 seems to be the most abundant cyclooxygenase metabolite in foetal rabbit lungs. It is only after the animal has been born that the lung gains the ability to biosynthesise PGI_2.

It is not clear precisely which cells in the lung are responsible for synthesising which prostanoid and since, the human lung comprises approximately 40 cell types, it may never be known. Most likely a large part of the PGI_2 formed in the lung is derived from endothelial cells lining the pulmonary blood vessels. Similarly, some of the TxA_2 which supposedly arises from bronchopulmonary tissue is actually produced by platelets

Table 7.1. *Conversion of PGH_2 to prostanoids by lung tissue from six different species*

Species	Prostanoid formed[a]				
	6-oxo-$PGF_{1\alpha}$	TxB_2	$PGF_{2\alpha}$	PGE_2	PGD_2
Man	17.3	17.9	12.9	7.4	5.1
Pig	28.2	21.7	5.1	6.4	3.8
Cow	26.0	24.1	4.0	3.4	< 1
Rat	23.0	18.2	6.4	12.6	5.0
Mouse	10.1	7.1	4.4	23.4	7.4
Guinea pig	2.6	46.0	1.8	3.0	< 1

[a] Figures show the percentage conversion of PGH_2 (4 nmol) to prostanoids following incubation for 10 min at 37 °C with lung parenchymal microsomes from each of the six species. Prostanoids were identified and assayed by high-pressure-liquid chromatography. PGI_2 and TxA_2 were measured as their stable hydrolysis products 6-oxo-$PGF_{1\alpha}$ and TxB_2 respectively. Data taken from Ally *et al.* (1982).

'trapped' in the lungs when the animal was killed. However, central (trachea, bronchi, bronchioles) and peripheral (parenchyma containing the alveoli) airways produce both PGI_2 and TxA_2 in their own right. Human and bronchial parenchymal fragments incubated in salt solution release 2–3 times more PGI_2 than TxA_2 (Schulman, Adkinson & Newball, 1982). In the lung parenchyma PGI_2 may be formed by the alveoli themselves since rat alveolar type II cells maintained in culture *in vitro* readily release large quantities of PGI_2.

Interpreting the results of experiments of this sort can be difficult. The pattern of prostanoids formed depends to a large extent upon the concentration of substrate PGH_2 or arachidonic acid. Since PGI_2 synthetase becomes saturated at relatively low concentrations of PGH_2 (greater than 10 μM) the major product of PGH_2 breakdown at high concentrations appears to be TxA_2. Thus, experiments *in vitro* of this sort can at best provide information only about the absolute activity of the individual enzymes involved in prostanoid biosynthesis. For more idea of the physiological formation of prostanoids it is necessary to carry out experiments using intact lungs.

Release of prostanoids from intact lungs

Spontaneous release. The spontaneous efflux of prostanoids has been assessed from intact trachea, bronchi and lung parenchyma of several species, from isolated lungs perfused with Krebs' solution via a cannula inserted into the pulmonary artery and from intact lungs in anaesthetised animals. In each of these cases unstimulated prostanoid release is low and reflects only a small fraction of the total amount of arachidonic acid products which may be released following mechanical, chemical or hormonal trauma (see below).

The identity and amount of prostanoids formed spontaneously by bronchopulmonary tissues varies according to the part of the lung used, species and the amount of damage done to the tissue during the experiment. For example, the isolated guinea pig trachea preparation exhibits spontaneous tone and releases both PGE_2 and $PGF_{2\alpha}$ into the bathing medium. Administration of indomethacin or mefanamic acid to block prostanoid biosynthesis succeeds in preventing prostaglandin efflux from the tissue and produces dose-dependent relaxations (Orohek *et al.*, 1973). Thus, in this tissue, sufficient prostanoids are released to regulate the natural tone of the preparation.

Many experiments have concentrated upon the measurement of prostanoid efflux from perfused lungs. Several species including rat, cat, dog, guinea pig, sheep and man have been studied in this way. In most species the major cyclooxygenase product released is PGI_2. An exception to this general finding is the guinea pig lung which releases approximately 2–4 times more TxA_2 than PGI_2 which agrees with the results of experiments *in vitro* using guinea pig lung homogenates described before. The spontaneous release of four prostanoids from the Krebs'-perfused rat lung *in vitro* is shown in Fig. 7.1. The rank order of pulmonary prostanoid release in this species is 6-oxo-$PGF_{1\alpha}$ > TxB_2 > PGE_2 = $PGF_{2\alpha}$. The amount of all prostanoids released normally increases the longer the lungs are perfused. Most likely the perfused lung *in vitro* becomes more and more oedematous as time goes on, causing tissue damage and thus prostanoid formation. In addition to PGI_2, TxA_2 and classical prostaglandins, perfused lungs also release trace amounts of PGD_2 and the 15-oxo and 13,14-dihydro-15-oxo metabolites of PGE_2, $PGF_{2\alpha}$ and TxB_2.

Although isolated perfused lungs readily release prostanoids into the perfusing medium *in vitro* it is unlikely that a similar process occurs in the lungs *in vivo*. Attempts to demonstrate significant pulmonary release of prostanoids in intact animals have generally proved unsuccessful. For example, Coker *et al.* (1982) measured the concentration of 6-oxo-$PGF_{1\alpha}$,

Fig. 7.1. Spontaneous release of four prostanoids from isolated, perfused rat lungs. Results show prostanoid output in ng/10 min period. The inset illustrates the rise in efflux of prostanoids from perfused rat lungs measured as release of all prostanoids over 10 min periods at 0, 30 and 60 min after setting up the lungs in cascade.

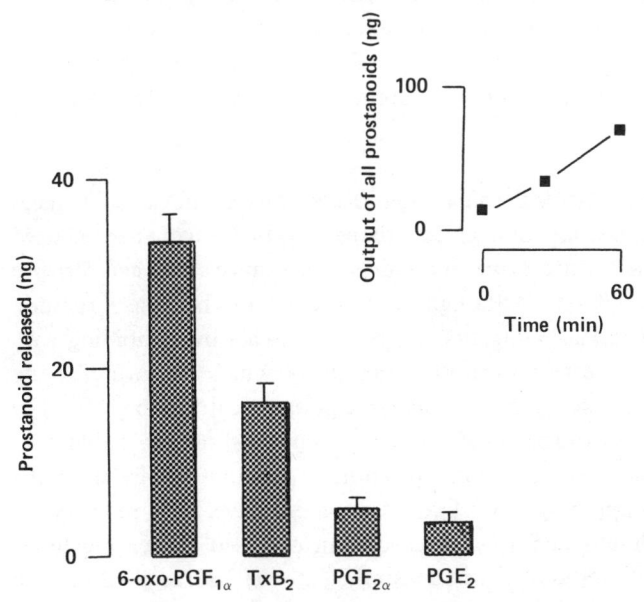

TxB_2, PGE_2 and $PGF_{2\alpha}$ in blood samples collected from the pulmonary vein and aortic arch of anaesthetised cats. Assuming prostanoids are not formed and/or released in the left heart, such samples represent blood immediately before and immediately after entering the lungs. If prostanoids are released from the lungs *in vivo* then they should be found in greater amount in aortic than in pulmonary venous blood. In fact in this study all four prostanoids had a similar plasma concentration irrespective of the vessel from which blood was collected. Subsequent experiments using a very sophisticated GC/MS technique have revealed that resting plasma 6-oxo-$PGF_{1\alpha}$ levels in man are much lower than was first thought, probably only 1–2 pg/ml and were not different in arterial or venous blood, implying that PGI_2 is not spontaneously 'leaked' into the blood stream by the lungs *in vivo*.

Release following mechanical stimulation. Chopped or minced lung tissue generate copious amounts of TxA_2, PGE_2 and $PGF_{2\alpha}$ (Yen, Mathe & Dugan, 1976). Colloidal suspensions of small (1–120 μm) particles of Sephadex infused into the pulmonary artery of isolated guinea pig lungs also causes prostanoid release, as does a variety or mechanical stimuli such as stroking, scratching or massaging the surface of the tissue. Even gentle ventilation of isolated perfused guinea pig lungs is a sufficient trauma to provoke prostanoid biosynthesis and release (Berry, Edmonds & Wyllie, 1971); hyperventilation in intact animals causes a flood of PGI_2 from the lungs into the circulation. Oedema induced by prolonged perfusion also causes release of prostanoid-like substances into the blood stream. These observations prompted the suggestion by Piper & Vane (1969) that cellular deformation is the common factor in stimulating prostanoid formation by the lung.

Release following chemical stimulation. Numerous chemical substances infused into isolated lungs or injected into the blood stream of anaesthetised animals elicit the release of prostanoid-like substances and leukotrienes into the venous effluent. Although very different in chemical structure, these substances (releasing agents) are believed to act by combining with and stimulating selective membrane receptors which in turn activate phospholipase enzymes to release free arachidonic acid for conversion to eicosanoids. A list of prostanoid-releasing agents is shown in Table 7.2.

In addition to substances which promote prostanoid release from the lungs, β-adrenoceptor agonists like isoprenaline have been reported to inhibit histamine-induced TxA_2 release from perfused guinea pig lungs (Folco *et al.*, 1977) possibly by increasing pulmonary intracellular cAMP concentration or by an action on Ca^{2+}-ion flux across the membrane. Other substances which in general have little or no effect on lung prostanoid output include histamine (acting on H_2 receptors), α-adrenoceptor agonists and cholinoceptor agonists.

Release following antigen–antibody interaction. Prostanoids are released from the lungs during anaphylactic shock. The anaphylactic reactions are classified as immediate hypersensitivity reactions. They occur following administration to man or animals of a foreign substance (the antigen) which stimulates the formation of a reaginic antibody which in turn becomes fixed to mast cells and basophils in the tissues. Experimentally, this procedure, called sensitisation, is usually performed in guinea pigs with ovalbumin as antigen. In man and most other species the reaginic antibodies are immunoglobulin Es (IgEs) whilst in guinea pigs they are immunoglobulin As (IgAs). Subsequent exposure to the antigen (the challenge) causes a reaction with the cell-fixed antibody, mast cell degranulation liberating large amounts of histamine and heparin and as we shall see prostanoid and leukotriene biosynthesis both from mast cells and from leucocytes.

In 1969, Piper & Vane carried out a series of studies in which they investigated the release of mediators from perfused lungs removed from guinea pigs sensitised a couple of weeks previously with ovalbumin (see Chapter 1). The lungs were challenged by injection of ovalbumin and the perfusate allowed to cascade over a bank of isolated pharmacological

Table 7.2. *Release of prostanoids from lungs by chemical substances*[a]

Chemical stimulus	Principal eicosanoid	Species
Arachidonic acid	TxA_2, PGI_2, PGE_2	Rabbit, guinea pig
PGH_2	TxA_2	Guinea pig
Bradykinin	TxA_2, PGI_2	Guinea pig
Angiotensin II	PGI_2, PGE_2, TxA_2	Guinea pig, rat, rabbit
Histamine (H_1)	TxA_2, PGE_2	Guinea pig
5-Hydroxytryptamine	TxA_2, PGE_2	Guinea pig
Adenosine/ATP	Not known	Guinea pig (trachea)
Leukotrienes B_4 C_4 D_4	TxA_2, PGE_2	Guinea pig
Platelet-activating factor	TxA_2, LtC_4, LtD_4	Rat
A23187	TxA_2, PGE_2, PGI_2	Rat
Escherichia coli endotoxin	PGE_2, TxA_2	Rat
	$PGF_{2\alpha}$, TxA_2	Cat

[a] Prepared from the following references: Greenberg *et al.* (1979), Alabaster (1980), Feuerstein & Ramwell (1981), Advenier *et al.* (1982), Coker *et al.* (1982), MacGlashan *et al.* (1982), Piper & Samhoun (1982), Voelkel *et al.* (1982).

preparations in order to detect the release of biologically active substances. A typical experiment of this sort is shown in Fig. 7.2. As expected, histamine was released in large amounts and caused a pronounced contraction of the cat terminal ileum preparation which is exquisitely sensitive to even small amounts of this amine. Histamine also produced a small contraction of the chick rectum preparation. Ovalbumin injection caused the release of slow-reacting substance of anaphylaxis (SRS-A) which produced a characteristically slow contraction of both the chick rectum and the guinea pig trachea. However, substances active on the rat stomach strip (now known to be mainly PGE_2 and $PGF_{2\alpha}$) and a second substance active on the rabbit aorta preparation (now identified as TxA_2)

Fig. 7.2. Ovalbumin-induced release of pharmacologically active substances from isolated perfused guinea pig lungs determined by superfusion cascade bioassay using six pharmacological preparations (RbA, rabbit aorta; CTI, cat terminal ileum; GPI, guinea pig ileum; CR, chick rectum; GPT, guinea pig trachea; and RSS, rat stomach strip).

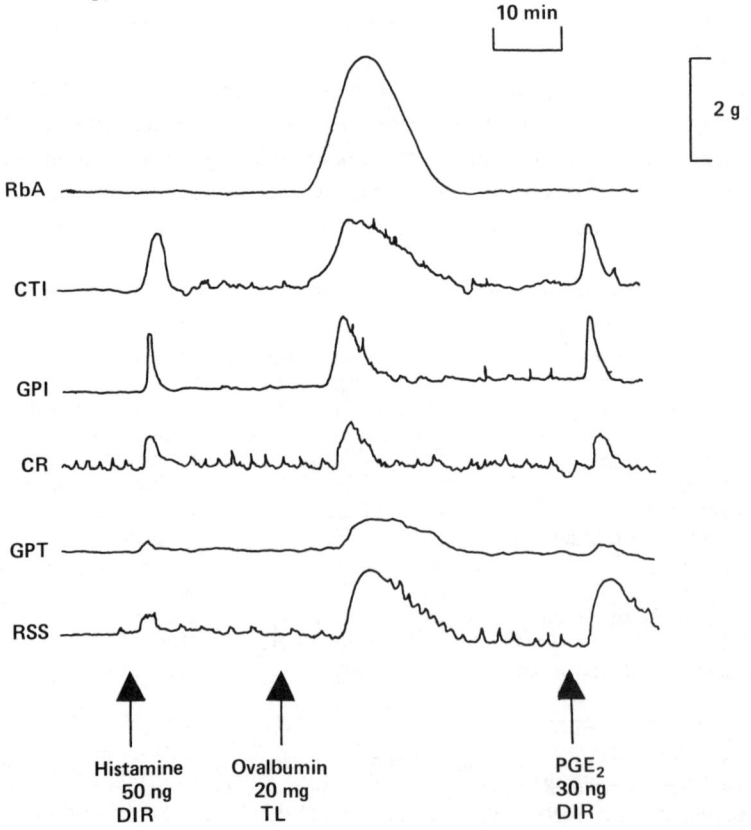

were also released during the anaphylactic shock reaction in the perfused guinea pig lung. In addition prostanoid metabolites are released from the lungs during anaphylaxis. Using a radioimmunoassay, Mathe & Levine (1973) found that the 15-oxo- and 13,14-dihydro-15-oxo-PG metabolites were released from guinea pig lungs during anaphylaxis. Thus anaphylactic shock *in vitro* causes the lungs to release a plethora of different prostanoids, leukotrienes and their metabolites.

The anaphylactic release of prostanoids from the lungs has also been demonstrated *in vivo*. For example, Burka & Eyre (1974) studied changes in the cardiovascular and respiratory systems following anaphylaxis in anaesthetised calves. In addition they measured the release of inflammatory mediators into the blood stream during anaphylaxis by superfusing carotid arterial blood over several bioassay preparations in cascade. These experiments demonstrated clearly that interaction of antigen with antibody *in vivo* as well as *in vitro* caused the pulmonary release of large amounts of E and F series prostaglandins. Some recent experiments using human lung samples removed from asthmatics revealed the release of large amounts of LtC_4 and LtD_4 along with 6-oxo-$PGF_{1\alpha}$, TxB_2, $PGF_{2\alpha}$, PGE_2 and PGD_2 following challenge with antigen which, in these cases, was birch pollen (Dahlen, 1983).

In conclusion, experimentally induced anaphylactic shock in animals is associated with the release of a battery of inflammatory mediators. Most likely, histamine released from lung mast cells is the 'primary' mediator and may in turn initiate the synthesis of prostanoids and leukotrienes which may therefore be considered to be 'secondary' mediators. The release of prostanoids from the lungs may also be of clinical importance when it comes to the aetiology of asthma, some cases of which are representative of anaphylactic shock.

Catabolism and uptake of prostanoids in the lung

As mentioned previously the lungs perform a vital service in the body by activating, inactivating or sometimes simply allowing free passage to biologically active substances passing through in the blood. The fate of some local hormones on transit across the pulmonary circulation is shown in Table 7.3.

The lungs are also responsible for the breakdown of whatever prostanoids and/or leukotrienes 'spill over' into the blood stream after synthesis in the peripheral tissues. Normally prostanoids are very efficiently catabolised in the tissues where they are formed and thus the lungs serve as a second line of defence ensuring that large amounts of prostanoids do not gain access to the arterial circulation. In some circumstances, however, the lungs do become an important site for prostanoid degradation in the body. During parturition, for example, the uterus releases massive quantities of prostanoids into the blood which are efficiently taken up and broken down in the lungs. This process is very effective. It has been calculated that 80–95% of PGE_2 and $PGF_{2\alpha}$ arriving at the lungs in the blood are catabolised to inactive products in species like the guinea pig, rat, rabbit, cat, dog and in humans. Since biological membranes are more or less impermeable to prostanoids the breakdown of circulating prostanoids in the lungs is almost certainly a two-step process involving firstly a carrier-mediated uptake into cells in the lung and thereafter prostanoid degradation by the sequential action of several catabolising enzymes (Fig. 7.3). Both of these processes have been described in general terms in Chapter 1. Only specific aspects of prostanoid uptake and catabolism in the lungs will be discussed here (see also Gillis & Roth, 1976).

Table 7.3. *Fate of biologically active substances in the lungs*

Biologically active substance	Effect of lungs	Nature of inactivation
Amines		
Noradrenaline	Inactivation	Taken into cells by uptake process which depends on Na^+-K^+-ATPase (inactivated by monoamine oxidase)
Adrenaline	Not affected	—
5-Hydroxytryptamine	Inactivation	As for noradrenaline
Histamine	Not affected	—
Peptides		
Bradykinin	Inactivation	Enzymatically catabolised by kininase on luminal surface of pulmonary endothelial cells – no uptake required
Angiotensin I	Activation (by conversion to AGT II)	Enzyme located on endothelial cells – no uptake required
Nucleotides		
ATP, ADP, AMP	Inactivated	Enzymatically catabolised by phosphate esterases on endothelial cells – no uptake required
Others		
Acetylcholine	Not affected	—

Prostanoid uptake

The evidence in favour of pulmonary, carrier-mediated uptake of prostanoids prior to enzymatic catabolism is summarised below.

(1) Lungs are capable of accumulating prostanoids against a concentration gradient. This may be demonstrated following incubation of lung pieces in salt solution containing radiolabelled PGE_1, PGE_2, $PGF_{2\alpha}$ or PGD_2. After 30–60 min at 37 °C the tissue: medium ratio of radioactive PG is greater than unity, indicating

Fig. 7.3. The catabolism of prostaglandins in the lungs. A, intact lungs; B, arrangement of blood vessels in a small section of pulmonary tissue; C, first step in prostaglandin catabolism is their transport across the blood vessel wall into lung cells; D, second step in prostaglandin catabolism is oxidation of transported prostaglandin by intracellular 15-PGDH.

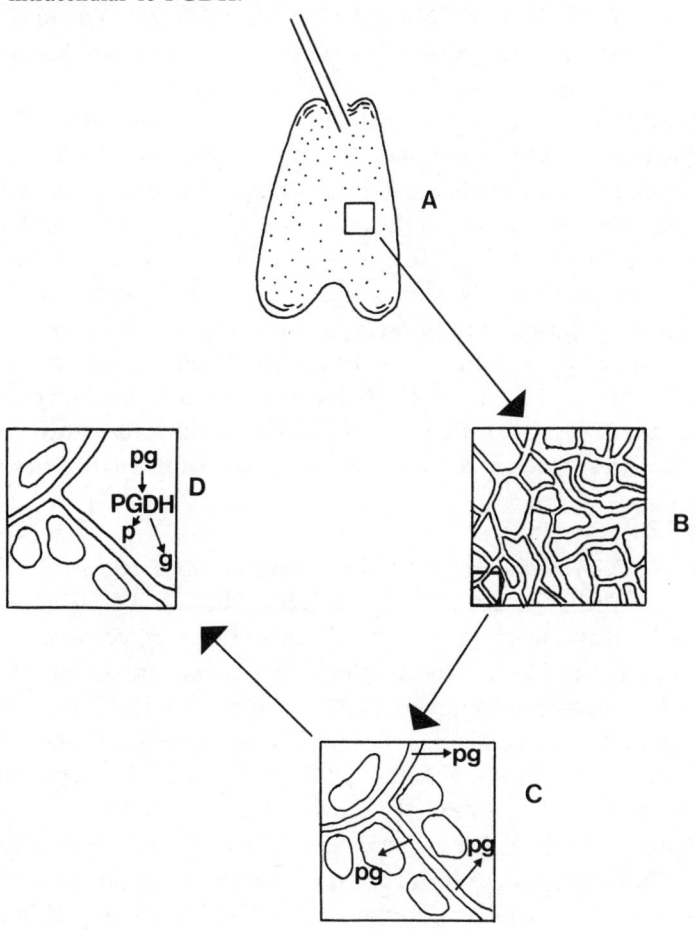

that the net movement of prostaglandin is from the medium across the cell membrane into the lung tissue.
(2) Compared with the appearance of an inert marker (e.g. blue dextran) which does not pass cell membranes, the appearance of tritiated PGs in the outflow of isolated, perfused rat lung is delayed suggesting that PGs must first be taken up prior to release.
(3) The accumulation and subsequent inactivation of PGs in perfused rabbit lung is reduced by pre-treating the animal with probenacid, a drug known to prevent carrier-mediated transport of other substances across biological membranes. In addition, reducing the temperature of the perfusing medium to 4 °C also reduced the ability of isolated perfused rabbit lungs to accumulate and catabolise PGE_2.
(4) Prostaglandin uptake into isolated perfused rat lung is unaffected if Na^+ ions are removed from the perfusing medium suggesting that carrier-mediated transport of prostanoids in the lungs is not dependent upon Na^+-K^+-ATPase activity.

Not all prostanoids are taken up in the lungs. Prostaglandins of the A series (which do not occur naturally other than as artifacts of PGE extraction), PGI_2 and its metabolite, 6-oxo-PGE_1 escape biological inactivation both in the perfused lung and in the pulmonary circulation *in vivo*. Since each of these prostaglandins is a substrate for 15-PGDH in broken-cell preparations, it has been concluded that they do not possess all the structural requirements to bind to the membrane-carrier molecule. Prostanoids which are substrates for the carrier system include: PGE_1, PGE_2, $PGF_{1\alpha}$, $PGF_{2\alpha}$, 15-oxo and 13,14-dihydro-15-oxo metabolites of the E and F series, PGD_2, 6-oxo-$PGF_{1\alpha}$, TxB_2 and 13,14-dihydro-15-oxo-TxB_2. It is not yet clear whether a similar uptake process for leukotrienes exists in the lungs.

Enzymatic breakdown of prostanoids and leukotrienes

The lungs contain large amounts of both 15-PGDH (NAD^+- and $NADP^+$-dependent: types I and II) and PG-Δ13R, consistent with the finding that 13,14-dihydro-15-oxo-PG metabolites are the most abundant cyclooxygenase products released from challenged perfused lungs. Precisely which of the many different cell types found in the lungs contain these two enzymes is not known although pulmonary endothelial cells maintained in culture have no 15-PGDH activity. Lung 15-PGDH utilises prostaglandins of the D, E, F and I series as well as TxB_2 as substrate. In common with the kidney and many other organs, lung homogenates have the ability to convert PGE_2 to $PGF_{2\alpha}$ by means of a PG-9KR enzyme. Homogenised guinea pig lungs also convert LtC_4 to LtD_4 and LtD_4 to LtE_4. It is not yet clear whether these enzymatic reactions are important for inactivating circulating leukotrienes.

Effect of classical prostaglandins on respiratory smooth muscle
Experiments in vitro

Much of the information about the bronchoconstrictor activity of classical prostaglandins has come from experiments on isolated airways smooth muscle either superfused with Krebs' solution or set up in the organ bath. With some exceptions such experiments *in vitro* have proved to be predictive of the situation in the intact animal. Central or peripheral airways smooth muscle may be studied *in vitro*. To study central airway smooth muscle the trachea from any of several species (commonly rat or guinea pig) may be removed and cut in one of several ways (e.g. spiral, zig-zag) to expose the circular smooth muscle, contraction or relaxation of which controls the aperture of the intact trachea. Smooth muscle from peripheral airways is provided by bronchi and small bronchioles. If the lungs are sufficiently large then bronchi may be dissected out, cut into a spiral and set up in an organ bath as before. With small animals this is obviously not possible. An alternative is to cut a lung parenchymal strip from a cat or guinea pig lung lower lobe which may be set up in the organ bath and made to contract or relax to the appropriate drugs. Although some of these drug effects may result from the contraction or relaxation of vascular smooth muscle, the major action appears to be on bronchial and bronchiolar smooth muscle (Lulich & Paterson, 1980). Figure 7.4 shows preparations of smooth muscle from major airways which have been used for prostaglandin studies.

Classical prostaglandins of the E, F and D series affect respiratory smooth muscle although, compared with their activity on smooth muscle in other parts of the body (e.g. the gut), they are only weakly active. As a general rule PGEs relax most preparations of airway smooth muscle from most species. This relaxation is seen either as a fall in the intrinsic tone of the preparation if it already has tone or, alternatively, by reducing contractions produced by bronchoconstrictor agents like acetylcholine, histamine, 5-hydroxytryptamine or even after anaphylactic shock in some species. PGEs relax smooth muscle from airways *in vitro* at concentrations of 10–50 ng/ml in an organ bath. In contrast, PGFs have little effect or produce only feeble contractions even in microgram amounts. Interestingly, $PGF_{2\beta}$, which is a stereoisomer of $PGF_{2\alpha}$ not naturally found in the body, potently relaxed guinea pig tracheal spiral preparations. PGD_2, the major metabolite of arachidonic acid formed when mast cells degranulate, contracts isolated smooth muscle from airways. Using the guinea pig trachea for comparison, PGD_2 is some 10 times more potent than $PGF_{2\alpha}$. Since the lungs may well be an important site of prostanoid catabolism, several authors have studied the effect of metabolites of classical prostaglandins on airways smooth muscle *in vitro*. 15-oxo metabolites of PGE_2

Fig. 7.4. The various types of isolated pharmacological preparation from lungs.

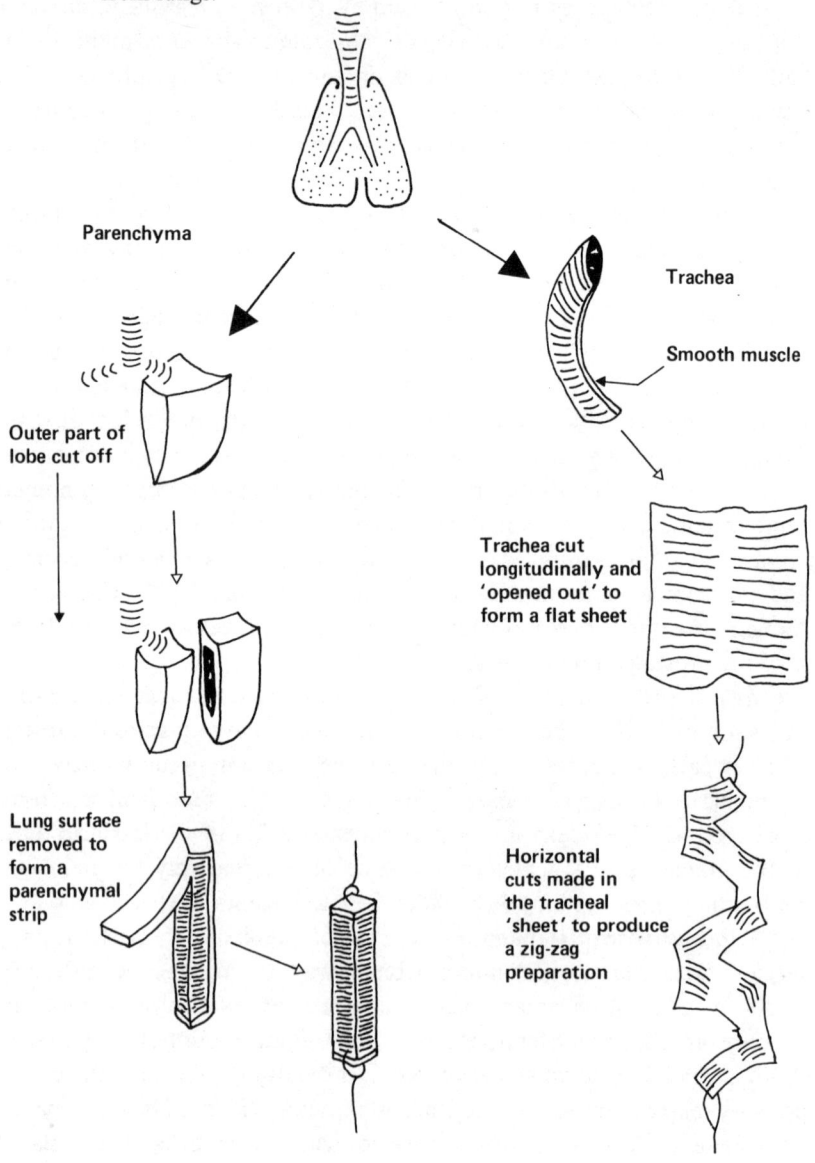

and $PGF_{2\alpha}$ have very similar biological activity on the airways to the parent molecules (Crutchley & Piper, 1975) while the 13,14-dihydro-15-oxo-PG metabolites have less than 10% the activity of the primary prostaglandins.

A great deal of interest has centred upon the bronchomotor effects of prostanoids in man. Although these experiments are limited by the availability of human tissue it is sometimes possible to obtain normal, healthy bronchi from patients undergoing surgery for lung cancer. In their pioneering studies, Sweatman & Collier (1968) demonstrated that spirally cut human bronchi had intrinsic tone and could be relaxed *in vitro* by PGE_2 and contracted by $PGF_{2\alpha}$. Later authors have confirmed that human bronchial muscle reacts to prostaglandins in the same way as animal tissues do although, in some preparations, PGE_2 may either cause a contraction or a relaxation (Gardiner, 1975).

Prostaglandins affect airways smooth muscle by first interacting with specific PG receptors on the cell membranes of the smooth muscle. Attempts to block the bronchomotor effect of PGs *in vitro* with a variety of antagonists (including phenoxybenzamine, atropine, propranolol, mepyramine, reserpine and the mast cell stabilising compound disodium chromoglycate) have all failed. The only drugs which do affect the action of prostaglandins on respiratory smooth muscle are the cyclooxygenase inhibitor mefanamic acid (Collier & Sweatman, 1968) and experimental prostanoid receptor blockers such as SC19220 and AH19437 (Kennedy *et al.*, 1982). Indeed, based upon their extensive studies of agonist-potency ratios these latter authors have characterised the PG receptors which mediate the relaxation of the cat trachea due to PGEs as being of the EP type. In contrast, $PGF_{2\alpha}$ appears to contract parenchymal strips of the guinea pig lung by acting as a weak agonist on TxA_2 receptors (termed TP receptors). Whichever type of receptor is involved, it seems likely that PGE-induced relaxation of trachea, bronchioles and parenchymal strips is caused by increasing intracellular formation of cAMP, i.e. the same post-receptor mechanism of action which underlies the effect of isoprenaline on airways smooth muscle.

Experiments in vitro

The respiratory effects of prostaglandins have been examined in experimental animals and man. There are several techniques available for studying the effect of drugs on the airways in man. A relatively crude index is the volume of air which can be forcibly expelled from the lungs in 1 s. This is the forced expiratory volume or FEV_1. The instrument which measures the volume of expired air is called a spirometer and is standard equipment in most departments of physiology. The FEV_1 is relatively insensitive and cannot detect small changes in airways tone. A more sensitive method for studying lung function is provided by the measurement of specific airways conductance or S_{GAW} using the technique of total body plethysmography. For this method the subject sits or stands in an air-tight box with a transparent perspex hood fitted with a mouthpiece through which he may breathe air from the outside or, by adjusting a valve, from the inside of the box. Changes in the volume of air in the box as the subject breathes may be monitored using a spirometer connected to the body plethysmograph.

Since animals are usually not sufficiently cooperative to blow into a spirometer when asked, alternative methods to measure changes in airways diameter in animals have been developed. Most frequently, ventilation in animals is measured by a method first described by Konzett & Rossler in 1940. In this technique an animal (cats, dogs, ferrets, rats etc., have been used but the guinea pig is most frequently employed) is anaesthetised sufficiently deeply to prevent spontaneous breathing and then artificially respirated by a pump through a cannula tied into the trachea. The lungs are inflated at a constant pressure and the air at each stroke left over after the lungs are completely filled is diverted to a pressure transducer attached to a side-arm of the tracheal cannula. Any increase in the resistance of the lungs to inflation as occurs, for example, following bronchoconstriction

Fig. 7.5. Apparatus for guinea pig Konzett–Rossler preparation. The animal is anaesthetised and positioned belly-up on an operating table. The skin over the throat is cut away, the trachea identified and a glass or plastic cannula tied into place. Air (6 ml/min) is pumped into the animal's lungs with a respirator and the resistance to flow provided by the airways is monitored continuously by a pressure transducer.

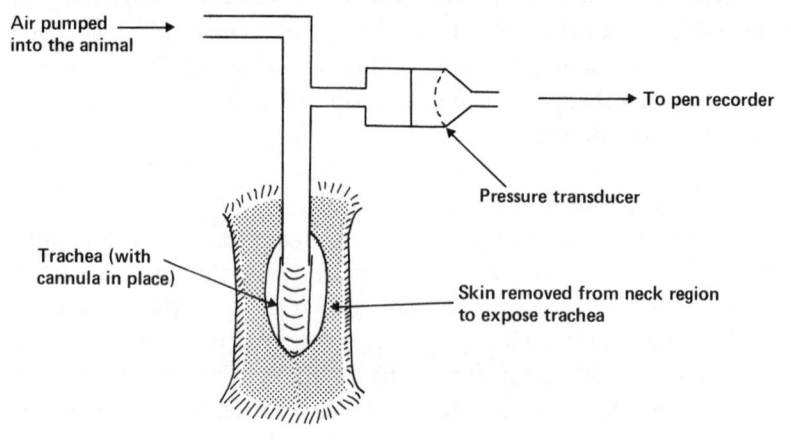

will increase the overflow volume of air pumped to the transducer and thus increase the pressure (called insufflation pressure) recorded. Conversely, if bronchodilatation occurs then more air passes into the lungs and less to the transducer, to produce a fall in insufflation pressure. This may become clearer by referring to Fig. 7.5 which shows a typical Konzett–Rossler set up. An added advantage of this technique is that the effect of the drug injected intravenously via either the jugular or femoral vein on both the airways smooth muscle and on arterial blood pressure (measured via a cannula inserted into the carotid artery) may be assessed at the same time.

The results of experiments in which prostaglandins have been administered to experimental animals have normally matched their effect on isolated airways preparations *in vitro*. In the anaesthetised guinea pig Konzett–Rossler preparation for example, intravenous PGE_2 (50–200 ng/kg upwards) greatly reduces or abolishes the bronchoconstrictor effect of agents such as acetylcholine, bradykinin and 5-hydroxytryptamine. PGE_2 also partly blocks the intense bronchoconstrictor effect which follows ovalbumin injection into previously sensitised guinea pigs (anaphylactic bronchoconstriction). $PGF_{2\alpha}$ produces bronchoconstriction only at high doses while 6-oxo-PGE_1 is also a potent bronchodilator. The results of a representative experiment are shown in Fig. 7.6.

Fig. 7.6. Effect of prostaglandins E_2 and $F_{2\alpha}$ on guinea pig airways resistance measured by the Konzett–Rossler procedure. An upward deflection of the pen indicates bronchoconstriction. H and A are standard doses (1 μg) of histamine and acetylcholine, injected intravenously. 0.1 μg PGE_2 (E) had no detectable effect on airways resistance but decreased the bronchoconstrictor response to both histamine and acetylcholine, indicating a bronchodilator effect. $PGF_{2\alpha}$ (F) (0.5 μg, 2 μg) caused bronchonconstriction. Notice that no decline in the bronchoconstrictor activity to histamine was observed when PGE_2 was injected 3 min before histamine. This is because PGE_2 is rapidly inactivated in the lungs, and other organs, following intravenous injection.

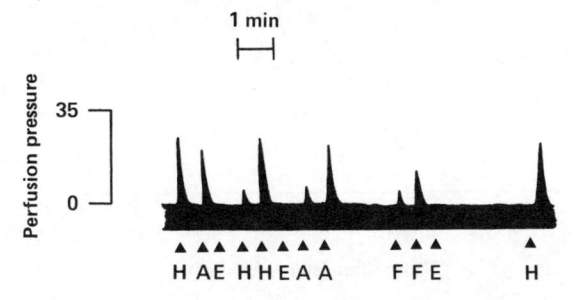

Species variations do occur and may complicate the interpretation of results. Whilst PGE_2 and $PGF_{2\alpha}$ have opposite effects on bronchomotor tone in guinea pigs, rabbits and monkeys, they both constrict the airways in cat Konzett–Rossler preparations and have little or no effect on dog airways. Despite these exceptions, PGEs are usually bronchodilator and PGFs usually bronchoconstrictor in animals.

The route of administration of PGE_2 is important since it governs the amount of bronchodilatation and also the incidence of side effects, which may include changes in blood pressure and heart rate. As can be seen from Fig. 7.6, the bronchodilator effect of PGE_2 in the guinea pig is short-lived. If the interval between injection of PGE_2 and histamine is increased from 1 min to 3 min, then no bronchodilatation is seen. Clearly after intravenous injection a lot of the PGE_2 is rapidly broken down not only in the lungs but also in other parts of the body which explains why PGE_2 has such a transient effect when administered by this route. One route of administration which ensures that large amounts of drug reach the airways without passing through the pulmonary circulation is inhalation. This may be achieved in man and experimental animals in the form of an aerosol. Inhalation of PGEs has proved to be a remarkably successful bronchodilator treatment. In the cat Konzett–Rossler preparation, in which a certain level

Fig. 7.7. Bronchodilator effect of intravenously injected PGE_2 (triangles) and aerosol-administered PGE_2 (squares) and isoprenaline (circles) in anaesthetised cats. Data from Rosenthale, Dervinis & Kassarich, 1974.

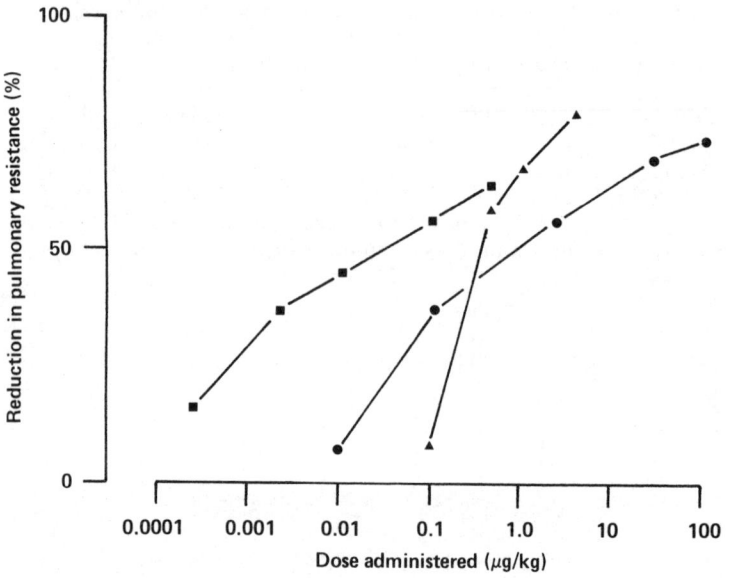

of bronchoconstriction is maintained by continuous intravenous infusion of the anti-cholinesterase drug neostigmine, the bronchodilator effect of inhaled E series prostaglandins is very marked, as shown in Fig. 7.7. The dose-response curves for PGE_2 given by aerosol and intravenously are not parallel and thus a strict estimate of their potency ratio by the two routes is not possible. However, it would appear that aerosol PGE_2 produces a more profound bronchodilator effect in this species than does intravenous PGE_2 or even inhaled isoprenaline. An added advantage is that PGE_2 administered by aerosol does not significantly affect blood pressure or heart rate in man or animals and thus may be a safer form of treatment than, for example, the intravenous route in which fairly large falls in blood pressure are sometimes reported.

Of the other prostaglandins which have been studied in experimental animals, $PGF_{2\beta}$ is an effective bronchodilator; it protects anaesthetised guinea pigs and cats against acetylcholine-induced bronchoconstriction. Surprisingly, and in contrast to other prostaglandins, $PGF_{2\beta}$ was considerably less effective given by aerosol than after intravenous injection. Why this should be so is unclear but may be related to an inability of inhaled $PGF_{2\beta}$ to cross biological membranes, thereby gaining only limited access to the target bronchial smooth muscle cells. As predicted from experiments *in vitro*, PGD_2 is a potent bronchoconstrictor agent in anaesthetised dogs and guinea pigs *in vivo*.

In man also PGEs administered either intravenously or by inhalation cause bronchodilatation whereas PGFs and PGD_2 are bronchoconstrictor. Again, smaller amounts of PGEs are required to produce the same amount of bronchodilatation when administered by aerosol and the incidence of side effects is lower. However, one problem of this route of administration which was not apparent with animal experiments was that all prostaglandins cause irritation of the mucous membranes of the upper respiratory tract, resulting in coughing and a sore throat. Such irritation was particularly apparent in early studies using prostaglandins as the free acid; it has been partially, but not completely, overcome by reformulation of prostaglandins as their neutral triethanolamine salts. It has been proposed that prostaglandins act on some form of 'irritant receptor' in the trachea and bronchi which, when stimulated, bring about bronchoconstriction by reflexly increasing nervous activity in parasympathetic nerves which innervate the airways. Evidence has come from studies in human volunteers in which pre-treatment with the muscarinic cholinoceptor antagonist Sch 1000 blocked some of the bronchoconstriction due to inhaled $PGF_{2\alpha}$. These considerations led Gardiner & Collier (1980) to propose an alternative classification for prostaglandin receptors in the lungs. According to this scheme, receptors which mediated bronchoconstrictor responses were χ, those causing bronchodilatation were ψ and those for irritancy or cough were ω. This simple classification may be useful for predicting the clinical value of PG analogues in the treatment of asthma. In this situation an analogue with high activity on ψ receptors and low activity on ω receptors would be desirable.

Of the other prostaglandins which have been administered to man, $PGF_{2\beta}$ has proved something of a disappointment. Although a good bronchodilator drug *in vitro* and in anaesthetised animals *in vivo*, $PGF_{2\beta}$ has little bronchomotor activity in man and – if it produces any response – usually causes bronchoconstriction. PGD_2 has not been studied for its effects in man. A summary of the major effects of classical prostaglandins on airways is shown in Table 7.4.

Effect of PG endoperoxides, TxA_2 and PGI_2 on respiratory smooth muscle

Perhaps because of their chemical instability, the bronchomotor effects of these prostanoids *in vitro* and *in vivo* has received less attention

Table 7.4. *Summary of the effects of prostaglandins on airways smooth muscle*

System studied	Effect[a]		
	PGEs	PGFs	PGD_2
Experiments in vitro			
Trachea	↓	↑ B↓	↑
Bronchi	↓	↑	—
Parenchyma	↓	↑	—
Experiments in vivo			
Animals (intravenous administration)	↓	↑ B↓	↑
Animals (aerosol administration)	↓↓	↑↑B↓	—
Man (intravenous administration)	↓	↑ B↓	—
Man (aerosol administration)	↓↓	↑ B↓	—

[a] ↑, Bronchoconstriction in intact animals and contraction of isolated tissues; ↓, bronchodilatation in intact animals and relaxation (or antagonism of the response to contractile drugs) in isolated tissues. — Indicates that the prostaglandin has little consistent effect. B Represents the response to $PGF_{2\beta}$.

than their classical prostaglandin counterparts. PGG_2 and PGH_2 contract the isolated guinea pig trachea. The potency ratio ($TxA_2:PGF_{2\alpha}$) in this preparation is about 45 (Svensson et al., 1977). PG endoperoxides administered by aerosol to guinea pigs produce a transient increase in insufflation pressure followed by a more prolonged bronchodilatation. The first effect is probably the direct action of the PG endoperoxide whereas the latter bronchodilatation results from the enzymatic transformation of the PG endoperoxide to PGI_2 or PGE_2 in the lungs. Since TxA_2 is not available for routine organ-bath studies, most authors have investigated the actions of the stable TxA_2-mimetic drug U46619 which potently contracts guinea pig tracheal and parenchymal smooth muscle. Further studies to characterise the receptors responsible for the effect of the PG endoperoxides and TxA_2 on the airways are needed.

There have been several attempts to characterise the effect of PGI_2 and its metabolites, 6-oxo-$PGF_{1\alpha}$ and 6-oxo-PGE_1, on airways smooth muscle. PGI_2 either weakly contracts the trachea or in some preparations actually causes a small relaxation *in vitro* (Omini, Moncada & Vane, 1977). PGI_2 relaxes the guinea pig lung parenchymal strip at high concentrations while 6-oxo-$PGF_{1\alpha}$ was virtually without effect on this tissue. In contrast,

Fig. 7.8. Relaxant effect of isoprenaline, 6-oxo-PGE_1 and PGI_2 on guinea pig lung parenchymal strips set up in an organ bath *in vitro*.

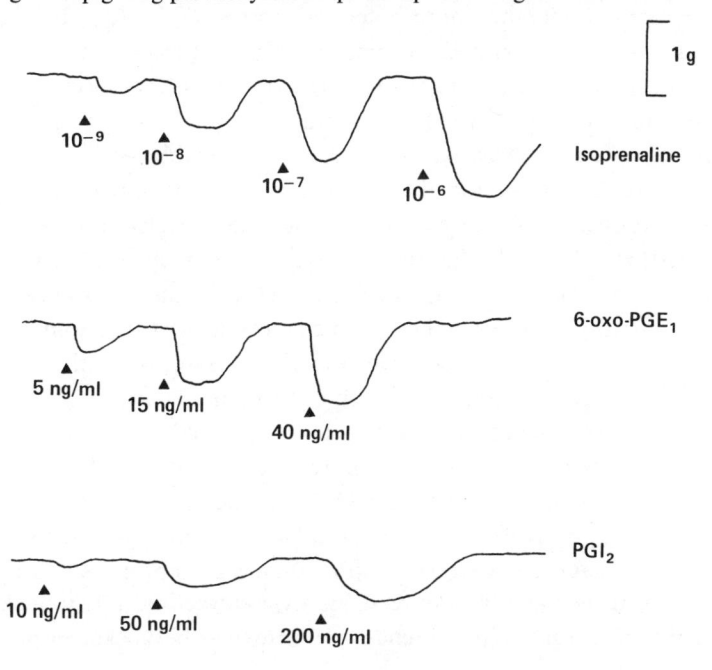

6-oxo-PGE$_1$ also relaxed guinea pig lung strips with a potency of up to 10 times that of PGI$_2$. Some typical results are shown in Fig. 7.8.

Administration of PGI$_2$ or its stable 20-methyl isomer by aerosol to healthy or asthmatic human volunteers had little effect on bronchomotor tone in these individuals. Both compounds did however produce some pharyngotracheal irritation and coughing suggesting that they do activate the ω irritant prostaglandin receptors (Szczklik et al., 1978a).

Effect of products of the lipoxygenase pathway on respiratory smooth muscle

The activity of slow reacting substance of anaphylaxis (SRS-A) on airways smooth muscle has been investigated for the last 40 years. Only in the last decade has SRS-A been shown to be comprised of a mixture of leukotrienes C$_4$, D$_4$ and E$_4$ which are formed by a lipoxygenase enzyme from arachidonic acid. Because of the possible relationship of SRS-A to human asthma a considerable amount of time and energy has been devoted to the study of leukotrienes on the ventilatory system both *in vitro* and *in vivo*.

Experiments in vitro

LtC$_4$ and LtD$_4$ potently contract guinea pig trachea and parenchyma strips and human isolated bronchial smooth muscle. In this respect they are considerably more potent than prostanoids (e.g. PGF$_{2\alpha}$, PGD$_2$ and TxA$_2$) and thousands of times more potent than histamine.

Like prostanoids, the response to leukotrienes is very species-dependent. LtC$_4$ and LtD$_4$ have little effect on tracheal preparations from the rat, cat or dog or on parenchymal strips from rats or rabbits.

A feature of this action of leukotrienes is the very slow contraction which they produce in the airways *in vitro* and when injected intravenously into anaesthetised animals. Frequently, contractions may take 3–5 min to peak and tachyphylaxis develops rapidly. Such slow-onset responses and the susceptibility to tachyphylaxis immediately suggest that some (or all) of the constrictor effect of leukotrienes may be secondary to the release from the tissue itself of substances with bronchomotor activity. There is now considerable evidence that LtC$_4$, LtD$_4$ and LtE$_4$ perfused through the isolated guinea pig lungs cause release of TxA$_2$ and other PG-like substances (Piper & Samhoun, 1982). Thus, the possibility that leukotrienes cause TxA$_2$ biosynthesis and release from the airways and that TxA$_2$ is responsible for the observed contraction has been proposed. Evidence for this has come from two sources. Firstly, contractions due to leukotrienes on superfused guinea pig parenchymal strips may be blocked by pre-treating the tissues with either indomethacin or the thromboxane synthesis inhibitor, carboxyheptylimidazole (Piper & Samhoun, 1982). Secondly, TxA_2 (measured by radioimmunoassay of TxB_2) is released into the organ bath when guinea pig lung parenchymal strips are challenged with LtC_4. However, if these tissues are pre-incubated with indomethacin then the contraction due to LtC_4 is reduced and at the same time the appearance of TxB_2 in the organ bath fluid is prevented.

These results strongly suggest that leukotrienes depend upon the formation of TxA_2 in order to contract airways smooth muscle *in vitro*. Unfortunately it is very difficult to lay down hard and fast rules since in other respiratory smooth muscle preparations, like the human bronchus, the response to leukotrienes is unaffected by indomethacin but blocked by the leukotriene receptor antagonist FPL 55712. Thus, in this tissue, the effect of leukotrienes seems to be independent of TxA_2 biosynthesis. The somewhat confusing picture which emerges of the role of cyclooxygenase products in the action of leukotrienes on the airways is depicted diagrammatically in Fig. 7.9.

Fig. 7.9. The mechanism of action of leukotrienes on pulmonary smooth muscle preparations *in vitro*. Notice that contractions of guinea pig lung parenchymal strips induced by leukotrienes are mediated by prostanoids, chiefly TxA_2.

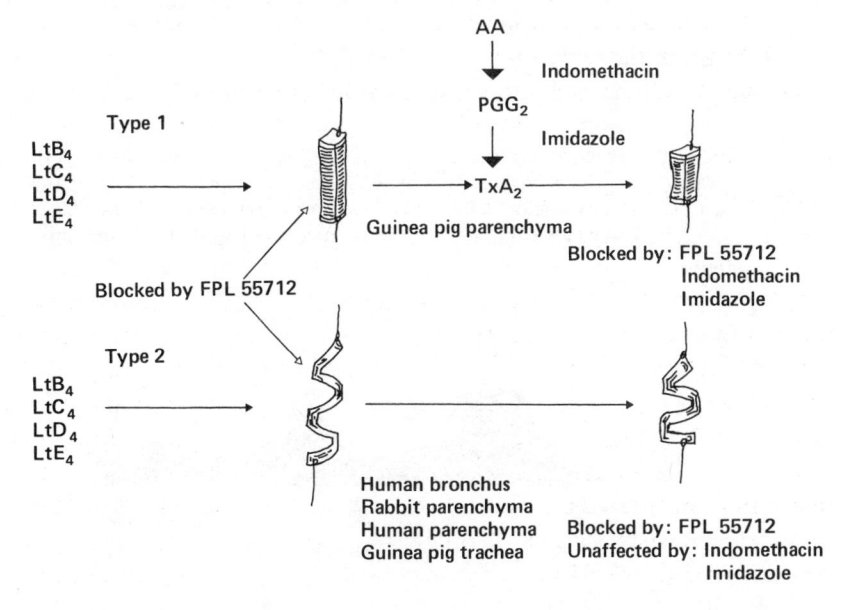

Experiments in vivo

LtC_4, LtD_4 and LtE_4 are potent bronchoconstrictors in the guinea pig Konzett–Rossler preparation *in vivo* (Fig. 7.10). Compared with histamine, LtC_4 injected intravenously was some 100 times more potent. Furthermore, LtC_4 produced a long-lasting increase in insufflation pressure compared with the transient effect of histamine. When given in the form of an aerosol, LtC_4 was even more potent than histamine with a potency ratio of 1000:1. The mechanism of action of leukotrienes varies according to the route of administration. When given intravenously, the bronchoconstrictor effect of LtC_4 is abolished by indomethacin pre-treatment but, when administered by aerosol, LtC_4 causes bronchoconstriction which is not blocked by NSAID (Weichmann & Tucker, 1982). Thus a cyclooxygenase metabolite (probably TxA_2) mediates the effect of leukotrienes given intravenously but not by inhalation. In man also leukotrienes are bronchoconstrictor (Holroyde *et al.*, 1981) causing wheezing and coughing in human volunteers.

Prostanoids and the pulmonary vasculature

The efficiency of oxygen and carbon dioxide transfer in the lungs depends not only upon the volume of air in the alveoli (determined by the calibre of the airways) but also on blood flow through the alveolar capillaries. If the amount of blood passing through the alveolar capillaries is reduced then the uptake of oxygen from air and expulsion of carbon dioxide from the blood is likely to be reduced. For this reason, the effect of prostanoids on the pulmonary vasculature may be of importance for gas exchange in the lungs.

Dilatation of the pulmonary vasculature is technically difficult to

Fig. 7.10. Effect of leukotrienes on guinea pig airways resistance measured by the Konzett–Rossler technique. Upward deflection of the pen indicates bronchoconstriction. Histamine (H), acetylcholine (A) and 5-hydroxytryptamine (5-HT) were injected (1 μg) into the jugular vein.

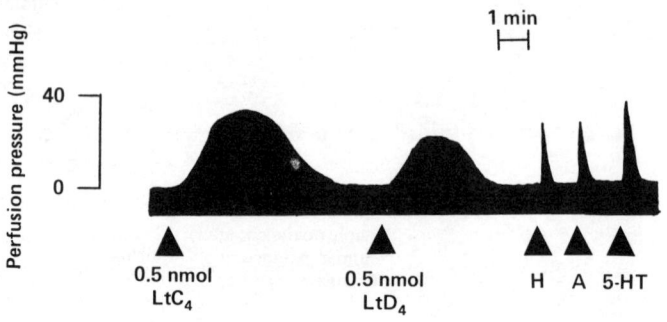

demonstrate experimentally since these vessels are already maximally dilated and normally will not dilate any further. To overcome this problem the effect of vasodilator prostanoids is assessed in animals in which pulmonary vasoconstriction has been induced by infusion of ADP to cause platelet aggregation and release of vasoconstrictor agents like TxA_2 and 5-hydroxytryptamine. In such animals, PGI_2 is a pulmonary vasodilator and in fact is the only metabolite of arachidonic acid which has this ability. The chemical hydrolysis product of PGI_2, 6-oxo-$PGF_{1\alpha}$, has negligible effects on the pulmonary vasculature while 6-oxo-PGE_1 retains some vasodilator activity (less potent than PGI_2) in this vascular bed. All other prostaglandins tested to date have pulmonary vasoconstrictor activity although the ratio of potency from one to the next depends upon the type of experiment performed and the species. $PGF_{2\alpha}$, for example, has remarkably good pulmonary vasoconstrictor activity in most species even though, when injected intravenously in anaesthetised animals, it produces only very small changes in systemic blood pressure and vascular resistance. Stable TxA_2-mimetic drugs, like U46619, also cause constriction of the pulmonary vascular bed as does PGD_2 and PGE_2 (which, interestingly, is a vasodilator in most other vascular beds). The order of potency is $TxA_2 > PGF_{2\alpha} = PGD_2 > PGE_2$.

Prostanoids, leukotrienes and bronchial secretion

Since the inflamed airway secretes mucus at an increased rate and asthma is associated with both bronchoconstriction and bronchial congestion it is obviously of interest to determine the effect of arachidonic acid metabolites on bronchopulmonary mucus formation. Relatively high doses of LtC_4 (10^{-6} M) increased mucus accumulation in the cat trachea *in vivo*, an effect partially blocked by FPL 55712 (Peatfield, Piper & Richardson, 1982). Similar doses of LtC_4 and LtD_4 also stimulated mucus secretion from strips of human bronchi *in vitro*. The effect of leukotrienes on mucus secretion is small compared with the effect of PGE_1 and $PGF_{2\alpha}$ (Richardson *et al.*, 1978). It is as yet unclear whether prostanoids or leukotrienes play any part in the massive bronchial secretion of mucus which occurs in human asthma.

Role of arachidonic acid metabolites in bronchial asthma

The role of chemical mediators in the aetiology of bronchial asthma in man is poorly understood. Histamine released from mast cells in the lung may not be of major importance in asthma, as was first thought, since anti-histamine drugs are largely of little therapeutic benefit in asthma sufferers. Similarly, it is unlikely that cyclooxygenase products of arachidonic acid are primary mediators of asthma since aspirin-like drugs are usually ineffective as treatment for this condition and, in a small number of patients, can actually provoke an extremely severe asthmatic attack. One explanation for this response is that NSAID may, by preventing cyclooxygenase activity, divert arachidonic acid metabolism along the lipoxygenase pathway resulting in the formation of leukotrienes which cause bronchoconstriction leading to asthma. Whether leukotrienes have an integral part to play in the genesis of bronchial asthma other than that induced by aspirin remains to be seen.

Although arachidonic acid metabolites may not play a primary role in asthma there is considerable evidence (summarised below) that both prostanoids and leukotrienes are formed locally in large amounts during asthma and that their presence may promote or exacerbate the asthmatic attack. Until such time that a more appropriate niche may be made for these substances in asthma, pulmonary prostanoids and leukotrienes should perhaps be considered as secondary mediators of this disease.

(1) Studies *in vitro* of IgE-mediated mast cell degranulation have shown that $PGF_{2\alpha}$, PGD_2, PGE_2, TxA_2, LtC_4 and LtD_4 are released in large quantities (Roberts *et al.*, 1979).

(2) Prostanoids (e.g. PGE_2, $PGF_{2\alpha}$, PGD_2, TxA_2) and leukotrienes (e.g. LtB_4, LtC_4, LtD_4) are released in copious amounts from isolated, perfused immunologically sensitised lungs following challenge with the appropriate antigen.

(3) Plasma levels of 13,14-dihydro-15-oxo-$PGF_{2\alpha}$ measured by GC–MS, in peripheral venous blood of asthmatic patients was increased 5–10-fold within 5 min of the induction of an asthmatic attack.

(4) Asthmatic patients are approximately 1000 times more sensitive to the bronchoconstrictor effect of inhaled $PGF_{2\alpha}$ than are normal, healthy subjects (Mathe *et al.*, 1973).

The role of arachidonic acid metabolites in other lung diseases remains largely unstudied. Brief reports have suggested that lung tissues from infants who have died as a result of sudden infant death syndrome (SIDS) generate less PGI_2 than tissue from infants killed in accidents or from some other known cause. More substantial experiments are required before definite conclusions linking deficient pulmonary PGI_2 biosynthesis with this clinical syndrome can be made. Other lung diseases which could alter the uptake and/or catabolism of prostanoids and leukotrienes from the lung include emphysema, pneumonia and sarcoidosis.

8
Prostanoids and the kidney

Introduction

The importance of prostanoids in renal function has been appreciated for many years but has only recently begun to be understood. Apart from a direct action on the nephron to stimulate water and electrolyte reabsorption, prostanoids also influence renal function indirectly by changing the distribution of blood flow within the kidney and by affecting the activity of renal sympathetic nerves. In addition, prostanoids interact in the kidney with local hormones and hormones such as angiotensin II, bradykinin and vasopressin (anti-diuretic hormone, ADH), each of which has very potent effects on renal water and ion transport. The kidney is anatomically and physiologically one of the most complex organs in the body and it is therefore no great surprise that the functions of renal prostanoids have proved difficult to unravel. Indeed a recent review on the subject compared the study of prostaglandins in the kidney with a 'trip down a rabbit hole' (Flamembaum & Kleinman, 1977). The aim of this chapter is to review the biochemistry and physiology of renal prostanoids in health and disease.

Synthesis and catabolism of prostanoids in the kidney

The mammalian kidney is an excellent source of cyclooxygenase and of the other enzymes which further convert prostaglandin endoperoxides. The most abundant prostanoids formed in the kidney are PGI_2 and PGE_2, with smaller amounts of TxA_2, $PGF_{2\alpha}$ and PGD_2. Despite much interest in the early 1960s there is no convincing evidence that PGA_2 is a normal end product of renal prostaglandin endoperoxide metabolism. Early studies suggested a gradient of PGE_2 biosynthesis from the cortex to the medulla and papilla (Anggard *et al.*, 1972) and these results have been confirmed by Dunn (1976) who found that the ratio of cyclooxygenase activity between rabbit renal papilla, medulla and cortex was about 100:10:1. Rabbit renal medulla cyclooxygenase has been extensively studied with respect to substrate, cofactor, pH and temperature requirements as well as susceptibility to inhibition by NSAID (Blackwell, Flower & Vane, 1975a) and is sometimes used as an alternative to bovine seminal vesicles as an inexpensive and readily available source of prostaglandin-synthesising enzymes for experiments *in vitro*.

Despite earlier claims to the contrary, the renal cortex can also synthesise small amounts of PGE_2 and $PGF_{2\alpha}$ (Pong & Levine, 1976) and in addition may be an important site of PGI_2 and TxA_2 formation. For example, microsomes from the rat renal cortex convert radiolabelled PGH_2 to a mixture of prostanoids including PGI_2 and TxA_2 (measured as 6-oxo-$PGF_{1\alpha}$ and TxB_2 respectively), PGE_2, $PGF_{2\alpha}$ and PGD_2 (Zenser, Herman & Gorman, 1977). In this study formation of PGI_2 and TxA_2, but not of classical prostaglandins, was actually greater in the cortex than in the medulla. Similarly, PGI_2 is reportedly the major metabolite of arachidonic acid in rabbit renal cortex, as determined by the gas-chromatographic–mass-spectrometric identification of 6-oxo-$PGF_{1\alpha}$ in renal cortex extracts (Whorton, Smigel & Oates, 1978). Studies in the last few years have revealed that prostacyclin may well be the major metabolite of arachidonic acid in the renal medulla as well as in the renal cortex. This certainly seems to be the case in human and rat renal medulla (Silberbauer & Sinzinger, 1978). It is hopefully clear even from this very brief account that there is still disagreement about the profile and amount of prostanoids formed by the renal cortex and medulla *in vitro*. The reasons for these discrepancies are probably several fold and include the use of different tissue fractions (i.e. homogenates or microsomes), different assay methods (gas-chromatography–mass-spectrometry, RIA, bioassay), the inclusion or non-inclusion of cofactors (GSH, adrenaline) coupled with natural species differences. To summarise, arachidonic acid is enzymatically converted to a variety of prostanoids in both the renal cortex and medulla *in vitro* in all species studied to date.

Anatomically the basic unit of the mammalian kidney is the nephron, which is responsible for filtering blood and producing urine of appropriate osmolality. Each nephron comprises a glomerulus, proximal convoluted tubule, loop of Henle, distal convoluted tubule and collecting duct which drains into the ureter (Fig. 8.1.). Two types of nephron occur in the kidney. The majority (about 85% in humans) are equipped with a glomerulus sited in the outer region of the cortex and a relatively short loop of Henle. These are the superficial cortical nephrons. The remaining nephrons (juxtamedullary) have glomeruli in the inner cortex and their loops of Henle protrude deep into the medulla. Since the osmotic gradient between the filtrate in the nephron and the interstitial fluid increases progressively on traversing the medulla from the cortex to the papilla, juxtamedullary nephrons are more efficient water reabsorbers and are thus responsible for disproportionately more reabsorption of electrolytes and water than are

Fig. 8.1. Biosynthesis of prostanoids in different parts of the nephron. The major prostanoid detected in experiments *in vitro* is shown in each case.

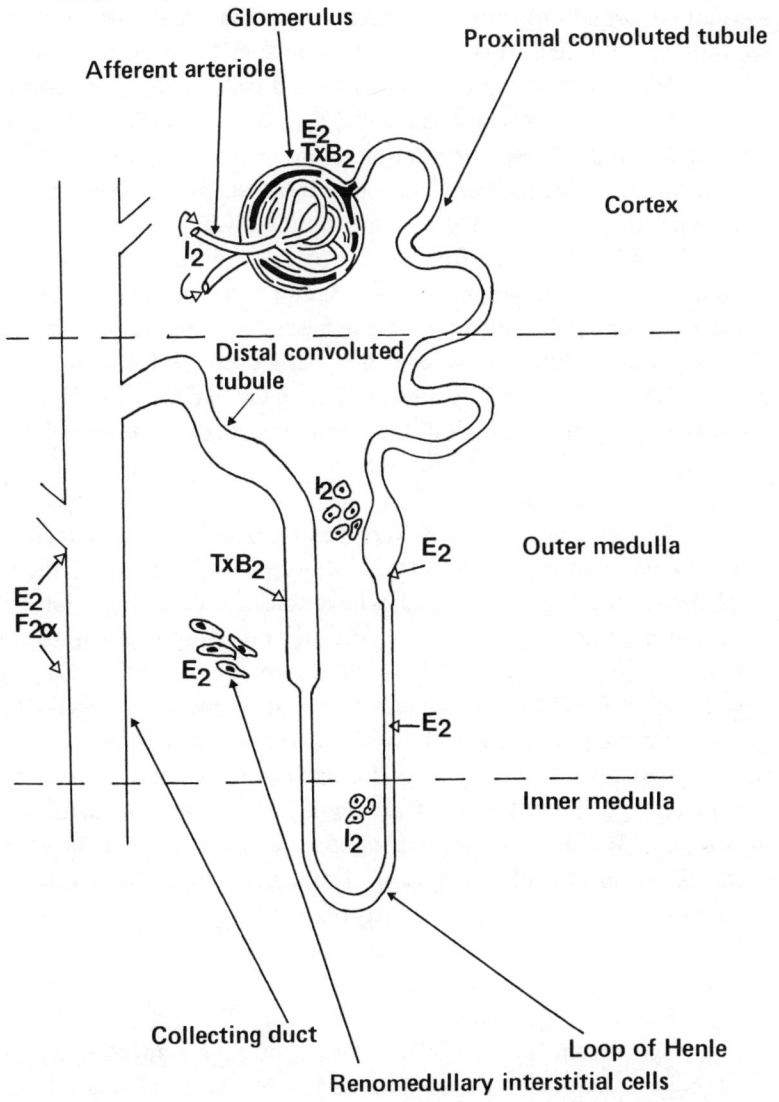

the superficial cortical nephrons. The physiological importance of this piece of renal anatomy will become clear later.

Bearing in mind the complex structure of the kidney, it is apparent that measuring renal cortical and medullary prostanoid biosynthesis is an extremely crude technique which provides little or no worthwhile information about which part of the nephron is responsible for producing which particular prostanoid. If such information were available, it might be possible to correlate the site of prostanoid formation with its function in that part of the nephron. Two approaches have been used to study prostanoid formation in individual parts of the nephron. Firstly, mechanical separation of the cells to be investigated can sometimes be achieved. This may be done by exposing collagen-dispersed cells of the renal cortex or medulla to procedures which selectively destroy all but the cells required. Grenier & Smith (1978), for example, were able to prepare a 97% pure suspension of rabbit papillary collecting tube cells by exposing dispersed cell preparations to hypotonic saline solution which osmotically lysed essentially all cells apart from those of the collecting tubule. As an alternative, the second approach is to maintain individual renal cell types in tissue culture and sample the culture medium for released prostanoids. Both techniques have provided some very useful information. From experiments of this sort it is now clear that the capacity for prostanoid formation is unequally distributed along the nephron as described below.

The glomerulus

The glomeruli of both superficial cortical and juxtamedullary nephrons occur in the renal cortex and represent about 5% of the total renal mass in the dog. Several studies have confirmed very high glomerular cyclooxygenase activity. Isolated rat cortical glomeruli convert arachidonic acid mainly to TxB_2 and PGE_2 with smaller amounts of $PGF_{2\alpha}$ and 6-oxo-$PGF_{1\alpha}$. Cyclooxygenase activity in rat glomeruli is several times greater than that in rat cortical distal convoluted tubules, another prime site of prostanoid biosynthesis in the nephron. In contrast to the rat, human renal glomeruli convert exogenous arachidonic acid mainly to prostacylin. Within the glomerulus, the major locus for prostanoid production is most likely the mesangial cells which (in cell culture) exhibit about twice the cyclooxygenase activity of cultured rat glomerular epithelial cells.

The collecting tubules

Apart from the glomerulus, the second most important region of prostanoid formation in the nephron is that of the collecting tubules in the medulla and papilla. Cultured rabbit renal papillary collecting tubule (RPCT) cells spontaneously release PGE_2, $PGF_{2\alpha}$, PGD_2, 6-oxo-$PGF_{1\alpha}$ and TxB_2 into the culture medium in the ratio 75:15:5:5:0. Inclusion of bradykinin increased the formation of all prostanoids in the same ratio (Grenier, Rollins & Smith, 1981). In accordance with the large amount of PGE_2 formed by RPCT cells in culture, homogenates of these cells contain very high enzyme activities for cyclooxygenase and prostaglandin endoperoxide E isomerase. High concentrations of cyclooxygenase enzyme in the collecting tubules of kidneys from several species have been confirmed using a highly specific immunofluorescence-staining method.

The medullary interstitial cells

A third site of prostanoid biosynthesis in the kidney are the interstitial cells, so-called because they lie between and 'pack out' the loops of Henle and their accompanying blood vessels (the vasa recta) in the medulla. Renomedullary interstitial cells were first shown to synthesise prostaglandins by Muirhead and his colleagues in 1972. These cells contain large amounts of arachidonic acid packed into densely stained granules and occur in small groups scattered liberally throughout the medulla and papilla. The granularity of these cells has been claimed to indicate the level of medullary PGE_2 formation *in vivo*. In experimental situations where renal PGE_2 formation is reduced below normal then arachidonic acid levels in these cells is high and renomedullary interstitial cell granularity is increased. Interestingly, the granularity of these cells is reduced in animals loaded with water or put on a high-salt diet (Nissen, 1968), both conditions known to augment renal prostanoid biosynthesis.

In tissue culture, renomedullary interstitial cells synthesise mainly PGE_2 and smaller amounts of $PGF_{2\alpha}$ (Muirhead *et al.*, 1972; Dunn, Staley & Harrison, 1976). Angiotensin II, bradykinin and arginine vasopressin increase PGE_2 biosynthesis by cultured rabbit renomedullary cells (Zusman & Keiser, 1977). These cells do not release prostacyclin. However, Smith *et al.* (1983) have localised interstitial cells in the medulla capable of forming PGI_2. In a series of elegant experiments these authors employed an immunofluorescence technique using mouse monoclonal antibodies raised to purified PGI_2 synthetase enzyme to demonstrate the occurrence of small groups of cells in the renal interstitium which were not identical to the cells first cultured by Muirhead's group. These cells were evenly distributed throughout the medulla, apparently associated with bundles of capillaries. As yet, these groups of cells have been given no name and have no known function. It is possible that medullary interstitial cells synthesise vasodilator prostaglandins (PGI_2 and PGE_2) which maintain the patency of the vasa recta and of the dense network of anastomosed blood vessels located between them. The role of prostanoids in the control of intrarenal blood flow is discussed later in the chapter.

Formation of prostanoids in the intact kidney

Although experiments *in vitro* can sometimes give a useful idea of the amount and type of prostanoids synthesised in the kidney, the results obtained do not necessarily reflect prostanoid formation in the intact kidney. Furthermore, experiments *in vitro* normally take no account of phospholipase activity which is probably the rate-limiting step in the biosynthesis of prostaglandins and leukotrienes, not only in the kidney but in all other organs as well. The intact kidney may be studied in several ways. For experiments *ex vivo*, the kidney is removed from a freshly killed animal (usually a rat or rabbit) and perfused at the rate of 5–15 ml/min with warmed, oxygenated Krebs' solution via a cannula inserted into the renal artery. The perfusate may be collected by means of cannulae in the renal vein or ureter or allowed to drip from the cut ends of these vessels over isolated pharmacological tissues for superfusion cascade bioassay. A slight modification to the apparatus allows a pressure transducer to be connected to a side arm of the arterial cannula to measure the perfusion pressure which represents renal vascular resistance. Experiments *ex vivo* of this sort have revealed the spontaneous release of small amounts of prostanoids from the isolated perfused rat and rabbit kidney. Most likely this prostanoid release is an artifact reflecting tissue damage during removal and preparation of the kidney for perfusion. Prostanoid release from the kidney may be increased several times above this basal efflux by injection or infusion into the renal artery of vasoactive substances such as bradykinin, angiotensin II and noradrenaline. A typical experiment of this sort showing prostaglandin release from the perfused rabbit kidney induced by another vasoactive substance, ATP, is shown in Fig. 8.2.

Enhanced prostanoid release from the kidney may reflect either stimulated prostanoid biosynthesis or reduced prostanoid catabolism or a combination of the two. All the currently available evidence suggests that vasoactive agents increase renal prostanoid formation by stimulating phospholipase activity. The released arachidonic acid is then rapidly converted by cyclooxygenase to prostaglandin endoperoxides which, in turn, are further catabolised to one or more of the various prostanoids. Precisely which prostanoid or combination of prostanoids is released by, for example, angiotensin II has not been fully established. However, we do know that in the dog kidney angiotensin II preferentially releases PGE_2 and PGI_2 (Shebuski & Aiken, 1980) while in kidneys from spontaneously hypertensive rats angiotensin II selectively releases TxA_2 (Shibouta, Inada & Terashita, 1979).

Renal prostanoid biosynthesis may also be studied *in vivo* either in conscious or in anaesthetised animals. In this case prostanoid concentrations are determined in samples of renal venous blood or in urine. If required, renal blood flow may be measured using an electromagnetic flow probe positioned around the exposed renal artery. Okahara, Imanishi & Yamamoto (1983) have measured the renal venous and renal arterial concentration of four important prostanoids in anaesthetised dogs. Renal venous PGE_2 (204 ± 76 pg/ml), $PGF_{2\alpha}$ (407 ± 73 pg/ml), 6-oxo-$PGF_{1\alpha}$ (378 ± 71 pg/ml) and TxB_2 (436 ± 66 pg/ml) levels were slightly higher than those found in renal arterial blood, suggesting that the kidney spontaneously synthesises and releases small amounts of prostanoids into the circulation. Indomethacin pre-treatment reduced renal venous prostanoid concentration while intravenous injection of angiotensin II and bradykinin stimulated renal prostanoid production in anaesthetised animals and increased the renal venous plasma concentration of PGI_2, PGE_2 and $PGF_{2\alpha}$ (Mullane & Moncada, 1980).

For several reasons renal venous prostanoid concentration is probably

Fig. 8.2. Release of prostanoids by ATP from isolated Krebs' perfused rabbit kidney *in vitro* determined biologically by passing the Krebs' solution emerging from the kidney over a rat stomach strip (RSS), chick rectum (CR) and rabbit coeliac artery (RbCA) in cascade. Injection of ATP (5 µmol) directly over the tissues (D) did not affect the bioassay tissues. ATP (1 and 5 µmol) injected into the renal artery (K) elicited the release of a substance (probably PGI_2) which contracted the RSS and CR but relaxed the RbCA.

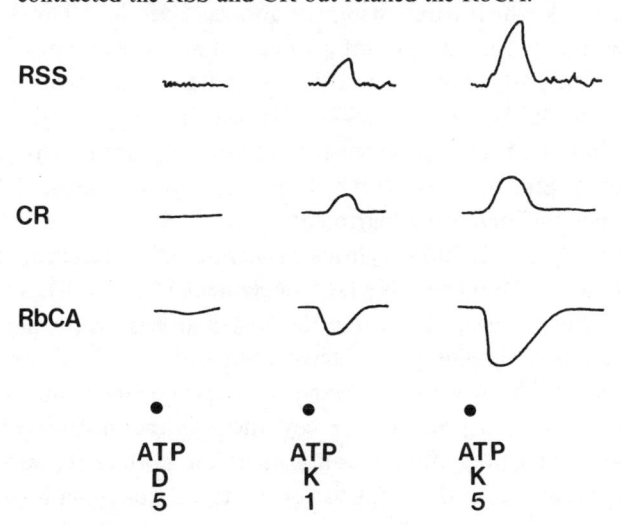

not a reliable indication of prostanoid synthesis in the kidney. Firstly, only very small amounts of prostaglandins are found in renal venous plasma necessitating the use of the most sensitive analytical methods. The majority of researchers have used radioimmunoassay since it is easy, cheap and normally reliable. However, it should be remembered that blood collected from the renal vein will contain not only prostanoids formed in the kidney but also large amounts of prostanoids and their metabolites made in organs all over the body. With the present state of knowledge it is more than likely that some of the metabolites appearing in the renal vein will not be known. Consequently there is a strong possibility that one or other of these uncharacterised prostanoid metabolites may cross-react with the antibodies to classical prostaglandins, TxB_2 and 6-oxo-$PGF_{1\alpha}$, thereby producing incorrect measures of these substances in renal venous blood. Another problem which frequently arises when blood or plasma is extracted prior to radioimmunoassay is that aggregating platelets release large amounts of some prostanoids (particularly TxB_2), thus producing erroneously high measurements in renal venous blood.

A widely used index of renal prostanoid formation is the concentration of unmetabolised prostaglandins in urine. PGE_2 and $PGF_{2\alpha}$ occur in human urine in both men and women. However only in women and in pre-pubertal males are unmetabolised prostaglandins in the urine believed to be of renal origin (Frolich, Wilson & Sweetman, 1975). In adult men, urine prostaglandin levels are frequently very high, probably the result of contamination with seminal fluid prostaglandins (Patrono, Wennmalm & Ciabottini, 1979). TxB_2 and 6-oxo-$PGF_{1\alpha}$ are also found in the urine.

Not surprisingly the mammalian kidney contains large amounts of the various enzymes which catabolise prostanoids. Cystolic 100000 g high-speed supernates prepared from rat, rabbit and monkey kidney have high 15-PGDH, PGΔ13R, PG-9KR and PG-9HDH enzyme activity (Katzen, Pong & Levine, 1975). Of these, 15-PGDH, which biologically inactivates prostaglandins, has been most extensively studied and purified from rabbit and pig kidney. Like the enzymes which synthesise prostanoids, 15-PGDH has a distinct intrarenal compartmentalisation. Renal, type I NAD^+-dependent 15-PGDH is 10 times more concentrated in the cortex than in the medulla while the type II, $NADP^+$-dependent 15-PGDH has twice as much activity in the medulla than in the cortex at least in the pig kidney (Lee et al., 1975). Similar results have been obtained in other species (Katzen et al., 1975). It would appear from experiments of this sort that the major site of renal prostanoid biosynthesis is the medulla while the major site of renal prostanoid catabolism is the cortex. However, it is incorrect to assume that the renal cortex is the only degradative site and the renal medulla the only synthetic site for prostanoids in the kidney. As previously discussed cortical prostanoid biosynthesis and medullary prostanoid catabolism does occur *in vitro*.

The half-life of renal 15-PGDH has been determined indirectly in experiments in which rats were killed at timed intervals after an intraperitoneal injection of the protein-synthesis blocker cycloheximide (Blackwell, Flower & Vane, 1975b). The rapid turnover of renal 15-PGDH ($t_\frac{1}{2}$ about 60 min) suggests that the activity of this enzyme in the kidney may be regulated by hormones which affect protein synthesis or transcription of DNA. For example, adrenalectomised or ovariectomised rats exhibit elevated lung and kidney 15-PGDH, suggesting that steroid hormones may regulate tissue concentrations of this enzyme *in vivo*. Renal 15-PGDH activity in the rat also undergoes age-dependent changes. The activity of this enzyme in rat kidney is high immediately after birth, increases to a peak (some 59 times the adult level) at day 19 and then declines to reach the minimum adult concentration by day 48. $PG\Delta13R$ shows a similar age-dependent profile in the developing rat kidney (Pace-Asciak, 1975). The physiological relevance of such high renal prostanoid-catabolising capacity in the early life of the rat (and sheep) is not clear. Pace-Asciak has suggested that the development of the kidney in the first few weeks of life may be retarded by the presence of vasoconstrictor prostaglandins (e.g. TxA_2, $PGF_{2\alpha}$) which interfere with the normal distribution of blood flow intrarenally, producing regions of isolated tissue ischaemia and preventing proper tissue development. This seems somewhat unlikely since the major renal prostanoids, PGI_2 and PGE_2, dilate the renal vascular bed and thus, according to this proposal, should be beneficial to the developing kidney. In addition to changes which occur with age, several groups have reported decreased renal 15-PGDH activity in spontaneously hypertensive rats (Pace-Asciak, 1976; Armstrong *et al.*, 1976).

PG-9KR, which converts PGE_2 to $PGF_{2\alpha}$, is also distributed unequally within the kidney. Rabbit kidney cortex contains 10 times more PG-9KR activity than kidney medulla (Stone & Hart, 1975). In contrast, PG-9KR is equally distributed between the cortex and medulla in the pig kidney (Lee *et al.*, 1975).

Rat kidney supernates contain a PG-9HDH enzyme which oxidises 13,14-dihydro-15-oxo-$PGF_{2\alpha}$ to the corresponding PGE_2 metabolite whilst human and rabbit kidney PG-9HDH preferentially convert $PGF_{2\alpha}$ directly to PGE_2 (Moore & Hoult, 1978; Hassid, Shebusky & Dunn, 1983). In the kidney, the oxidation of $PGF_{2\alpha}$ to PGE_2 represents a biological activation since PGE_2 (but not $PGF_{2\alpha}$) has potent effects on renal blood flow and electrolyte and fluid excretion. Human, rat, pig and rabbit kidney

100000 g supernates also convert PGI_2 to 6-oxo-PGE_1 by means of a PG-9HDH enzyme (Moore & Griffiths, 1983b).

Biological effects of prostanoids in the kidney
Modulation of renal blood flow

Intrarenal infusion of PGE_2, PGD_2 and PGI_2 dilates renal blood vessels and increases blood flow through the kidneys of most species (Fulgraff & Brandenbush, 1974; Bolger, Eisner & Ramwell, 1978). An exception is the rat kidney in which PGE_2 may be either vasodilator or vasoconstrictor depending upon the pre-existing renal vascular tone. Thus, when renal vascular resistance is low, PGE_2 is a vasoconstrictor but if renal vascular resistance is raised (for example, by infusion of vasopressin) then PGE_2 exerts a dilator effect (Pace-Asciak & Rosenthale, 1982). In contrast PGI_2 and its stable metabolite 6-oxo-PGE_1 are potent vasodilator agents in the kidneys of all species, including the rat. $PGF_{2\alpha}$ which is also formed in small amounts in the kidneys of most species has little or no effect on the renal vasculature. PG endoperoxides have been shown to constrict the canine renal vasculature (Gerber, Ellis & Nies, 1977). Similarly, intrarenal infusion of arachidonic acid in the perfused rat kidney *in vivo* decreases renal blood flow, an action reduced by pre-treatment with a thromboxane synthetase inhibitor. This result provides indirect evidence that TxA_2, like the PG endoperoxides, constricts renal blood vessels.

Prostanoids also influence the distribution of blood flow inside the kidney. In the blood-perfused dog kidney, close arterial injection of PGE_2 increased the amount of blood flowing through the inner cortex at the expense of blood flow to the outer cortex (Itskovitz, Terragno & McGiff, 1974). Increasing endogenous renal prostanoid concentrations by intra-arterial infusion of arachidonic acid in the dog kidney has the same effect of shifting blood flow from the outer cortex to the inner cortex and medulla. Presumably in this experiment arachidonic acid is converted to PGE_2 or PGI_2 in glomerular or interstitial medullary cells, which then dilate medullary and inner cortical blood vessels to increase blood flow to the juxtamedullary nephrons.

From this very brief discussion it is clear that administration of exogenous prostanoids or of arachidonic acid normally dilate blood vessels in the kidney. Studies with renal homogenates and microsomes have shown that the kidney contains all the enzymes necessary to synthesise the full range of prostanoids and the detection of unmetabolised prostanoids in renal venous blood and urine suggests a continuous low-level formation of prostanoids in the kidneys. These observations raise two important questions. Firstly, do endogenous prostanoids made in the kidney also dilate the renal vasculature and, if they do, are these prostanoids involved in the physiological control of renal blood flow?

One way to determine the role of prostanoids in the regulation of blood distribution within the kidney is to study the effect of aspirin and similar NSAID on the renal circulation. Unfortunately experiments of this sort have produced varying results. Indomethacin or sodium meclofenamate administered to anaesthetised, surgically operated animals did reduce renal blood flow, suggesting that prostanoids made in the kidney exert a continuous dilator effect on renal blood vessels (Lonigro et al., 1973). However, indomethacin does not affect renal blood flow in conscious dogs with indwelling cannulae or in normal, healthy humans. Furthermore, endogenous prostanoids are not responsible for autoregulating renal blood flow and glomerular filtration rate in dogs and rats. This discrepancy between the effect of indomethacin on intact animals and on operated animals may have a logical explanation. As discussed earlier some vasoactive substances e.g. angiotensin II and noradrenaline, stimulate prostanoid biosynthesis in the kidney. The PGE_2 and PGI_2 which are released produce dilatation of the renal vascular bed which opposes the constrictor effect of angiotensin II and noradrenaline. Thus, indomethacin and other NSAID cause renal vasoconstriction and reduce renal blood flow only in those experimental conditions in which angiotensin II and/or noradrenaline concentrations in the kidney are raised as occurs, for example, after surgical operations. In conditions in which angiotensin II and noradrenaline are not artificially increased (e.g. in healthy human volunteers) NSAID do not cause vasoconstriction. These experiments serve to highlight the important interaction which occurs in the kidney between prostanoids and other local hormones for the control of renal blood flow. The next two sections are devoted to a discussion of the role of arachidonic acid metabolites in the renal actions of angiotensin II and bradykinin.

Prostanoids and the renin angiotensin system. Renin is an enzyme synthesised and released from the kidney in response to a reduction in Na^+ ion concentration in the filtrate in the proximal convoluted tubule. Renin acts on a plasma protein to form a decapeptide, angiotension I, which is subsequently converted to an octapeptide, angiotensin II, by a converting enzyme located chiefly in the lung. Angiotensin II is a potent vasoconstrictor in all vascular beds and also releases the sodium-retaining hormone, aldosterone, from the adrenal cortex.

Plasma renin activity (PRA) and plasma angiotensin II concentration vary with the amount of sodium in the diet. For example, dogs maintained for 5–7 d on a sodium-enriched diet (100 mEq/d) have low PRA. These animals are said to be sodium replete. Alternatively, if dietary intake of sodium is restricted and frusemide is administered to promote loss of Na^+ ions into the urine, then PRA is increased. Such animals are said to be sodium deprived. As pointed out already, if endogenous prostanoids do dilate the renal vasculature then cyclooxygenase inhibitors should produce renal vasoconstriction. In fact sodium meclofenamate causes renal vasoconstriction in sodium-deprived dogs but not in normal or sodium replete dogs (Blasingham, Shade & Share, 1980). As shown in Fig. 8.3, the ability of sodium meclofenamate to constrict renal blood vessels is correlated with PRA in dogs; this strongly suggests that angiotensin II, formed by the action of renin, elevates renal PGE_2/PGI_2 biosynthesis *in vivo* which then dilate the renal vasculature. In the absence of this compensatory prostanoid formation, as for example after treatment with sodium meclofenamate or some other NSAID, then the constrictor action of angiotensin II is left unopposed. The results of these and other similar studies strongly suggest that, when animals are either sodium replete or in normal sodium balance, the spontaneous formation of prostanoids in the kidney does not produce a background vasodilator 'tone' in the kidney. Only when prostanoid biosynthesis is stimulated in the presence of angiotensin II do endogenous dilator prostanoids play a definite role by opposing renal vasoconstriction.

Fig. 8.3. Relationship between plasma renin activity (PRA) and sodium meclofenamate-induced renal vasoconstriction in unanaesthetised, sodium-deplete (open squares) and sodium-replete (filled squares) dogs. Data from Blasingham *et al.* (1980).

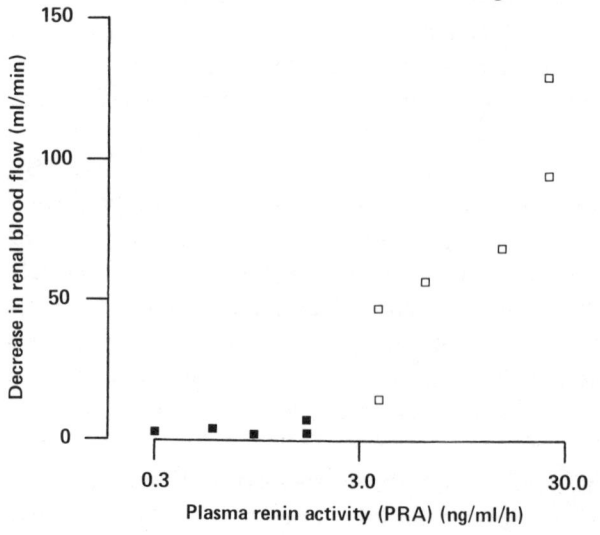

This interaction between angiotensin II and arachidonic acid metabolites has been confirmed and extended by the studies of Blasingham & Nasjletti (1980). In conscious dogs maintained on a low-sodium diet (less than 5 mEq/d) and thus in a sodium-deprived state, sodium meclofenamate produced renal vasoconstriction as expected. However, if animals were pre-treated with the angiotensin-II-receptor blocker, saralasin, then meclofenamate-induced vasoconstriction was abolished. Clearly, angiotensin II must first combine with a specific receptor before stimulating phospholipase A_2 activity and thereby increasing renal prostanoid biosynthesis.

Prostanoids made locally in the kidney oppose the constrictor effect of angiotensin II and of renal nerve stimulation not only in animals artificially depleted of sodium ions but also in a number of pathophysiological conditions which occur clinically. This list includes renal ischaemia, surgical stress, heart failure and cirrhosis of the liver. In each of these conditions treatment with NSAID would be expected to bring about renal vasoconstriction.

Prostanoids and the kallikrein–kinin system. As well as opposing the renal vasoconstrictor effect of the renin–angiotensin II system and sympathetic nerve stimulation, prostanoids also mediate the renal vasodilatation produced by the kallikrein–kinin system. Like renin, kallikrein is an enzyme occurring in urine, plasma and some glands which acts on a plasma protein precursor to form bradykinin. Kallikrein is constantly released from the kidney into the urine. Bradykinin formed within the kidney dilates renal blood vessels and increase renal blood flow. Like angiotensin II, bradykinin stimulates prostanoid biosynthesis in several perfused organs (including the kidney) by activating membrane phospholipase A_2 activity. Many of the pharmacological effects of bradykinin in the body may be altered in character or abolished completely by NSAID, indicating that they are mediated by prostanoids. In the dog kidney, indomethacin significantly reduced the dilatation induced by bradykinin. Furthermore, infusion of bradykinin into the dog renal artery attenuated the renal vasoconstrictor effect both of angiotensin II and of noradrenaline. In each case this action was associated with increased intrarenal prostanoid biosynthesis (Susic, Nasjletti & Malik, 1981). Thus it seems likely that renal vasodilatation by endogenous bradykinin is actually secondary to formation and release of vasodilator PGI_2 and/or PGE_2.

Prostanoids appear to play a pivotal role in the control of blood flow within the kidney interacting with and providing the 'fine tuning' for two opposing systems, the vasoconstrictor renin–angiotensin–noradrenaline system and the vasodilator kallikrein-kinin system. Fig. 8.4 illustrates the interaction of prostanoids with these two systems in the mammalian kidney.

Modulation of fluid and electrolyte excretion

Prostaglandins of the E, A and I series administered to man or experimental animals are natriuretic i.e. they increased urine volume and urinary output of Na^+ and Cl^- ions (Fig. 8.5, Johnstone, Herzog & Lauler, 1967; Bolger et al., 1978). In contrast, $PGF_{2\alpha}$ has little effect on water or electrolyte excretion and the action of PG endoperoxides and TxA_2 has yet to be determined. There is still controversy whether endogenous prostanoids synthesised locally within the kidney exert a similar natriuretic effect of physiological importance. The major reason for the uncertainty is that most studies are by necessity carried out in isolated kidneys or in surgically prepared animals in which renal prostanoid biosynthesis is artificially raised as a result of tissue trauma combined with the effect of circulating angiotensin II and noradrenaline. In each case the result is increased deacylation of membrane phospholipids and release of free arachidonic acid for conversion to prostanoids. Attempts to assess the function of renal prostanoids in the regulation of sodium excretion by blocking endogenous cyclooxygenase with aspirin-like drugs have produced conflicting results. Acute administration of indomethacin to rats or dogs

Fig. 8.4. Control of renal blood flow: interaction of prostanoids, angiotensin II (AGT II) and bradykinin (BK). Notice that both BK (formed from a plasma protein precursor called kininogen by the enzyme kallikrein) and AGT II (formed from a plasma precursor called angiotensinogen by the sequential action of two enzymes renin and converting enzyme) stimulate conversion of arachidonic acid and thereby increase renal prostanoid biosynthesis. Renal prostanoids probably mediate the vasodilator action of BK and oppose or modulate the vasoconstrictor action of AGT II.

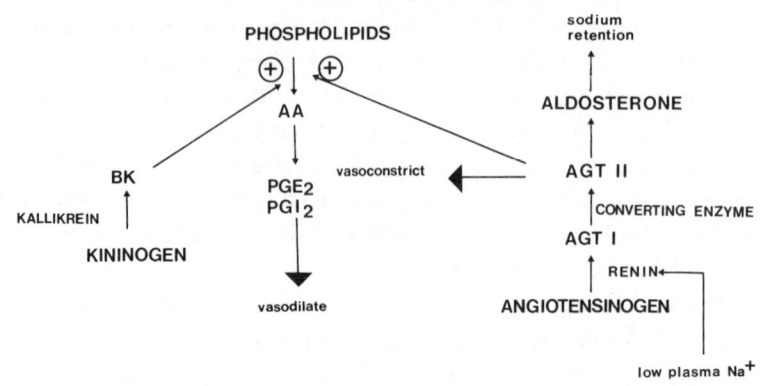

reduced urine volume and sodium excretion (Altsheler *et al.*, 1978) whereas, in other studies, indomethacin either had no effect on sodium excretion in anaesthetised dogs or actually increased sodium excretion in conscious dogs. Similarly in man indomethacin treatment has been reported either to reduce sodium excretion or to have no effect on urine sodium concentration. The effect of indomethacin on salt and water homeostasis in man has been ably reviewed by Brater (1979).

A second approach which has been used to assess the role of prostanoids in the control of electrolyte excretion has centred upon the use of diuretic drugs. Evidence has accumulated in recent years that the so-called loop diuretics such as frusemide and ethacrynic acid increase renal blood flow

Fig. 8.5. Natriuretic action of PGE_1 in anaesthetised dogs. Urine was collected from indwelling cannulae positioned in the ureters from each kidney before (control, C) and after (test, T) intrarenal infusion of PGE_1 or $PGF_{2\alpha}$ (2 μg/min). Histograms show urinary output of Na^+ and K^+. PGE_1 significantly increased excretion of both ions whereas $PGF_{2\alpha}$ was without effect. Asterisks indicate statistical significance of results as determined by Student's *t* test, $P < 0.05$. Data from Gross & Barrter (1973).

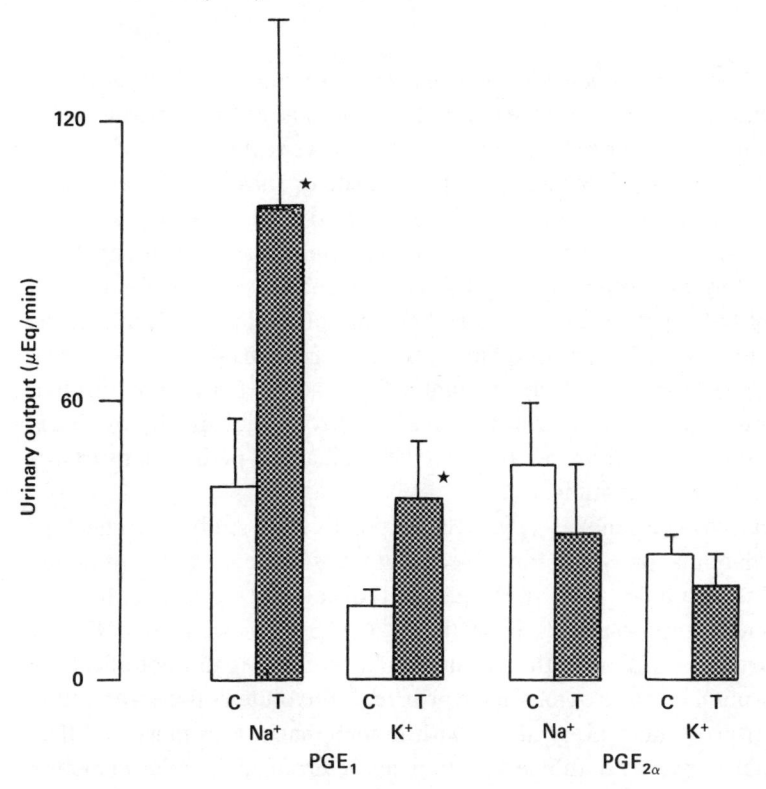

in experimental animals by stimulating the biosynthesis of renal prostanoids. In these experiments renal vasodilatation was prevented by pre-treating the animals with indomethacin without any consistent effect on frusemide-induced natriuresis in man or in dogs (Weber, Scherer & Larsson, 1977).

Although it is still not clear whether endogenous prostanoids made locally within the kidney affect electrolyte excretion into the urine, there is no doubting the very potent natriuretic action of administered prostanoids. The mechanism of this natriuretic effect has been the subject of intensive research. It is now widely believed that prostaglandins are natriuretic by one or more of the following mechanisms:

(1) by inhibiting the active reabsorption of Na^+ and Cl^- ions in the thick descending limb of the loop of Henle and in the collecting tubule;
(2) by antagonising the fluid-retaining effect of vasopressin in the collecting tubule;
(3) by dilating renal medullary blood vessels so producing 'washout' of the medullary osmotic gradient which is responsible for passive fluid absorption across the thin ascending limb of the loop of Henle.

Effect of prostanoids on active sodium chloride reabsorption in the nephron. In recent years a considerable amount of effort has been directed towards determining whether prostanoids affect the active reabsorption of sodium chloride in the nephron and if so at which site or sites. Experiments have usually been of two types: (*a*) micropuncture studies in rats or rabbits in which fluid is withdrawn at intervals along a single nephron and analysed for electrolyte concentration; and (*b*) isolation and perfusion of individual parts of the nephron, e.g. the ascending limb of the loop of Henle or the collecting tubule. Absorption of ions can be determined either by measuring the ionic composition of the medium before and after perfusion through the portion of nephron or alternatively electrophysiologically by determining the voltage difference between the luminal and peritubular surfaces of the tubule under study.

With these techniques it has been possible to study the effect of prostaglandins on water and electrolyte absorption in the proximal convoluted tubule, loop of Henle and distal and collecting tubules. Intravenous infusion of PGE_2 in the rat (Fulgraff & Meiforth, 1971) or renal arterial infusion of this prostaglandin in the dog did not affect the reabsorption of water or ions in the proximal convoluted tubule. Inhibition of renal prostanoid biosynthesis with indomethacin also failed to affect proximal convoluted tubule function as determined in micropuncture studies in rats. These experiments suggest that the site of altered sodium chloride and fluid reabsorption following prostaglandin treatment must be either the loop of Henle, the distal convoluted tubule or the collecting duct of the nephron. Of these, most research points towards the loop of Henle as the major locus of prostaglandin action. Stokes (1979) has studied the effect of PGE_2 on sodium chloride transport using isolated segments of thick ascending limbs of the loop of Henle dissected from rabbit kidneys. The method used involved measuring changes in the transepithelial voltage in perfused ascending limb of Henle's loop before and after PGE_2 treatment. After a period of equilibration the transepithelial voltage stabilised at 2–4 mV. PGE_2 rapidly reduced the transepithelial voltage to 1–2 mV, indicating a reduction in net sodium chloride absorption (Fig. 8.6). The effect of PGE_2 reversed within 2–3 h. Most likely, PGE_2 decreases the transepithelial voltage in these experiments by blocking activity of the Na^+-K^+-ATPase in the cells lining the loop of Henle thereby preventing active uptake of sodium ions.

Other authors using micropuncture studies in the rat and rabbit kidney nephron have confirmed that some prostanoids inhibit sodium chloride absorption in the loop of Henle. Higashihara, DuBose & Kokko (1978) sampled fluid at several points along individual nephrons in rats and found that indomethacin or meclofenamate reduced the delivery of chloride ions to the distal tubule, indicating that medullary prostanoids may also inhibit

Fig. 8.6. The inhibitory effect of PGE_2 (2 μM) on transepithelial voltage across the rabbit isolated, perfused medullary ascending limb of the loop of Henle. A fall in transepithelial voltage reflects a decrease in net sodium chloride absorption. Data from Stokes (1979).

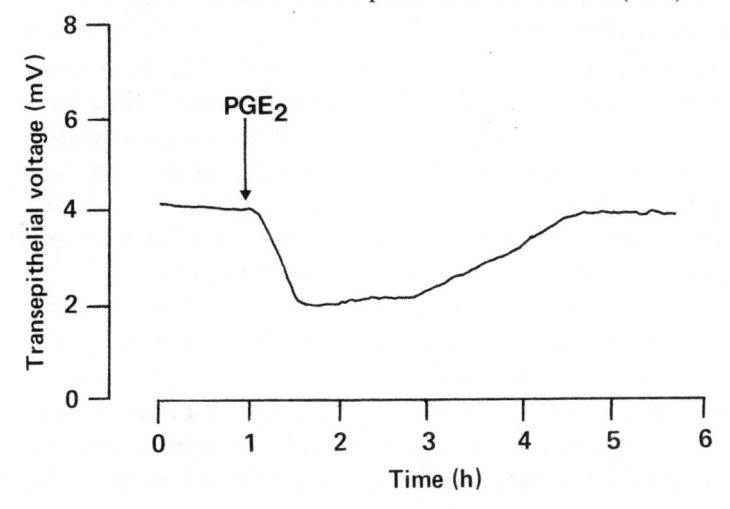

sodium chloride absorption in the loop of Henle. As further support for a role of endogenous prostanoids, Ganguli, Tobian & Azar (1977) have reported that cyclooxygenase inhibitors increase the sodium chloride content of the renal papilla in the rat, presumably by preventing prostanoid formation and thus reducing the output of Na^+ ions into the urine. As a consequence, sodium chloride builds up in the renal papilla.

In addition to the loop of Henle, prostanoids also inhibit absorption of sodium chloride in the latter part of the nephron. For example, PGE_2 reduced the absorption of Na^+ ions on passage along the rat distal convoluted tubule (Fulgraff & Meiforth, 1971). PGE_2, PGI_2 and $PGF_{2\alpha}$ reduced the absorption of sodium ions across the isolated perfused rabbit cortical collecting tubule (Stokes, 1981). It is not clear whether these effects on the distal and collecting tubules are quantitatively more or less important in determining urine electrolyte concentration and volume than the effect on the loop of Henle. However, simply by being 'downstream' of the loop of Henle, an effect on the distal and/or collecting tubule could not be so easily countered. In addition to a direct action on sodium transport in the distal and collecting tubules, prostanoids also have an indirect effect on water and electrolyte absorption in this part of the nephron be affecting the release and action of vasopressin.

Vasopressin release and its effect on the collecting tubule. The interaction of vasopressin and prostanoids in the kidney has been extensively studied ever since the original report by Orloff, Handler & Bergström (1965) that PGE_1 reduced the effect of vasopressin on water transport in the toad bladder. Subsequently, a similar action was shown to occur in the perfused rabbit cortical collecting tubule. Since indomethacin, meclofenamate and aspirin enhance the anti-diuretic effect of vasopressin in several species, including man, it may be assumed that endogenous renal prostanoids also antagonise the effect of vasopressin in the kidney. The collecting tubule is not necessarily the only intrarenal site where prostaglandins antagonise the effect of vasopressin. In some species vasopressin additionally stimulates sodium chloride absorption in the loop of Henle; PGE_2 or PGI_2 possibly formed within the renomedullary interstitial cells may thus block this action of vasopressin as well as its action in the collecting tubule.

Like angiotensin II and bradykinin which, as we have seen, rely to some extent upon stimulating prostanoid biosynthesis in the kidney for their renal effects, there is now convincing evidence that vasopressin also stimulates renal prostanoid formation. For example, vasopressin enhances prostanoid synthesis and release by cultured rabbit renomedullary interstitial cells, by rabbit kidney medulla slices, in anaesthetised rabbits and in the Brattleboro rat (a genetic strain of rat deficient in vasopressin; see Dunn et al., 1983 for further information and references). Thus PGE_2 may have an important role to play as a negative-feedback modulator of the renal effects of vasopressin. However, the situation may be complicated by the fact that vasopressin also stimulates TxA_2 biosynthesis in some organs. Unlike PGE_2, stable TxA_2 analogues stimulate osmotic water flow in the toad bladder. If TxA_2 has a similar action in the mammalian kidney *in vivo* then it would be expected to potentiate and not inhibit the action of vasopressin. Certainly thromboxane synthetase inhibitors have been reported to inhibit vasopressin-stimulated water absorption, implying that vasopressin does stimulate TxA_2 production in the kidney and that TxA_2 also promotes the diuretic effect of vasopressin. For a detailed review the reader is referred to the excellent article by Zusman (1981).

Exactly how prostaglandins block the action of vasopressin in the kidney has also received attention. There now seems little doubt that vasopressin exerts its physiological effect on the collecting tubules to reabsorb fluid back into the body by stimulating collecting tubule cell adenylate cyclase activity thereby increasing the concentration of cAMP within these cells. Prostaglandins, on the other hand, inhibit both the basal and vasopressin-stimulated adenylate cyclase activity, effectively lowering the amount of cAMP within collecting tubule cells. Fig. 8.7 is an attempt to illustrate the complex interaction of vasopressin with arachidonic acid metabolites in the kidney.

Fig. 8.7. Regulation of fluid and electrolyte absorption in the kidney collecting duct: interaction of vasopressin and prostanoids.

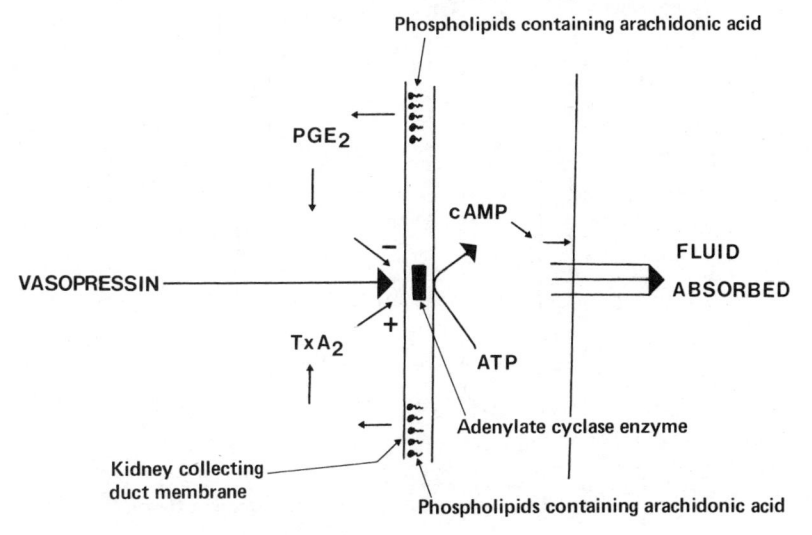

Intrarenal blood flow distribution. Prostanoids also affect sodium chloride and water excretion by an influence on medullary blood flow. Evidence that prostaglandins of the E series and PGI_2 divert blood away from the outer and towards the inner cortex and medulla has already been discussed in the section on renal blood flow. To understand how this circulatory change within the kidney may cause natriuresis it is necessary to look at how sodium chloride and water are reabsorbed in the loop of Henle.

The ascending limb of the loop actively transports sodium chloride from the ultrafiltrate passing through the nephron into the interstitial space but does not allow the free flow of water in the same direction. The result is that the salt concentration in the ultrafiltrate decreases while the salt concentration in the interstitial space increases, so setting up an osmotic

Fig. 8.8. The countercurrent exchange mechanism for reabsorption of fluid and electrolytes in the loop of Henle. Sodium chloride is actively reabsorbed in the thick ascending limb of the loop of Henle; the now hyperosmotic interstitium (region between the loop of Henle and the vasa recta) 'sucks' water out of the thin water-permeable descending limb. In this way an osmotic gradient develops in the interstitium. Osmotic pressure (measured in mOsmols/kg water) is lowest at the junction of the cortex with the medulla (about 300 mOsmol/kg) and increases to a maximum of about 1200 mOsmol/kg at the tip of the loop of Henle in the medulla. Formation of prostanoids in the renal medulla dilates the vasa recta blood vessels thereby 'washing out' the osmotic pressure gradient and allowing fluid to flood through the nephron without being reabsorbed.

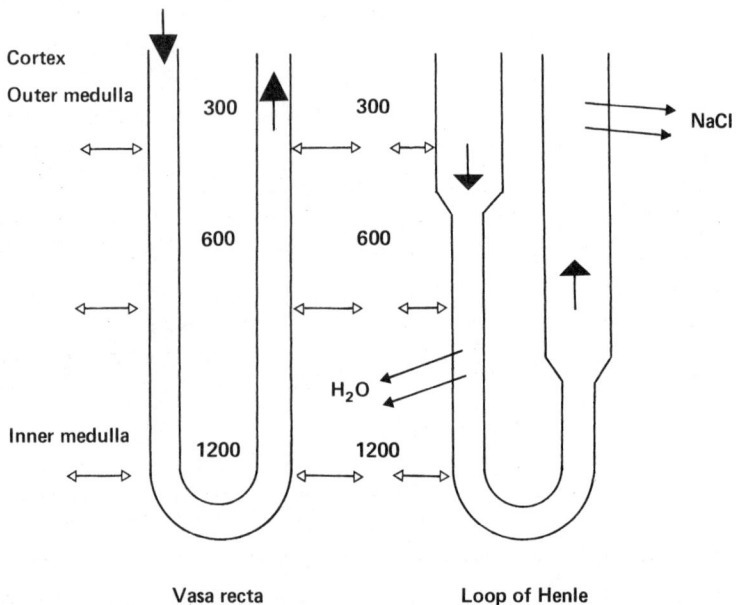

gradient between the contents of the loop and the interstitium. The walls of the vasa recta blood vessels are freely permeable both to sodium chloride and to water and thus an equilibrium is also set up between the interstitial fluid and blood. The osmotic gradient between the ultrafiltrate and interstitial fluid gradually increases to reach a maximum at the tip of the loop. The walls of the descending limb of the loop of Henle are freely permeable to water which flows out of the nephron and into the interstitium because of the osmotic gradient. This process is that of a normal countercurrent-multiplier mechanism which very efficiently ensures the reabsorption of fluid and electrolytes in this part of the nephron. In the presence of medullary vasodilator prostanoids, blood flow through the vasa recta is increased. The result of this redistribution of blood flow inside the kidney is that the various equilibria set up firstly between the ultrafiltrate in the loop and the interstitial fluid and then between the interstitial fluid and blood in the vasa recta are removed. This process is sometimes referred to as a 'washout' of the osmotic gradient. One consequence of this 'washout' phenomenon is that sodium chloride and water reabsorption in the loop of Henle is reduced or abolished altogether, resulting in natriuresis. This solute washout effect reverses when the vasodilator prostanoids are enzymatically broken down (PGE_2) or spontaneously degrade (PGI_2) and vasa recta blood flow returns to normal. This process is shown diagrammatically in Fig. 8.8.

Effect of prostanoids on renin and aldosterone release

Earlier in this chapter the role of prostanoids in modifying the renal vascular effects of angiotensin II was discussed. In the mid-1970s the emphasis placed on the interplay between the renin–angiotensin system and prostanoids was shifted with the discovery that prostaglandins affected not only the vascular effects of angiotensin II but also influenced renin release from the juxtaglomerular apparatus (Larsson, Weber & Anggard, 1974). In these pioneering experiments, arachidonic acid was found to increase while indomethacin decreased PRA in rabbits suggesting that one (or more) cyclooxygenase products stimulate renin release *in vivo*. This effect of indomethacin was not restricted to rabbits. Later experiments revealed that NSAID also decreased PRA in man and in other experimental animals. It was concluded from these studies that prostanoids formed continuously within the kidney exerted a 'tonic' stimulation of renin release. The identity of the prostanoid responsible is not established. PGE_2, PGI_2, PGG_2 and 6-oxo-PGE_1 stimulate while $PGF_{2\alpha}$ inhibits renin release from rabbit kidney slices *in vitro* (Fig. 8.9), an effect correlated with the ability of these prostaglandins to stimulate adenylate cyclase activity and increase intracellular cAMP levels within the kidney (Weber et al., 1976). The weight of available evidence suggests that PGI_2 is the endogenous prostanoid responsible for stimulating renin release from the juxtaglomerular apparatus *in vivo*. However, the possibility that PGI_2 owes some of its renin secretagogue activity to its enzymatic conversion to 6-oxo-PGE_1 has been suggested and should not be ignored. Certainly rabbit renal cortex contains high PG-9HDH activity, converting PGI_2 to 6-oxo-PGE_1 which is as (or more) potent than PGI_2 as a renin secretagogue.

Evidence has also been presented that prostanoids are responsible for the process of tuberoglomerular feedback. This is the mechanism whereby renin released from the juxtaglomerular apparatus of an individual nephron regulates the glomerular filtration rate of that particular nephron. In this way sodium chloride reabsorption from each nephron is maximised. Exactly how this process operates is not known but inhibition of renal prostanoid biosynthesis to about 10% of its control value attenuates this feedback mechanism which may then be restored to full efficiency by

Fig. 8.9. Effect of 6-oxo-PGE_1 (columns 1–4), PGI_2 (column 5), 6-oxo-$PGF_{1\alpha}$ (column 6) and isoprenaline (column 7) on renin release from rabbit renal cortical slices incubated in Krebs'–Ringer physiological salt solution. Asterisks indicate a statistically significant (Student's paired t test, $P < 0.05$) increase in renin release compared with a control incubation period with no drugs present. Data from McGiff, Spokas & Wang (1982).

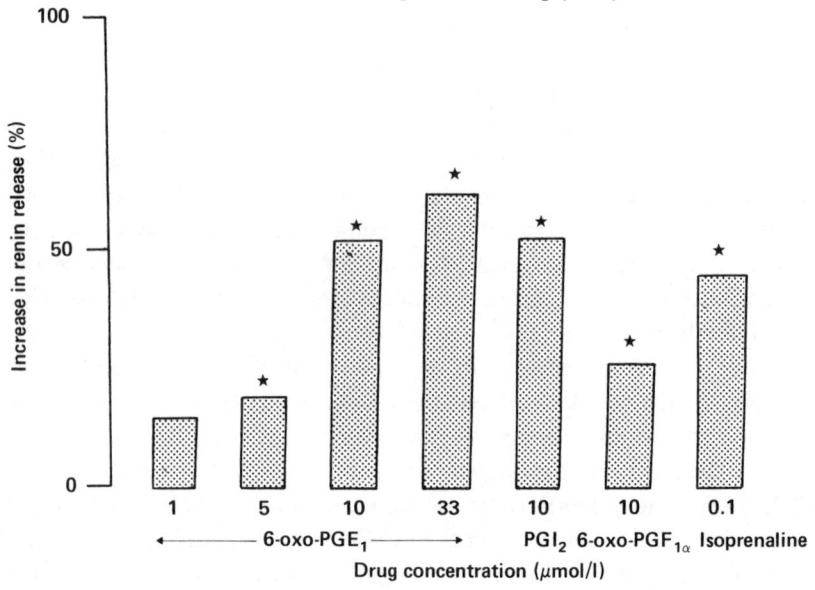

infusing PGI_2 into the renal artery. Presumably PGI_2 causes renin release to restore tuberoglomerular feedback.

Prostaglandins of the E and I series increase renin release from the kidney by a direct action on the juxtaglomerular apparatus. This is probably not the only way in which prostanoids may affect renin release *in vivo*. Some prostaglandins, particularly those of the E series, modulate the pre-junctional release and post-junctional actions of noradrenaline released from stimulated sympathetic nerves (see Chapter 9). Stimulation of renal sympathetic nerves causes renin to be released following activation of β-receptors. Prostaglandins are not only released from isolated perfused kidneys when the renal nerves are electrically stimulated *in vitro* but may also potentiate the effect of noradrenaline on renal β-receptors (thus increasing renin release) or alternatively inhibit noradrenaline efflux from these nerves (thus reducing renin release). Whether either of these effects plays an important part in the action of prostanoids on renin release *in vivo* is unknown. However, it should be stressed that whenever arachidonic acid or prostanoids have been administered to whole animals, the effect is to enhance the release of renin. For this reason it is most unlikely that inhibition of noradrenaline release from the renal nerves contributes significantly to the renin secretagogue action of prostanoids.

The effect of drugs on the kidney

Many drugs affect prostanoid synthesis and catabolism (see Chapter 2) but only a few of these have a specific action in the kidney. Among these are some diuretic drugs which rely, at least in part, for their clinical effect on the ability to stimulate renal prostanoid biosynthesis. Hydrochlorothiazide, for example, given to human subjects produces a mild diuresis affecting primarily the loop of Henle and, additionally, significantly increases the appearance of PGE_2 in the urine. Since PGE_2 is a potent natriuretic agent it may contribute to the diuretic activity of this drug. However, pre-treatment with NSAID failed to modify the diuretic activity of hydrochlorothiazide either in man or in experimental animals, which would argue against a direct role for prostanoids in the action of this drug.

A better case can be made for the intermediacy of prostanoids in the action of loop diuretics such as frusemide, ethacrynic acid and bumetanide. Aspirin-like drugs reduce the diuretic effect of all three of these drugs in dog, rabbit and man by anything up to 80%. In addition, frusemide increased urinary excretion of PGE_2, $PGF_{2\alpha}$, TxB_2 and 6-oxo-$PGF_{1\alpha}$ in man (Ciabottini, Puglisi & Cinotti, 1980) and elevated dog renal venous plasma PGE_2 concentration. These observations suggest that the loop diuretics stimulate medullary PGI_2/PGE_2 formation which then exerts a natriuretic effect on the loop of Henle by the mechanisms discussed earlier. *In vitro*, frusemide and ethacrynic acid are potent inhibitors of renal 15-PGDH and PG-9KR which may explain their ability to increase kidney PGE_2 content (Stone & Hart, 1976). The aldosterone antagonist, spironolactone, as well as amiloride and triamterene all increase urine PGE_2 output upon repeated use in humans (Kramer, Dusing & Stinnisbeck, 1980) suggesting that they too enhance renal prostanoid turnover. However, cyclooxygenase inhibitors do not blunt the diuretic effect of these three drugs which again casts some doubt upon the part that prostanoids play in their action. The anti-hypertensive drug clonidine also produces a characteristic water diuresis in dogs which is inhibited by indomethacin pre-treatment and is associated with increased PGE_2 excretion into the urine (Olsen, 1976).

All of the studies linking diuretic drugs with renal prostanoids depend upon two types of experiment. First, measurement of PGE_2 or another prostanoid or combination of prostanoids in the urine before and after drug treatment. The unreliability of such measurements has been dealt with earlier in this chapter. Second, the demonstration of a reduced effect after treatment with drugs which prevent prostanoid biosynthesis. Unfortunately such drugs frequently have effects, other than inhibition of cyclooxygenase, which could alter the effectiveness of diuretics. Anti-inflammatory drugs like aspirin, for example, are known to compete for the same organic acid transport process responsible for the uptake of frusemide and bumetanide into the loop of Henle. Thus aspirin may well reduce the diuretic activity of these two drugs simply by preventing access to their site of action. Until such time as truly specific inhibitors of cyclooxygenase become available, the newcomer to prostanoid research should perhaps be wary of over-enthusiastic interpretation of results from studies using NSAID unless good corroborative evidence is also presented.

Renal disease

Several renal disease states are associated with an imbalance in the biosynthesis and/or catabolism of prostanoids. In most cases this results in clinical symptoms which do not directly affect the kidney but are manifested elsewhere in the body. Reduced formation of vasodilator prostaglandins in the kidney, for example, may predispose to hypertension as discussed in Chapter 3. On the other hand, renal diseases such as uraemia, haemolytic uraemic syndrome and nephrosis are associated with changes in platelet reactivity and thrombosis. These diseases and the involvement of prostanoids in their aetiology are discussed in Chapter 4.

Only disease states in which renal prostanoids play a part and which involve altered renal function will be discussed here.

Barrter's syndrome

In 1962, Barrter, Pronove & Gill described a novel form of hyperaldosteronism characterised clinically by muscle weakness and cramps, nocturia, high PRA and hypokalaemia. Later studies of this condition, renamed Barrter's syndrome after its discoverer, suggested that the primary disturbance was a reduced ability to reabsorb sodium chloride from the ascending limb of the loop of Henle in the kidney. Since this time, much attention has been focussed upon the role of prostanoids in the pathophysiology of this condition. Evidence has accumulated that elevated renal prostanoid formation plays a part in the aetiology of Barrter's syndrome. This evidence is summarised below.

(1) Urinary excretion of PGE_2, PGD_2 and 6-oxo-$PGF_{1\alpha}$ is markedly increased in patients with Barrter's syndrome but urinary excretion of 7α-hydroxy-5,11-dioxotetranorprostane-1,16-dioic acid (the major metabolite of PGE_2 in the body, and thus reflecting extrarenal PGE_2 formation) was not altered. Thus an overproduction of prostanoids is restricted to the kidney in this clinical condition.

(2) Indomethacin treatment of patients with Barrter's syndrome does bring about some clinical improvement with a reduction in PRA and plasma aldosterone concentration as well as correction of the hypokalaemia.

(3) As described previously in this chapter, prostaglandins of the E series inhibit chloride reabsorption in the ascending limb of the loop of Henle in rats and rabbits (Stokes, 1979). A similar action in the human kidney may underlie the renal chloride leak in patients with Barrter's syndrome.

(4) Prostaglandins of the E and I series potently stimulate renin release from the kidney and therefore may be responsible for the much elevated plasma renin and aldosterone concentrations observed in this disease.

(5) Patients with Barrter's syndrome exhibit a defect in platelet aggregability associated with increased plasma concentration of anti-aggregatory 6-oxo-PGE_1.

A good case can therefore be made for increased biosynthesis of prostanoids in the kidneys of patients with Barrter's syndrome. Whether PGE_2, PGI_2 or some other prostaglandin is responsible and in which part of the kidney this prostaglandin is made has yet to be fully established. However, it may be relevant to point out that renal biopsies of patients with Barrter's syndrome invariably reveal hyperplasia of the PGE_2-secreting renomedullary interstitial cells.

Hepatorenal disease

Functional changes in renal haemodynamics and sodium chloride excretion is a common finding in patients with severe liver disease. In some cases such abnormalities may lead to renal failure characterised by oliguria, sodium and fluid retention and ischaemia of the renal cortex. This is the so-called hepatorenal syndrome. Renal failure is a fairly common result of cirrhosis of the liver and is partly responsible for the ascites (accumulation of fluid in the peritoneal cavity) of hepatic cirrhosis. Over the years abnormalities of several vasoactive systems in the kidney have been observed in patients with hepatorenal syndrome and suggested to play a part in the aetiology of the disease. More recently prostanoids have been implicated in hepatorenal disease. The conclusions of some relevant studies are listed below:

(1) urine PGE_2 concentration over a 12 h collection period is decreased and TxB_2 excretion increased in patients with hepatorenal disease (Zipser *et al.*, 1982);
(2) plasma PGE_2 is elevated in patients with hepatorenal disease;
(3) indomethacin and other cyclooxygenase blockers produce clinical deterioration in patients who have hepatorenal disease with or without ascites (Boyer, Zia & Reynolds, 1979);
(4) in an animal model of hepatorenal disease in man (chronic ligation of the bile duct in dogs to make the animals cirrhotic), Zambraski & Dunn (1981) demonstrated that over half of the animals used developed ascites (in such animals urinary excretion of PGE_2 was greatly increased);
(5) in another model of the kidney during renal failure (induced in rabbits by tying off either the ureter or the renal vein), synthesis of vasoconstrictor TxA_2 was markedly increased.

These apparently disparate pieces of evidence can be brought together to produce a rational explanation of the role of prostanoids in hepatorenal disease. According to this scheme, elevated activity of the kidney renin angiotensin system is the root cause of renal disease associated with hepatic cirrhosis. Aldosterone, present in large amounts in the blood of such patients, promotes excessive sodium and water retention and thus contributes to the ascites. Angiotensin II causes renal cortical vasoconstriction leading to renal ischaemia. Vasodilator prostaglandins are released in the kidney at this time, possibly because of the presence of angiotensin II which stimulates phospholipase A_2 activity, and partially restore renal blood flow. Increased renal PGE_2 turnover is reflected in raised plasma PGE_2 and urinary PGE_2 excretion while inhibiting prostaglandin biosynthesis (with, for example, indomethacin) allows the vasoconstrictor action of angiotensin II to go unopposed, thereby explaining the deleterious effect of such drugs. While elevated PGE_2 and PGI_2 concentrations intrarenally would have a beneficial effect because they are both vasodilator, any increase in TxA_2 formation in the kidney would have the opposite effect causing vasoconstriction and predisposing to renal failure.

All the available evidence suggests that arachidonic acid metabolites have an important function in several hormonal, haemodynamic and electrolyte abnormalities in the kidney of patients with liver disease. The precise role of individual prostanoids has yet to be established. Certainly one approach to unravel the part played by, for example, TxA_2 in the hepatorenal syndrome is to determine the way in which selective thromboxane synthetase blockers alter the course of the disease in man. Such studies are awaited with interest especially now that selective and potent as well as safe thromboxane synthetase inhibitors are available clinically (see Chapter 2).

9
Prostaglandins and the nervous system

Central nervous system
Biosynthesis and catabolism of prostanoids in the brain
In contrast to many other parts of the body, the brain has only feeble ability to convert exogenous arachidonic acid to prostanoids. For this reason most experimenters have concentrated upon measuring endogenous prostanoid concentrations in brain homogenates or in the cerebrospinal fluid (CSF). Prostanoid precursors other than arachidonic acid e.g. eicosatri- and eicosapentaenoic acids are not found in brain neuronal membranes and thus prostanoids containing 1 or 3 double bonds do not occur in the CNS. A common problem of such research is that prostanoid concentrations in the brain increase several fold in the first few minutes after death. Possibly the combination of brain ischaemia coupled with the physical trauma of actually removing the brain from the skull provokes de-esterification of membrane phospholipids to release free arachidonic acid and thereby produce erroneously high brain prostanoid concentrations. For this reason great care must be taken to estimate prostanoids in the brain as soon after the death of the animal as possible. For example, cerebral cortex $PGF_{2\alpha}$ levels in rats (about 1 ng/100 g tissue) sacrificed almost instantaneously by exposure to microwave radiation is considerably less than the value obtained when rats were killed by a blow to the head and then exsanguinated (about 10 ng/100 g tissue). The difference between these two figures presumably represents formation of $PGF_{2\alpha}$ from endogenous arachidonic acid during tissue removal, preparation and work-up.

Bearing in mind these technical problems, several prostanoids have been found in brains of different species. The first to be discovered, $PGF_{2\alpha}$, was detected by gas-chromatography–mass-spectrometry in ox-brain homogenate. Subsequently both PGE_2 and $PGF_{2\alpha}$ were found in rat brain cortex homogenates and slices and later in guinea pig, cat and human brain samples. The range of prostanoids found in the brain depends to a large extent on the species studied. Quantitatively the major prostanoid in rat brain is PGD_2 (Abdel-Halim et al., 1977) but this prostaglandin does not occur in human brain or CSF (Abdel-Halim, Ekstedt & Anggard, 1979). Not surprisingly rat brain contains high PG-endoperoxide-D-isomerase activity, particularly in the hypothalamus, while this enzyme has not been reported in human brain. TxA_2 is formed by guinea pig, rat and cat but not human brain. Small amounts of 6-oxo-$PGF_{1\alpha}$ occur in extracts of brains from several species, including humans, where the site of synthesis is more likely to be the cerebral blood vessels rather than the neurones. Lipoxygenase is present centrally with 12-HETE being detected in extracts of gerbil brain (Sautebin et al., 1978). Whether leukotriene formation also takes place centrally is not yet known.

Different parts of the brain synthesise different amounts of prostanoids. In rat and cat brain the major areas of PGE_2 and $PGF_{2\alpha}$ biosynthesis are the cerebral cortex with lesser amounts formed in the cerebellum, caudate nucleus and hypothalamus (Wolfe, Pappius & Marion, 1976). The regional distribution of prostanoids throughout the brain is not correlated with the localisation of any known neurotransmitter substance.

Prostanoids are not rapidly broken down centrally. In general, the brain is a very poor source of 15-PGDH and PG-Δ13R although both monkey and pigeon brain contain high PG-9KR activity, which normally converts PGE_2 to $PGF_{2\alpha}$. This may help explain the high ratio of $PGF_{2\alpha}$ to PGE_2 (sometimes as much as 3:1) which occurs in the brain. Despite the obvious fact that brain as a whole only poorly catabolises prostanoids this does not exclude the possibility of high 15-PGDH activity associated with particular nerve tracts. In any case a reasonably efficient mechanism for removing prostanoids from the brain does exist in the form of an active transport system from the brain across the choroid plexus into the CSF. This transport process is blocked by probenacid and takes place against a concentration gradient (Bito, Davson & Hollingsworth, 1976). Since CSF can neither synthesise nor break down prostanoids, it provides a useful means of estimating central prostanoid turnover in humans. Many researchers have estimated CSF prostanoid levels, with the first studies reported over 20 years ago. In normal subjects CSF contains between 50 and 100 pg/ml $PGF_{2\alpha}$ and little or no PGE_2, TxB_2 or 6-oxo-$PGF_{1\alpha}$. However, very much higher amounts of $PGF_{2\alpha}$ (greater than 200 pg/ml) occur in patients in a variety of clinical conditions affecting the brain, e.g. epilepsy, stroke and subarachnoid haemorrhage (see Egg, Herold, Rumpl & Gunther, 1980) and PGE and 6-oxo-$PGF_{1\alpha}$ occur in CSF of patients with schizophrenia and essential hypertension respectively. Further studies are necessary to determine whether prostanoids are involved in the aetiology of these conditions or occur in CSF as a consequence of the disease state.

Central effects of prostanoids
Behaviour

Shortly after the discovery of $PGF_{2\alpha}$ in the brain, several authors reported that prostaglandins of the E and F series (administered by different routes in several species) produced characteristic changes in animal behaviour such as sedation, loss of righting reflexes, diminished motor activity and ptosis (closure of the eye). Much of this early work has been reviewed by Holmes & Horton (1968). As a general rule threshold behavioural effects were observed after parenteral administration (intravenous, subcutaneous, intraperitoneal) of PGEs or PGFs at doses of 0.5–1.0 mg/kg with the effect lasting from 15–60 min, depending upon the species studied. The relatively large doses used in such studies almost certainly produced unphysiological prostaglandin concentrations in the brain. However, it must be remembered that parenterally administered prostaglandins are extensively broken down in the lung, liver and other organs and so the amount actually reaching active sites in the brain may well have been very small. One thing which is clear is that the amount of prostaglandin penetrating the brain after parenteral administration will vary considerably from one species to the next and (more than likely) from one animal to the next.

Another way of assessing the behavioural effects of PGs which at the very least allows an accurate idea of the concentration reached in the brain is injection of a fixed amount through a chronically implanted cannula into the lateral ventricle of an unanaesthetised animal. In such experiments in cats much smaller doses of PGE_1 or PGE_2 (3–10 µg/kg) produce behavioural effects (e.g. reduced movement, stupor, catatonia) which lasted for up to 48 h. Most published reports suggest that $PGF_{1\alpha}$ and $PGF_{2\alpha}$ have less behavioural effects than the corresponding E series compounds. Although PGI_2 infusion occasionally causes headache in human volunteers, no pronounced central effects on behaviour have been reported with this prostaglandin.

Since the amount of prostaglandins administered to produce behavioural effects are several orders of magnitude greater than the endogenous concentration of these substances in the brain, it is likely that the effects which I have described are pharmacological rather than physiological. Furthermore, aspirin-like drugs do not cause behavioural excitation or elevate motor activity in animals or man, which implies that brain prostanoids do not normally exert significant central actions on behaviour.

Motor effects and convulsions

The motor effects of prostaglandins have been extensively studied not least because of their potent anti-convulsant action (reviewed by Coceani, 1974). Intraventricularly injected PGE_1 and PGE_2 dose-dependently antagonise convulsions in mice elicited by a range of drugs including pentylenetetrazole, picrotoxin, strychnine or isoniazid and partially protect against electroshock-induced seizures. The anti-convulsant effect of prostaglandins may be secondary to stimulation of central turnover of gamma amino butyric acid (GABA), an inhibitory amino acid neurotransmitter in the cerebellum which normally dampens down the activity of Purkinje cells and thus prevents spontaneous convulsions. PGE_2 has been shown to increase mouse brain GABA concentration by up to 50% (Rosenkranz, Thayer & Killam, 1980). Interestingly, PGE_2 prevents the pentylenetetrazole-induced increase in cerebellar cGMP in mice which may provide a clue to the molecular mechanism underlying the anti-convulsant effect of this prostaglandin. Alternatively prostaglandins may constrict cerebral blood vessels, to cause a fall in blood flow to the brainstem and thereby a reduction in the rate of firing of neurones in this area and prevention of convulsions. Whatever the precise mechanism of action it is likely that, as with the effects on behaviour noted earlier, the anti-convulsant activity of prostaglandins is of pharmacological and not of physiological interest. The 'acid test' is that aspirin-like drugs do not cause convulsions and assuming that they do inhibit brain-prostanoid biosynthesis it can be reasoned that endogenous prostanoids do not normally exert an anti-convulsant action in the body.

Temperature control

Prostanoids synthesised within the brain do not affect the control of body temperature (thermoregulation) in normal, healthy non-febrile (i.e. without fever) animals. Several pieces of evidence support this conclusion. Firstly, aspirin-like drugs do not lower normal body temperature in non-febrile rabbits, cats and humans (Rosendorff & Cranston, 1968). Secondly, these drugs do not affect the temperature of animals maintained in a cold environment to raise their body temperature to levels similar to those which occur in fever of bacterial origin. Finally the amount of prostaglandin-like activity in the CSF collected from the cisterna magna did not change in cats exposed to a hot climate or in the same animals subsequently exposed to a cold environment. Thus prostanoids play no part in the regulation of temperature in hot or cold climates.

In contrast, there is considerable evidence that some prostaglandins mediate the increase in body temperature (fever) caused by bacterial infection (reviewed by Milton, 1982). To understand how prostaglandins are involved in fever one must first return to the sequence of events which follow invasion of the body by bacteria. Fig. 9.1 indicates what happens when a Gram-negative bacillus (an endotoxin) gets into the blood stream.

Fig. 9.1. Steps involved in the production of fever after bacterial infection.

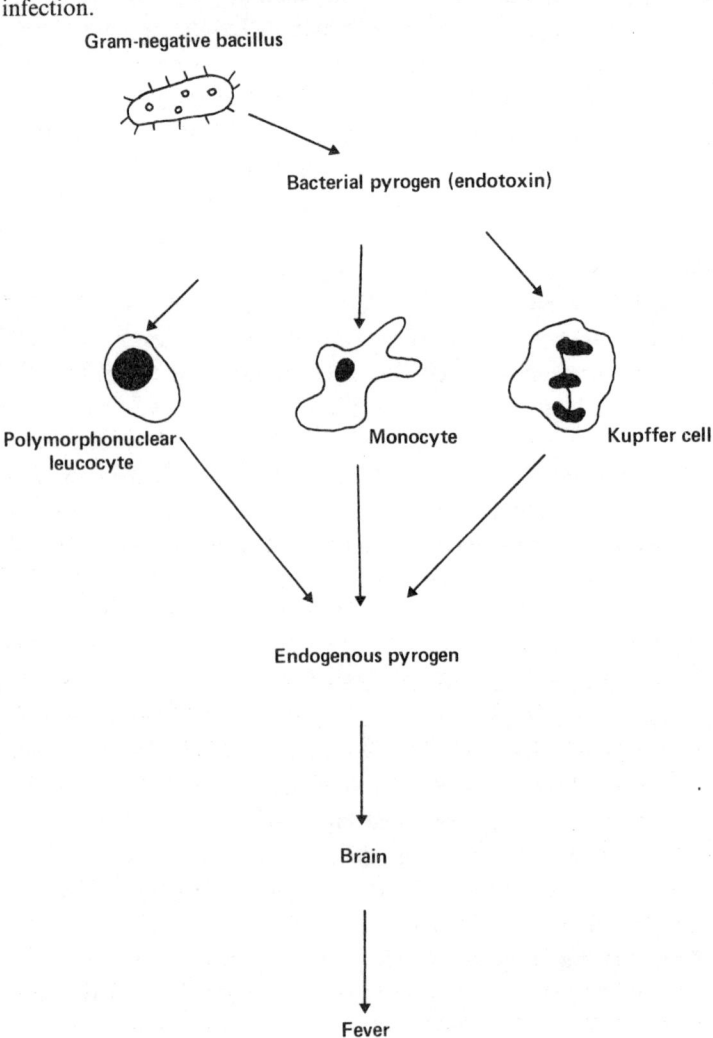

The same events occur if the body is invaded by a variety of viruses, fungi or even yeasts. The first step is activation of any of a number of normal cells in the body to synthesise and release an endogenous pyrogen which is a small protein of molecular weight 15 000. The endogenous pyrogen is the final common mediator of all substances which cause fever. It crosses the blood–brain barrier and acts within the pre-optic area of the anterior hypothalamus (AH/POA) to increase the set point around which body temperature is regulated. This may be likened to turning up the thermostatic control of a central heating system. The body retains its ability to thermoregulate in response to changes in the environment but this is now achieved at a temperature 2–3 deg F greater than normal.

Several pieces of evidence support the obligatory role of prostanoids in fever of bacterial origin. Firstly, injection of prostaglandins of the E series directly into the brain of unanaesthetised animals produces a rapid increase in body temperature. This hyperthermia occurs even at doses as low as 100 pg, similar to concentrations found locally in some parts of the brain. Indeed, on a weight-for-weight basis, prostaglandins are probably the most potent hyperthermia-producing substances yet discovered. Prostaglandins of the F series are devoid of such activity while PGI_2 causes only mild hyperthermia in the cat. Although it is impossible to carry out properly controlled studies in man, an increase in body temperature is frequently observed in women receiving prostaglandins for the induction of labour or for abortion. Mapping experiments have shown that the only part of the brain sensitive to direct ionophoretic application of prostaglandins was the AH/POA. Similarly, this region was the only location in which pyrogenic substances also caused an increase in temperature. From this it would appear that prostaglandins may mediate pyrogen-induced fever by an action in the AH/POA. Furthermore, cerebrospinal fluid collected from the third ventricle of cats made febrile by intravenous injection of *Shigella* endotoxin contained 2–4 times more prostaglandin-like biological activity (estimated as contractions of the isolated rat stomach strip preparation) than did samples obtained from animals with a normal body temperature. Administration of paracetamol, an anti-pyretic and inhibitor of brain cyclooxygenase, lowered body temperature to a normal level and decreased CSF prostaglandin levels to values observed in non-febrile animals (Fig. 9.2).

Similar experiments with almost identical results have been carried out in rabbits and a specific radioimmunoassay has been used to demonstrate that the prostaglandin-like biological activity in cat CSF after *Shigella* treatment is PGE_2. In man too, CSF PGE-like activity was raised in pyrexic patients with bacterial or viral infections. Thus there is considerable evidence that CSF-prostaglandin levels are raised during hyperthermia, suggesting a generalised increase in brain-prostaglandin biosynthesis. Whether the concentration of prostaglandin in the brain area of interest, the AH/POA, also increases to a similar extent cannot be determined simply by measuring prostaglandins in the CSF.

Not all of the experimental evidence supports the intermediacy of PGE_2 in pyrogen-induced fever. Indeed, several objections to such a mechanism have been put forward (Veale, Cooper & Pittman, 1977). Probably the

Fig. 9.2. Appearance of prostaglandins in the cerebrospinal fluid (CSF) of cats made feverish by injection of *Shigella* endotoxin. CSF collected from the third ventricle was bioassayed on the rat stomach strip preparation (upper panel), (*a*) before injection of the endotoxin, (*b*) after injection of the endotoxin, and (*c*) after intraperitoneal injection of the NSAID, paracetamol. Notice that endotoxin injection leads to the appearance in the CSF of a rat stomach strip contracting substance (almost certainly a prostaglandin) and an increase in body temperature (lower panel). Paracetamol treatment rapidly lowers body temperature and the prostaglandin content of ventricular CSF. Data from Feldberg & Gupta (1973).

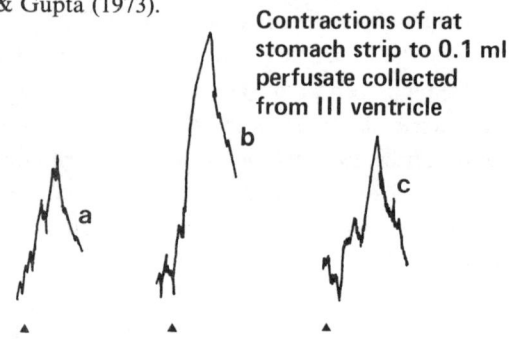

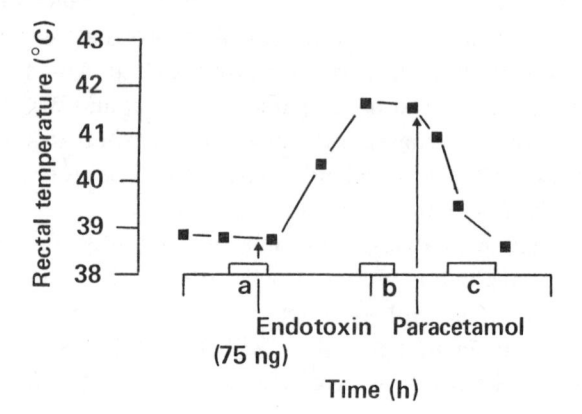

most damaging of these objections is the demonstration by Cranston and his colleagues (1975) that the increase in PGE_2 detected in cat CSF during fever may well be a side effect of pyrogen administration which is not related to the fever. These authors managed to obtain a dose regimen of sodium salicylate which, given intravenously, prevented the increase in CSF PGE_2 in endotoxin-treated cats but did not normalise body temperature; this suggested that the hyperthemia was unrelated to the increase in prostaglandins centrally. Other difficulties with the theory have arisen from experiments with the prostaglandin receptor antagonist SC19220 which given intraventricularly to cats blocked fever due to PGE_2 but had no effect on that caused by intravenous pyrogen, again implying that the pyrogen does not necessarily require intervention by PGE_2 in order to bring about an increase in body temperature. Furthermore, PGE_1 or PGE_2 injected directly into the lateral ventricles of anaesthetised newborn lambs did not raise body temperature, although intravenous pyrogen did cause a hyperthermia as expected.

A fair appraisal of the role of prostaglandins in pyrogen-induced fever would probably be one of 'non proven'. Further studies to measure prostaglandin release from the AH/POA in animals with and without fever are necessary in order to overcome the problems of relating CSF prostaglandin levels, which actually reflect whole-brain prostaglandin turnover, to the formation of prostaglandins in a small part of the hypothalamus. Other studies to determine the role of PGI_2, TxA_2 and leukotrienes in fever are also important.

Food intake

Several studies suggest that prostaglandins exert an action on the hypothalamus to influence food intake. Unfortunately the reports are contradictory and confusing. Subcutaneous or intrahypothalamic injection of PGEs (but not PGFs) decreased food intake in rats and sheep in one study but PGE_1 increased eating in another study when injected directly into the anterior hypothalamus of sheep. One thing is clear about the role of prostaglandins in the control of food intake. The doses of prostaglandins used in these studies are huge, sometimes several times the maximal prostaglandin-synthesising capacity of the whole brain. Consequently, it is most unlikely that prostaglandins formed naturally in the brain have a physiological function in the control of appetite or food ingestion.

Hypothalamus–pituitary axis

Prostaglandins may be involved in the regulation of hormone release from the pituitary. The effect of prostaglandins on pituitary release of gonadotrophins is discussed in detail in Chapter 5. PGE_1, PGE_2, $PGF_{1\alpha}$ and $PGF_{2\alpha}$ injected by any of a number of routes (e.g. subcutaneous, intravenous, intrapituitary, intrahypothalamic) in a number of different species (e.g. rat, sheep, pigeon, human) have been reported to enhance pituitary release of LH, FSH, prolactin, growth hormone (GH), thyroid stimulating hormone (TSH) and adrenocorticotrophic hormone (ACTH) (see Kenimer, Goldberg & Blecher, 1977 for review). Studies *in vitro* using pituitary preparations have in general confirmed that PGEs stimulate pituitary hormone release. This is most directly demonstrated by injecting prostaglandins either into the median eminence of the hypothalamus or into the pituitary. Only when injected into the hypothalamus does PGE_2 increase blood ACTH levels indicating that the important site of action of prostaglandins is somewhere in the hypothalamus. There is other evidence that prostaglandins affect the release of other pituitary hormones by an effect within the hypothalamus. In some elegant experiments Drouin *et al.* (1976) first showed that intravenous PGE_2 raised plasma LH levels in female rats and then proved that this effect was completely abolished after an injection of anti-LHRH serum, indicating that the effect of PGE_2 must have occurred prior to release of LHRH from the hypothalamus and its subsequent effect in the pituitary. Experiments of this type in which large, unphysiological concentrations of prostaglandins are injected directly into the brain, must obviously be treated with caution.

Much interest has been focussed upon the molecular mechanisms of these effects and, not surprisingly, cAMP appears to play a central role. For example, prostaglandins stimulate cAMP accumulation in rat pituitary glands *in vitro* as does arachidonic acid and dihomo-γ-linolenic acid. The effect of these fatty acid precursors, but not of the prostaglandins themselves, was abolished by treatment with NSAID. Furthermore, the time course of PGE_1-induced enhancement of GH release from the anterior pituitary *in vitro* was paralleled by a rise in pituitary cAMP. Both effects (GH release and increase in pituitary cAMP content) were prevented by the prostaglandin receptor blocker 7-oxo-13-prostynoic acid (Ratner, Wilson & Peake, 1973). It is also well known that cAMP stimulates the release of all adenohypophysial hormones. Thus, it is reasonable to conclude that prostaglandins first stimulate cAMP formation in the pituitary and hypothalamus and thereafter cAMP, possibly by a mechanism which involves other neurotransmitters in the hypothalamus–pituitary axis, brings about the release of pituitary hormones.

Water intake

In Chapter 8 the interaction of ADH with prostanoids in the collecting duct of the kidney nephron was discussed. Prostanoids also affect intake of fluid by a direct action on the so-called thirst sensor in the hypothalamus. As in the kidney, the effect of prostanoids centrally is under the control of angiotensin II. The brain contains all the enzymes necessary to synthesise angiotensin II, the major function of which is to act on neurones in the thirst sensor region to stimulate the feeling of thirst and to cause release of ADH from the posterior pituitary gland. PGE$_1$ and PGE$_2$ mimic angiotensin II in both of these actions, stimulating ADH release when injected into the common carotid artery of cats and potentiating the thirst-producing activity of prostaglandin E$_2$ in goats. Again whether this represents a physiological function of central prostaglandins remains to be seen.

Autonomic nervous system
Biosynthesis and catabolism of prostanoids

Prostanoids are released following stimulation of autonomic nerves or administration of catecholamines in a number of tissues including spleen, heart, blood vessels, kidney, adipose tissue and vas deferens of several species. In most of these tissues the major prostanoid formed is probably PGE$_2$ with lesser amounts of PGI$_2$ and PGF$_{2\alpha}$.

Effects of prostanoids on autonomic neurotransmission

Sympathetic nervous system. Stimulation of adrenergic nerves innervating a peripheral organ or tissue causes the release of noradrenaline from sympathetic nerve terminals in that organ. The first step in this process is the invasion of the nerve terminal by a propagating action potential followed by the influx of Ca^{2+} ions across the pre-junctional nerve membrane. Noradrenaline release may be modified by a variety of chemically dissimilar substances acting on receptors located on the pre-junctional nerve membrane. Substances known to influence noradrenaline release in this way include noradrenaline itself (α-receptor stimulation inhibits while β-receptor stimulation potentiates noradrenaline release), dopamine (inhibitory), 5-hydroxytryptamine (inhibitory or facilitatory), histamine (inhibitory), GABA (inhibitory or facilitatory), adenosine and acetylcholine (both inhibitory), opiates (inhibitory) and angiotensin II (facilitatory). To this list we may also add prostaglandins of the E series which, in many organs innervated by the sympathetic nervous system, potently reduce noradrenaline release after nerve stimulation.

Most experimenters who have studied the way in which prostaglandins affect noradrenaline release have employed classical pharmacological techniques using isolated organs perfused or superfused with physiological salt solution or alternatively biological tissues mounted in an organ bath. Few, if any, attempts have been made to look at this phenomenon in the intact animal. Three different approaches have been used to assess noradrenaline release *in vitro*:

(1) recording of the mechanical (or, more rarely, the electrical) response of the tissue to nerve stimulation;

(2) measurement of the release of endogenous noradrenaline, its metabolites and/or the enzyme dopamine-β-hydroxylase from the tissue before, during and after nerve stimulation;

(3) measurement of the release of exogenous noradrenaline after pre-labelling stores in the synaptic vesicles by incubating the tissue with radiolabelled ([^{3}H] or [^{14}C]) noradrenaline.

Some of the early experiments to investigate the action of prostaglandins on sympathetic neurotransmission using only the first method were plagued with inconsistent results not only from one tissue to the next but also from one research group to the next. It was later realised that prostaglandins interfere not only with the pre-junctional release of noradrenaline but also with its effect post-junctionally. This aspect of the prostaglandin–noradrenaline interaction will be dealt with later in the chapter. The introduction of sensitive assay methods for catecholamines and the use of [^{3}H]noradrenaline of high specific activity have resulted in the widespread application of the second and third methods which provide a rapid and reliable way of quantifying the effect of prostaglandins on noradrenaline efflux, separate from any action they may have on the post-junctional response to noradrenaline.

Most investigators agree that PGE_1 and/or PGE_2 inhibit noradrenaline release from adrenergic nerves innervating the heart, spleen, blood vessels, vas deferens, intestines, iris and kidney from several species (see Starke, 1977 for review). Some tissues are remarkably sensitive to very low doses of PGE_2. For example, the guinea pig mesenteric artery preparation in which noradrenaline release following perivascular nerve stimulation is reduced by as little as 10 pg/ml PGE_2 (Kuriyama & Makita, 1982). In striking contrast noradrenaline release from some tissues is not affected by PGE_2 even at very high concentrations (above 1 μg/ml). Examples of such 'prostaglandin-resistant' preparations include the rat vas deferens and the dog spleen.

Other prostaglandins have not been so widely investigated. PGI_2 inhibits contractions of the field-stimulated guinea pig vas deferens although very large concentrations are required. Clearly the chemical instability of PGI_2 at 37 °C in physiological solution may contribute to its lack of potency in this respect. A more reasonable experiment would be to determine the effect of prostacyclin synthetase inhibitors on noradrenergic transmission to determine whether transmission is affected when endogenous PGI_2 biosynthesis is prevented. Such an experimental approach has not yet been carried out. Similar problems of chemical instability make it difficult to interpret experiments in which prostaglandin endoperoxides were reported to inhibit noradrenaline release from the guinea pig vas deferens (Hedqvist, 1976). In such experiments there is the additional problem that prostaglandin endoperoxides may also break down chemically into E series prostaglandins which reduce noradrenaline release in their own right. As yet there have been no attempts to determine the effects (if any) of TxA_2 and leukotrienes on sympathetic neurotransmission. Of the other prostaglandins tested, $PGF_{2\alpha}$ potentiates the mechanical response to sympathetic nerve stimulation without affecting the response to added noradrenaline. This effect also occurs in several pharmacological preparations but direct evidence that $PGF_{2\alpha}$ promotes noradrenaline release pre-junctionally has not been obtained. In the isolated superfused rat heart pre-loaded with [^{3}H]noradrenaline, $PGF_{2\alpha}$ (up to 10^{-6} M) did not affect nerve-stimulation-evoked noradrenaline efflux (Hedqvist & Wennmalm, 1971). PGD_2 appears to have a similar effect to PGE_1 and PGE_2 in many tissues but is much less potent. The prostacyclin metabolite, 6-oxo-PGE_1, inhibits contractions of the field-stimulated guinea pig vas deferens without affecting the contractile response of this tissue to added noradrenaline suggesting a pre-junctional site of action.

Mechanism of action. At concentrations which prevent noradrenaline release following nerve stimulation, prostaglandins of the E series have no effect on action potential generation in adrenergic nerves, enzymatic catabolism of noradrenaline by monoamine oxidase (MAO) or catechol O-methyl transferase (COMT), neuronal/extraneuronal uptake of noradrenaline or post-junctional membrane potential. Additionally, PGEs do not affect noradrenaline release from isolated adrenergic storage vesicles prepared from bovine splenic nerve (Hedqvist, 1970) which suggests that the inhibitory effect of prostaglandins on sympathetic neurotransmission requires an intact nerve terminal to operate. Since Ca^{2+} ions have an important role to play in excitation–secretion coupling and, since prostaglandins affect Ca^{2+} ion utilisation in other tissues (e.g. platelets – see chapter 4), it is reasonable to propose that prostaglandins reduce noradrenaline release by uncoupling calcium-mediated excitation-secretion coupling. Evidence for such an uncoupling mechanism of action includes:

(1) Some authors have found that the inhibitory effect of prostaglandins on noradrenaline release (e.g. in guinea pig vas deferens) varies according to the Ca^{2+} ion concentration in the surrounding bathing medium. The potency of PGE_2 was greatest at low concentrations of this ion and declined as the amount of Ca^{2+} in the bathing medium was raised. Such dependence on Ca^{2+} ions does not occur in all organs in which prostaglandins have been tested. In the perfused spleen, for example, PGE_2 inhibits noradrenaline release over a very wide range of extracellular Ca^{2+} ion concentrations with little if any change in potency.

(2) The inhibitory effect of PGEs on sympathetic neurotransmission in several tissues is inversely related to the frequency of nerve stimulation and also to the length of time that each electric shock is applied (i.e. the pulse width). Since increasing the frequency and/or pulse width of sympathetic nerve stimulation promotes noradrenaline release by increasing the uptake and/or utilisation of Ca^{2+} intraneuronally, these results are compatible with the theory that PGEs prevent noradrenaline release by inhibiting the

Fig. 9.3. A, Release of noradrenaline from sympathetic nerve endings following electrical stimulation of the pre-junctional nerve. B, The same process in the presence of prostaglandins of the E series. Notice that prostaglandins prevent influx of Ca^{2+} ions through calcium ionophores to reduce the efflux of noradrenaline.

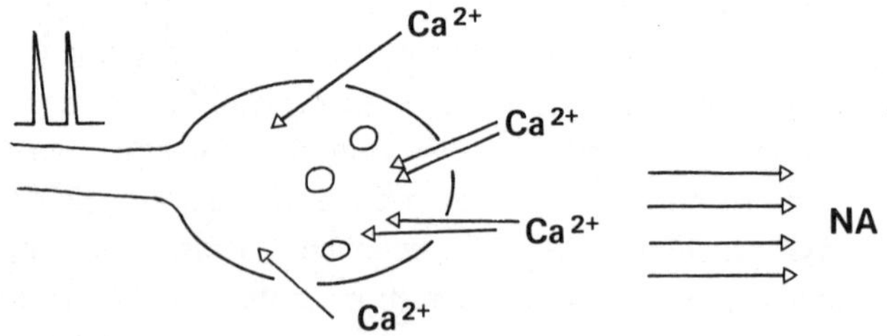

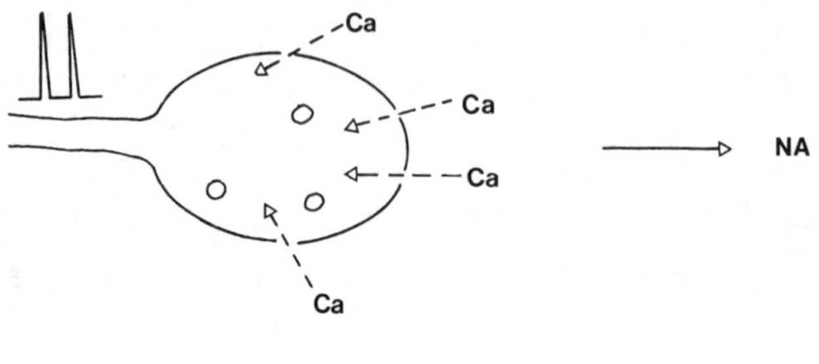

influx of Ca^{2+} ions into the pre-junctional nerve terminal and thus blocking excitation–secretion coupling.

(3) Release of noradrenaline by indirect-acting sympathomimetic drugs (e.g. tyramine) which is brought about by a mechanism independent of Ca^{2+} ions is unaffected by prostaglandins.

Several models have been put forward to explain how PGEs and Ca^{2+} ions interact in the neurone pre-junctionally. To date, no single model has been universally accepted. One possible interpretation of the evidence outlined above is shown in Fig. 9.3. In this model the arrival of an action potential at the nerve terminal results in the influx of extracellular Ca^{2+} ions or the release of this ion from membrane stores. The resulting elevation in intraneuronal Ca^{2+} ion concentration sets in motion the sequence of events culminating in the release of noradrenaline into the synaptic cleft. The action of prostaglandins may be either to decrease the influx of Ca^{2+} ions, possibly by closing the membrane Ca^{2+} ionophore, or alternatively to promote uptake of intraneuronal Ca^{2+} ions into storage vesicles. The resulting reduction and sequestration of calcium within the nerve terminal prevents excitation–secretion coupling and blocks release of noradrenaline.

The physiological role of prostanoids in noradrenergic transmission – the Hedqvist hypothesis

Hedqvist (1969) proposed that endogenous prostaglandins (particularly PGE_2) formed locally in adrenergically innervated organs have a physiological function to regulate noradrenaline release from adrenergic nerves. Although considerable evidence has been put forward in support of the so-called Hedqvist hypothesis, conflicting evidence has also been reported (reviewed by Dubocovich & Langer, 1975; Westfall, 1977; Starjne, 1979; Langer, 1981). The following section deals with the 'pros' and 'cons' of an involvement of endogenous PGE_2 in sympathetic neurotransmission.

1. Exogenous PGEs act pre-synaptically to prevent noradrenaline (NA) release.

For
Very small amounts of PGE_1 and PGE_2 depress NA release from adrenergic nerves in many tissues.

Against
a. Only prostaglandins of the E series influence NA release at concentrations which may be achieved *in vivo*.
b. Some adrenergic nerves are resistant to the inhibitory effect of PGs.

2. Sympathetic nerve stimulation causes PGE release

For
a. Stimulation of sympathetic nerves to many organs causes release of PGE_2, PGI_2, $PGF_{2\alpha}$ and PGD_2 in amounts which block NA release.

Against
a. PGs are released from adrenergically innervated organs by a host of other stimuli e.g. bradykinin, ATP, ischaemia, hypoxia, mechanical trauma etc.

Against
b. Nerve stimulation releases large amounts of PGEs from some organs (dog spleen) in which the adrenergic nerves are resistant to the inhibitory effect of exogenous PGEs.
c. The amounts of PGEs released from isolated perfused organs freshly prepared is very low, insufficient to affect NA release. Only when organs have been perfused for some time does PGE release increase to levels which will decrease NA release. PGE release at this stage of the experiment is likely the result of non-specific tissue damage.

3. Inhibitors of prostaglandin formation increase NA release

For
a. Cyclooxygenase inhibitors (e.g. indomethacin) increase NA release following nerve stimulation in a large number of peripheral tissues.
b. Indomethacin administration to rats *in vivo* increases urinary excretion of NA, suggesting that inhibition of PGE biosynthesis promotes sympathetic neurotransmission *in vivo* as well as *in vitro*.

Against
a. In some organs (e.g. cat spleen, dog adipose tissue) indomethacin or meclofenamate does not alter NA release following nerve stimulation.

4. Increased PG formation reduces NA release.

For
a. Administration of arachidonic acid reduces the release of NA by nerve stimulation in the rabbit kidney. This action is prevented by indomethacin.

The origin of the released prostaglandins has also been hotly disputed. Since the output of PGEs following sympathetic nerve stimulation may be prevented by administration of α- and/or β-adrenoceptor antagonists to block the response of the tissue (Gilmore, Vane & Wyllie, 1968) it seems likely that, if endogenous prostaglandins do influence noradrenaline release *in vivo*, then they are probably synthesised and released by the target tissue and not by the pre-junctional nerve terminal. If this is the case then prostaglandins act as chemical regulators of a unique form of 'trans-synaptic' negative-feedback control of noradrenaline release (Fig. 9.4). However, release of modulator PGEs pre-junctionally cannot be ruled out. If the guinea pig vas deferens is electrically (field) stimulated (5 Hz) in the presence of low extracellular Ca^{2+} ion concentration, no contractile response occurs. However, treatment with the cyclooxygenase-inhibiting drug, 5,8,11,14-eicosatetraynoic acid, still enhanced nerve-stimulated noradrenaline release (Starjne, 1973, 1979) indicating that prostaglandins are still released and inhibit noradrenaline output when no physical contraction of the tissue occurs. In this case the origin of the released prostaglandin would appear to be pre-junctional.

Interaction between α-adrenoceptor and PG-mediated inhibition of noradrenaline release

There is good evidence that the release of noradrenaline from adrenergic nerves is under the control of two feedback processes, one mediated by PGEs and the other mediated by the action of noradrenaline on pre-junctional α_2-adrenoceptors. Since PGE synthesis and release occurs when perfused tissues are challenged with noradrenaline it is clearly important to establish whether these two processes operate independently or whether inhibition of noradrenaline release after stimulation of α_2-adrenoceptors is dependent upon the formation of PGEs, as suggested by scattered reports in the literature.

The increase in noradrenaline overflow which occurs when pre-junctional α_2-adrenoceptors are blocked by drugs like phenoxybenzamine is unaffected by inhibition of prostaglandin biosynthesis with either indomethacin or 5,8,11,14-eicosatetraynoic acid. Similarly, inhibition of noradrenaline

Fig. 9.4. The origin of E series prostaglandins formed during sympathetic neurotransmission, (a) PGEs synthesised by the target organ when it contracts feed back across the synaptic gap to block noradrenaline output i.e. trans-synaptic regulation. (b) PGEs made pre-junctionally inhibit noradrenaline release i.e. pre-synaptic regulation.

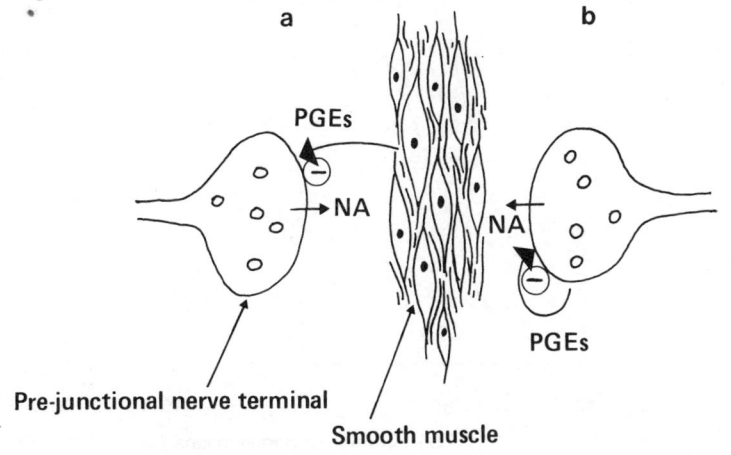

release by selective α_2-adrenoceptor agonists like clonidine is either unaffected or only marginally reduced by aspirin-like drugs. Both of these results strongly suggest that activation of pre-junctional α_2-receptors by noradrenaline and other agonists blocks noradrenaline efflux by a process which does not involve prostaglandins.

Post-junctional effects of prostaglandins on sympathetic neurotransmission

In addition to their pre-junctional effect, prostaglandins may also affect sympathetic neurotransmission by increasing or decreasing the post-junctional sensitivity to released noradrenaline. Prostaglandins of the E series (PGE_1, PGE_2, 6-oxo-PGE_1) enhance the response to exogenous catecholamines in both the guinea pig and rabbit vas deferens preparations set up in an organ bath (Fig. 9.5). In this tissue the concentrations required for post-junctional potentiation of exogenous noradrenaline are very similar to the amounts needed to inhibit noradrenaline release pre-junctionally. For example, PGE_1 (0.8 ng/ml) produced 50% inhibition of the contractile response following field stimulation of the isolated guinea pig vas deferens while the response to exogenous noradrenaline was potentiated several fold by 0.5 ng/ml PGE_1 (Griffiths & Moore, 1983). This

Fig. 9.5. Potentiation of the contractile response to noradrenaline (filled squares) on the isolated guinea pig vas deferens preparation in the presence of 0.5 ng/ml PGE_1 (open squares) and 1.5 ng/ml 6-oxo-PGE_1 (triangles).

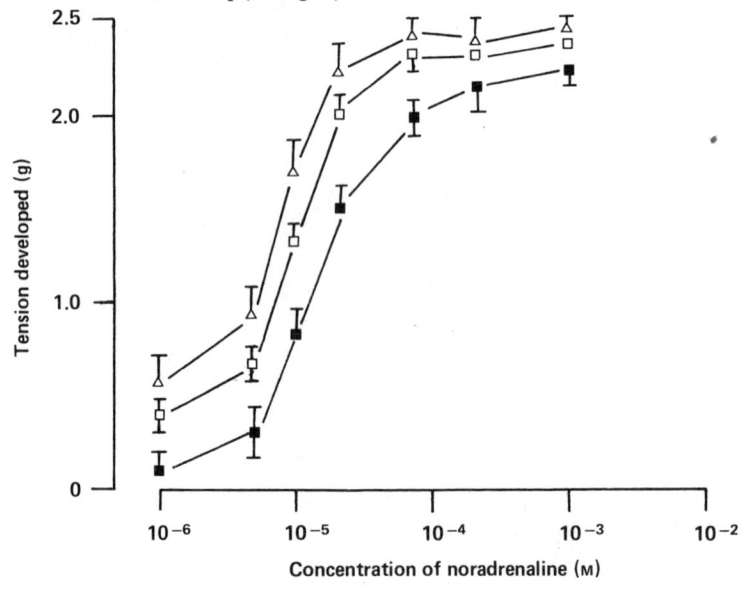

clearly illustrates the multifactorial nature of the effect of prostaglandins on sympathetic neurotransmission and highlights the necessity of measuring noradrenaline release directly if the pre-junctional effects of prostaglandins are to be unambiguously determined. The post-synaptic effects of prostaglandins of the E and F series on several isolated pharmacological preparations are shown in Table 9.1. Other prostanoids such as PGI_2, TxA_2 and prostaglandin endoperoxides have not been studied.

Parasympathetic neuroeffector transmission

Prostaglandins have very complicated effects on cholinergic neuroeffector transmission. Various permutations of pre-junctional and post-junctional inhibition/potentiation have been reported using isolated tissues *in vitro*. The precise result depends upon the choice of prostaglandin, tissue, species and experimental method adopted. Examples of some of the ways in which prostaglandins may interfere with acetylcholine release from parasympathetic nerve endings and post-junctional response to acetylcholine are discussed below.

In the majority of tissues, prostaglandins of the E series enhance cholinergic transmission. Examples include rabbit and guinea pig ileum, bovine iris and rabbit rectococcygeus. Early experiments suggested that PGEs promoted neurotransmission in these tissues by increasing the release of acetylcholine following nerve stimulation. This conclusion was

Table 9.1. *Post-junctional actions of prostaglandins*[a]

Species and tissue	PG	Response to NS	Response to CA
Dog			
Spleen	E_2	↑	↑
	$F_{2\alpha}$	↑	0
Paw	E_2, $F_{2\alpha}$	↑	↑
Heart, kidney, uterine artery	E_1, E_2, $F_{2\alpha}$	↑	↑
Rabbit			
Aorta	E_1, E_2	N.D.	↑
Mesenteric artery	E_1	0	↑
Kidney, heart	E_2, $F_{2\alpha}$	0	↑
Rat			
Heart, kidney	E_2	↑	↑
Mesenteric artery	E_2, $F_{2\alpha}$	↑	↑

[a] NS, nerve stimulation; CA, catecholamine; N.D., not determined.

based upon studies in which PGE_1 and PGE_2 were shown to enhance the contractile response of the isolated guinea pig ileum preparation to field stimulation without affecting the response of the tissue to exogenous acetylcholine added to the bath (Ehrenpreis, Greenberg & Belman, 1973).

Other authors have questioned this result, finding that PGEs potentiated the response of the guinea pig ileum to field stimulation and applied acetylcholine almost equally (Botting & Salzman, 1974). More recent experiments in which acetylcholine efflux from electrically stimulated guinea pig ileum has been measured directly using a sensitive physicochemical technique (GC–MS) have revealed no effect of prostaglandins on acetylcholine release even at concentrations higher than required to potentiate the mechanical response of the tissue (Gustafsson, Hedqvist & Lundgren, 1980). Thus, in these tissues prostaglandins have a predominantly post-junctional potentiating effect, in all probability involving an increase in membrane permeability of the effector cell to Ca^{2+} ions.

In some parasympathetically innervated organs (e.g. rabbit and mouse heart) PGE_1 blocked the fall in heart rate due to stimulation of the vagus. However, the negative chronotropic effect of acetylcholine was unaffected by PGE_1. Together with the observation that prostaglandins may be released from the intact heart either by vagal nerve stimulation or by acetylcholine, these results have led to the proposal that PGEs may regulate acetylcholine release and cholinergic transmission in the heart in a manner similar to that proposed for the noradrenergic system (Wennmalm, 1971). This theory has not been widely accepted. A major problem with this proposal are the findings that PGE_1 blocked the negative chronotropic effect both of acetylcholine and cholinergic nerve stimulation in the isolated guinea pig atria preparation and, furthermore, that PGE_2 had no effect on the response to acetylcholine of the rabbit sino-atrial node preparation. The reasons why different experimenters obtain diametrically opposite results in this way are not clear. One major reason which seems to be widely ignored by most researchers is the lack of uniformity when it comes to the choice of experimental animal in these studies.

Since the data linking prostaglandins with parasympathetic neurotransmission are so conflicting it is not reasonable to draw any general conclusions on the subject. A more careful characterisation of the pre- and post-junctional activity of prostanoids (particularly those like PGD_2, $PGF_{2\alpha}$, PGI_2 and TxA_2 which have so far gone largely unstudied) is clearly required.

Effect of prostaglandins on chemical neurotransmission at other sites

What little information which is available suggests that prostaglandins have little or no effect on peripheral ganglionic neurotransmission. To cite one example, PGE_1 and PGE_2 did not affect postganglionic action potentials elicited by pre-ganglionic stimulation of the cat cervical sympathetic chain. In the central nervous system, PGE_2 reportedly increased [^{3}H]noradrenaline release from pre-labelled rat brain slices. Whether this effect was mediated by an action on pre-synaptic receptors located on noradrenergic terminals or from post-synaptic stimulation of other neurones which in turn acted upon and inhibited noradrenaline release is not known.

10

Inflammation

Introduction

Inflammation is a defensive reaction of the body to injury. It is normally a desirable, physiological process which ensures that the tissues are protected against microorganisms and that any tissue damage which does occur is swiftly repaired. Occasionally the inflammatory response is uncontrolled and persists after the damaged tissues have been repaired. In this case damage to the host's tissues can occur causing the development of chronic inflammatory disease conditions like arthritis.

An inflammatory reaction may be evoked by mechanical trauma (e.g. cutting, crushing), exposure to corrosive chemicals, radiation, antigen–antibody reactions or invasion of the body by disease-producing organisms called pathogens. In the last case pathogens may be bacterial, viral or fungal.

The main features of the acute inflammatory response were first described by Celsus in the first century AD and include:

(1) rubor (redness) due to vasodilatation of blood vessels in the inflamed area increasing the supply of white blood cells to the damaged tissues;

(2) tumor (swelling) caused by an increase in the permeability of blood vessels in the inflamed area and leakage of fluid and white blood cells to the site of damage;

(3) calor (heat) due to increased blood flow to the inflamed area;

(4) dolor (pain) caused by stimulation of sensory nerve endings in the damaged tissues and serving to alert the whole organism of the danger.

Each of these components of the inflammatory reaction are brought about by the regulated synthesis and release or alternatively release from pre-formed stores of tissue mediators. Numerous inflammatory mediators occur naturally in the body. Examples include low molecular weight biogenic amines (e.g. 5-hydroxytryptamine and histamine), peptides (e.g. kinins) and lipids (e.g. platelet activating factor, PAF) as well as higher molecular weight proteins such as eosinophil chemotactic factor (ECF) and components of complement.

Before any substance can be identified as an inflammatory mediator it should satisfy certain conditions based loosely on the criteria for peripheral neurotransmitters first suggested by Sir Henry Dale in the 1920s. These criteria include:
(1) the proposed mediator must occur in an appropriate concentration at the site of inflammation, and enzymes for its synthesis and breakdown must also be present;
(2) the proposed mediator must reproduce all of the biological features observed during inflammation;
(3) pharmacological manipulation of the synthesis, release and/or actions of the proposed mediator should produce an appropriate alleviation (or worsening) of the symptoms of inflammation.

A considerable amount of experimental evidence suggests that both cyclooxygenase and lipoxygenase metabolites of arachidonic acid are mediators or modulate the activity of other mediators in acute inflammation. There is less evidence of a role for these metabolites in chronic inflammatory disease states although even here prostanoids may modify the immune response by affecting the release of lymphokines (see later).

Evidence that prostanoids are mediators of inflammation
Occurrence at the inflammatory site

Virtually all cell types so far studied have the capacity to synthesise prostaglandins. The trigger which switches on the deesterification of membrane phospholipids and the subsequent conversion of free arachidonic acid to prostanoids may well be the distortion or trauma of the cell membrane. Since acute inflammation, by definition, must be preceded by tissue damage it is therefore not surprising that prostanoids are always found at the site of inflammation.

Stimuli which are known to cause inflammation in man and which also provoke tissue prostanoid biosynthesis range from allergic reactions and autoimmune diseases to any type of thermal, mechanical, chemical- or radiation-induced tissue damage. PGE_2, $PGF_{2\alpha}$, PGI_2 and TxA_2 for example have been detected in inflammatory exudates induced by the implantation of carrageenan-impregnated polyester sponges just under the skin of the flank of rats (Chang, Murota & Tsurufuji, 1977). In such experiments classical prostaglandins and PGI_2 do not appear in the exudate until at least 6–8 h have elapsed and TxA_2 (measured as TxB_2) can be detected only after 8 d. Prostaglandins also occur in synovial fluid removed from the paws of rats with *Streptococcus*-extract-, carrageenan- or adjuvant-induced arthritis, as well as from the joints of human rheumatoid or osteoarthritis sufferers in which PGI_2 is the predominant cyclooxygenase product (Brodie *et al.*, 1980). Other chronic inflammatory states in which prostanoids have been detected are allergic contact eczema, anterior uveitis, psoriasis, burn damage and ulcerative colitis.

There is little doubt that prostanoids occur in abundance in and around the inflamed tissue. What is in doubt is the source of these prostanoids. Certainly in acute inflammation the immediate source of prostanoids will be the damaged tissues themselves e.g. skin, smooth muscle, blood vessels. As the inflammatory response develops and phagocytic leucocytes are attracted to the area of damage then the leucocytes themselves are very capable of synthesising and releasing large amounts of both prostanoids and leukotrienes. In chronic inflammation leucocytes are presumed to be the major source of arachidonic acid metabolites. Many studies have been performed to identify which is the most prominent prostanoid-producing leucocytes and to determine how prostanoid biosynthesis by these cells is triggered. As a result of these studies it seems likely that virtually all inflammatory cell types, with the exception of lymphocytes, are capable of responding to an appropriate stimulus with prostanoid biosynthesis. However, the macrophage is probably the major leucocyte source of prostanoids during the inflammatory reaction. To demonstrate the superiority of the macrophage in this respect Gemsa and his colleagues (1981) exposed a fixed number of different leucocyte cell types (eosinophils, neutrophils, macrophages and lymphocytes) to a standard concentration of zymosan and assayed the amount of PGE_2 released. Whilst 5×10^6 eosinophils and neutrophils released only 1 ng and 1.5 ng PGE_2 respectively the same number of macrophages released 4 ng PGE_2. In contrast, lymphocytes produced little or no PGE_2. More recent experiments have revealed that human macrophages challenged with zymosan particles also synthesise PGI_2 and TxA_2 as well as PGE_2 (Humes *et al.*, 1977).

An important stimulus which turns on leucocyte prostanoid biosynthesis *in vitro* is the process of phagocytosis. Zymosan and antigen–antibody complexes are both phagocytosed by macrophages and provoke these cells into forming large amounts of prostanoids. Since a primary function of leucocytes at the site of inflammation is the phagocytosis of cell debris, foreign particles etc. it seems likely that this is the process which provokes leucocyte prostanoid formation *in vivo* as well as *in vitro*. Other stimuli which cause leucocyte prostanoid production include agents which provoke

Inflammatory properties of prostanoids

Vascular effects. A characteristic feature of the acute inflammatory response is vasodilatation of blood vessels in the immediate vicinity. Prostaglandins of the E series and PGI_2 are very powerful vasodilators of the microcirculation (Nakano & McCurdy, 1967; Moncada et al., 1976). Depending upon the blood vessel under study and the circumstances of the experiment PGI_2 is some 2–50 times more potent than PGE_2. Indeed PGI_2 is 80–90 times more potent than PGE_2 as a vasodilator of the precapillary arterioles in the epithelial tissue of the hamster cheek pouch (Higgs et al., 1979a). In addition to a direct vasodilator effect, prostaglandins also oppose the vasoconstrictor action of noradrenaline which further ensures that blood vessels within the inflamed area remain dilated. Most importantly the concentration of PGE_2 and PGI_2 found in inflammatory exudates is sufficiently high to exert a pronounced vasodilator effect on the microvasculature.

Another feature of inflammation which affects the vascular system is the increase in vascular permeability resulting in oedema. Exogenous prostanoids are poor inducers of oedema in man and experimental animals. Subplantar injections of PGI_2 (up to 100 ng) or PGE_2 (up to 50 ng) dose-dependently increased paw volume in rats (Higgs, Moncada & Vane, 1978a) but lower doses (10–50 ng) significantly potentiated the oedema due to carrageenan injection. Prostaglandins also potentiate oedema caused by histamine and bradykinin at doses which do not produce oedema alone. It is likely that potentiation of oedema by prostanoids reflects the vasodilator activity of prostaglandins of the E and I series and is not a

Fig. 10.1. Interaction of bradykinin and prostaglandins in the control of fluid leakage from blood vessels.

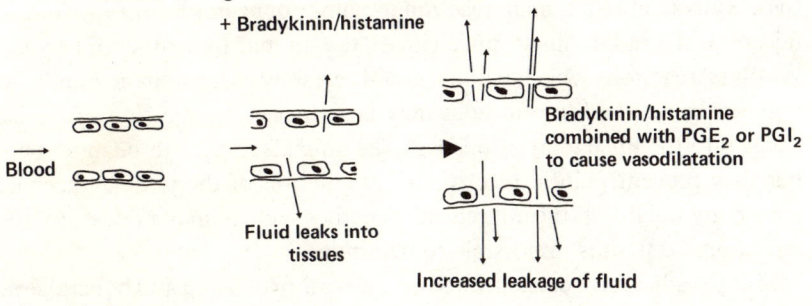

direct action on the endothelial cells which control passage of fluid and cells from the blood into the tissues (see Fig. 10.1).

It is now widely believed that for oedema to occur two processes must take place. Firstly, contraction of the endothelial cells lining post-capillary and collecting venules so that they move part and allow fluid through and, secondly, an increase in blood flow. As shown in Fig. 10.1 bradykinin and histamine increase vascular permeability by contracting venular endothelial cells. In contrast, PGI_2 and PGE_2 vasodilate blood vessels but have little or no ability to contract the endothelial cells. This would explain why prostanoids are poor inducers of oedema on their own but potentiate plasma exudation caused by bradykinin and histamine (see Williams, 1983).

In this respect prostanoids act not as primary mediators of the inflammatory response but as modulators of the activity of other inflammatory mediators. As this chapter progresses, it will become increasingly clear that prostanoids are almost certainly not primary mediators of the acute inflammatory response but secondary mediators promoting the activity of other agents like 5-hydroxytryptamine, histamine and complement.

Pain-producing and hyperalgesic effects. Acute inflammation in man is associated with pain. The inflamed region may also be tender and sensitive to touch. This latter condition is called hyperalgesia and may be defined as a state in which pain can be inflicted by normally painless stimuli (be they mechanical, chemical or by some other process). Since pain and hyperalgesia are subjective phenomena the effect of drugs which cause pain (algesics) or alleviate pain (analgesics) can really be studied only in humans. Such experiments carried out in human beings after informed consent include application of algesics onto blister bases on the skin. Alternatively, the efficacy of analgesics can be assessed before and after electrical stimulation of the dental pulp as the pain-producing stimulus.

Several animal assays have been developed in which pain is assumed to have been felt by the animal when its reaction to a particular stimulus seems to be equivalent to a human reaction to pain. For example, intraperitoneal injection of bradykinin to mice causes the animal to writhe and hyperventilate, reactions which, in man, could certainly indicate pain. Similarly, a hyperalgesic condition in dogs may be achieved by injection of algesic drugs into the knee joint of one leg. The animals appear to be in no pain but they prevent painful mechanical stimulation of the joint by keeping their body weight off the injured leg. Whether such animal models are true reflections of pain is impossible to determine.

Many studies have concentrated on the pain-producing and hyperalgesic actions of prostanoids and this area has been reviewed by Moncada, Ferreira & Vane (1979). A general conclusion from all of these studies is that at concentrations likely to occur at the site of the inflammatory reaction, prostaglandins of the E, F and I series do not cause pain in the majority of human and animal tests. At concentrations 1-2 orders of magnitude higher than those found in the damaged tissues during inflammation prostaglandins are algesic. In contrast, low doses of PGE_2 and PGI_2 produce hyperalgesia and strongly augment the pain-producing ability of other inflammatory mediators like bradykinin and histamine. The hyperalgesia which follows subplantar injection of PGE_2 into the rat paw is long-lasting, suggesting that the sensitisation of afferent pain receptors by prostaglandins persists for a considerable period of time. Although PGI_2 is some 5 times more potent than PGE_2 as a hyperalgesic in the rat paw model, the effect diminishes rapidly (presumably as PGI_2 is chemically degraded into 6-oxo-$PGF_{1\alpha}$).

In summary, prostanoids potentiate algesia by other inflammatory mediators, highlighting once again their status as regulators and not primary mediators of the inflammatory reaction.

Effect on leucocytes. Before discussing the involvement of prostanoids in leucocyte function it seems worthwhile to provide some background information about the different types of leucocytes which occur in the body and their role in inflammation.

Some of the inflammatory white cells which play a part in the acute inflammatory response are already found in the tissues. These are the mast cells which, when degranulated following tissue damage, release large quantities of histamine and heparin which produce vasodilatation/increase vascular permeability and prevent blood coagulation. Other white cells enter the inflammatory site from the blood stream. These are the leucocytes and are of two types:

(1) polymorphonuclear cells (PMNs) which have lobed nuclei and are subdivided into the neutrophils, eosinophils and basophils;
(2) mononuclear cells which have a single non-lobed nucleus and are subdivided into monocytes and lymphocytes.

Following tissue damage, the PMNs are the first leucocytes onto the scene of the inflammatory reaction. They first adhere to the blood vessel wall and then migrate between the endothelial cells. PMNs phagocytose microorganisms and destroy them by means of a variety of digestive enzymes contained within their lysosomal granules. It is these enzymes which can cause excessive damage to the host's tissues causing the setting up of a chronic inflammatory state. The monocytes arrive on the scene many hours after the PMNs and are transformed in the tissues into macrophages which are also phagocytic. These have a bigger 'appetite' than PMNs and digest not only microorganisms but also cell debris.

It is now known that PMNs and other leucocytes are attracted to the inflammatory region by specific chemical substances which are described as having either chemotactic or chemokinetic properties. Chemotaxis is defined as the process in which the direction of movement of cells is influenced by chemicals. It may be either positive (cells move towards the chemotactic chemical) or negative (cells move away from the chemotactic chemical). Chemokinesis is the process whereby a chemical substance either increases (positively chemokinetic) or decreases (negatively chemokinetic) the speed of cell locomotion.

Some prostanoids are chemotactic for leucocytes. PGE_1 and TxA_2, for example, are chemotactic for rabbit and mouse PMNs but have no such activity when leucocytes from other species including man are used. Generally, high concentrations of these prostanoids are required which make it unlikely that they are chemotactic agents under physiological circumstances. In direct contrast, PGI_2 prevents PMN chemotaxis *in vitro* and, when infused intravenously, blocks adhesion of leucocytes in inflamed hamster cheek blood vessels (Higgs *et al.*, 1978*b*). Most likely, PGI_2 protects the vascular endothelial wall against adhesion not only of platelets but also of white blood cells. By preventing leucocyte adhesion to the endothelium of inflamed blood vessel walls PGI_2 will also stop leucocyte migration into the damaged tissues i.e. it will exert an anti-inflammatory action. However, the concentration of PGI_2 which is required to prevent leucocyte adhesion to vascular walls (about 10–100 ng/ml) is greater than the concentration of this prostanoid found at the inflammatory site. Thus this effect of prostacyclin is probably of more pharmacological than physiological interest.

Prostanoids may be involved physiologically in cellular immune reactions. This possibility is raised by the observation that modest doses of PGE_1, PGE_2, PGD_2 and PGI_2 inhibit the activation of lymphocytes and prevent them from secreting lymphokines. These are a heterogenous group of substances of uncertain structure released from lymphocytes which (among other things) promote lymphocyte cell division, activate macrophages and neutrophils, are chemotactic for PMNs and cause bone resorption, oedema and vasodilatation (for review see Morley, Hanson & Rumjanek, 1984). In other words lymphokines are potent inflammatory mediators. It has been suggested that PGE_2 synthesised by activated macrophages is involved in a negative-feedback control of lymphocyte activation (Gordon, Bray & Morley, 1976). This interaction is depicted diagrammatically in Fig. 10.2.

In this proposal, activated macrophages at the site of inflammation synthesise and release PGE_2 which acts on lymphocytes to block activation and lymphokine secretion. The fall in lymphokine concentration in the inflamed tissue reduces activation of macrophages and so completes the feedback loop. Interestingly, macrophages also synthesise TxA_2 which promotes lymphocyte activation and thus lymphokine release. In this way the relative amounts of PGE_2 and TxA_2 synthesised by the macrophage would be critically important in regulating lymphocyte activation and the immune response. Although the precise details of the way in which arachidonic acid metabolites affect lymphocyte function have still to be worked out, it is known that the concentration of PGE_2 required to prevent lymphocyte activation *in vitro* does occur in inflammatory fluids (range 3–300 nmol/l). Thus, PGE_2 at least could act as an immune regulator.

Effect of anti-inflammatory drugs on prostanoid biosynthesis

The third criterion for an inflammatory mediator is that drugs which prevent the synthesis and/or effects of the mediator should be anti-inflammatory. The two groups of drugs which are most useful clinically to treat inflammation (NSAID and corticosteroids) prevent prostanoid biosynthesis by affecting different parts of the arachidonic acid cascade (see Chapter 2). Furthermore, a good correlation has been demonstrated between the ability of a series of NSAID to inhibit cyclo-oxygenase activity *in vitro* and their ability to suppress inflammation

Fig. 10.2. Control of lymphokine release by prostanoids.

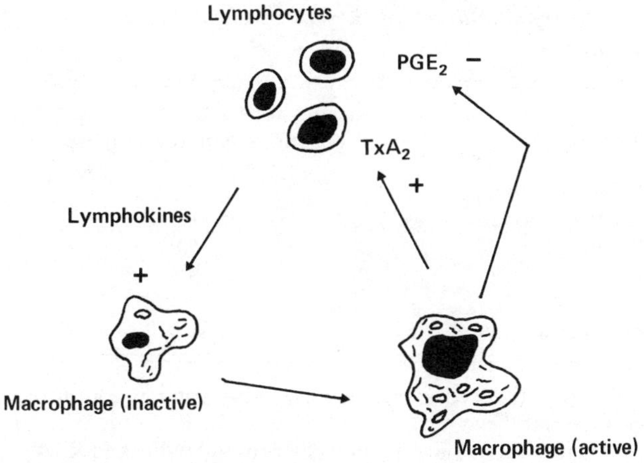

caused by injection of carrageenan into the rat paw (Table 10.1). Some convincing evidence that aspirin and other NSAID reduce inflammation by lowering local prostanoid concentrations has come from experiments using optical isomers of different NSAID (Tomlinson et al., 1972). To quote just one example the (+) isomer of a structural analogue of indomethacin blocks cyclooxygenase activity *in vitro* and prevents rat paw oedema following carrageenan injection whereas the (−) isomer had neither cyclooxygenase blocking nor anti-inflammatory activity (Ham et al., 1972). This topic has been dealt with in Chapter 2.

Table 10.1. *Rank order of potency of NSAID as anti-inflammatory agents and cyclooxygenase inhibitors*

NSAID	Suppression of oedema[a]	Inhibition of cyclooxygenase[b]
Sodium salicylate	0.9	0.3
Aspirin	1	1
Phenylbutazone	6.5	0.6
Ibuprofen	7.5	3
Naproxen	9	14
Indomethacin	50	78
Ketoprofen	80	920
Flurbiprofen	250	490

[a] Oedema produced in rat paw by subplantar injection of carrageenan.
[b] Inhibition of prostanoid biosynthesis *in vivo* in the rat.

Table 10.2. *Rank order of potency of corticosteroids as anti-inflammatory agents and phospholipase inhibitors*

Steroid	Suppression of oedema[a]	Inhibition of phospholipase[b]
Hydrocortisone	1	1
Corticosterone	0.4	1.3
Prednisolone	4	6.8
Fludrocortisone	10	8.5
Triamcinolone	5	10.4
Dexamethasone	25	31.2

[a] Oedema induced in rat paw by sub-plantar injection of carrageenan.
[b] Measured as ability to prevent prostaglandin and TxA_2 release from isolated perfused guinea pig lungs.

In addition to inhibiting the production of oedema, NSAID also inhibit other aspects of the inflammatory reaction. Aspirin is widely used, for example, to prevent pain and to reduce body temperature during fever which are both characteristics of acute inflammation.

The anti-inflammatory activity of steroid drugs is also well correlated with their ability to reduce prostanoid biosynthesis. As discussed in Chapter 2, these drugs do not inhibit cyclooxygenase at clinically obtained concentrations but do lower prostanoid levels at the site of inflammation by blocking the phospholipase step in the arachidonic acid cascade. The anti-phospholipase activity of several steroid drugs has been assessed by their ability to prevent release of prostanoids from perfused guinea pig lungs induced by a variety of 'releasing factors' like angiotensin II or bradykinin (see Chapter 2). The anti-phospholipase effect of six steroids is correlated with their anti-inflammatory action in Table 10.2.

Steroid anti-inflammatory drugs inhibit almost all of the manifestations of acute inflammation including vasodilatation, pain and oedema and they also prevent the movement of PMNs and monocytes from the blood across the vessel wall to the damaged tissues. All types of inflammation (e.g. induced by pathogens like bacteria, chemical or physical stimuli, or disordered immune reactions like those which occur in autoimmune disease) are prevented or reduced by steroid drugs. It is likely that some of the anti-inflammatory activity of steroids are not related to block of prostanoid biosynthesis. Other mechanisms of action e.g. stabilisation of lysosomal vesicle membranes, may also contribute to the anti-inflammatory effect of steroids. Nevertheless, there is convincing evidence that steroids to a large extent are effective in the treatment of inflammatory diseases by preventing prostanoid and leukotriene formation in the damaged region.

Evidence that leukotrienes are mediators of inflammation
Occurrence at the inflammatory site

Leukotrienes were originally isolated from arachidonic acid challenged rabbit PMNs by Borgeat & Samuelsson in 1979 and indeed were named after these leucocytes. Since that time leucocytes of several species have been shown to be a good source of lipoxygenase enzymes which convert arachidonic acid by an action at C-5 to yield leukotrienes and by oxygenation at positions C-8, C-9, C-11 and C-12 to yield a variety of mono- and dihydroxyeicosatetraenoic acids (HETEs) and the corresponding mono- and dihydroperoxyeicosatetraenoic acids (HPETEs). All lipoxygenase metabolites of arachidonic acid are biologically active. In general the rank order of potency is leukotrienes > HPETEs > HETEs. Of the leukotrienes, LtB_4 has the most potent chemotactic and chemokinetic activity as discussed in the next section. A summary of the various cultured cell lines which release LtB_4 is shown in Table 10.3.

Lipoxygenase products have also been detected in chronic inflammatory disease states in humans. LtB_4, 5-HETE and 12,15-diHETE occur in synovial fluid of patients with rheumatoid arthritis. LtB_4 and 12-HETE have been detected in the epidermis of patients with psoriasis, and LtB_4-like activity has been measured in the sputum of cystic fibrosis sufferers.

In summary, HPETEs, HETEs and leukotrienes (particularly LtB_4) can be generated from a wide variety of cells following an equally wide variety of stimuli. Furthermore, the concentration of LtB_4 synthesised and released from white cells at the inflammatory site *in vivo* is almost certainly adequate to cause many of the inflammatory activities described in the next section.

Inflammatory properties of leukotrienes

Vascular effects. Unlike prostaglandins of the E and I series, leukotrienes C_4 and D_4 are vasoconstrictor in the coronary and renal vascular beds and also in the microvasculature. Direct observation of the small blood vessels of the hamster cheek pouch by means of a microscope with attached camera have shown the major vasoconstrictor action of leukotrienes to be on the terminal arterioles. In these experiments the vasoconstriction is short-lived and is followed almost immediately by leaking of fluid out of the blood vessels into the tissues. LtB_4, LtC_4 and LtD_4 similarly increase vascular permeability when injected just under the skin of experimental animals like, for example, the guinea pig. Because leukotrienes are vasoconstrictor this blunts their ability to promote

Table 10.3. *Biosynthesis of LtB_4 in vitro*

Cell	Species	Stimulus
PMNs	Rat	
	Rabbit	Arachidonic acid
	Human	A23187
	Pig	Zymosan
Eosinophils	Human	Arachidonic acid
	Mouse	A23187
	Horse	Complement
T-lymphocytes	Human	A23187
Macrophages	Human	A23187
	Rat	Complement

vascular permeability. Even so LtB_4 causes oedema in these experiments at low doses (less than 1 µg) and is even more potent when combined with a vasodilator like PGE_2 or histamine (Williams & Piper, 1980; Williams, 1983). Quite possibly therefore the synthesis of leukotrienes contributes to oedema production in inflammation.

Effect on leucocytes. LtB_4 is an extremely potent chemokinetic and chemotactic agent for PMNs, eosinophils, monocytes and macrophages harvested from peritoneal fluid and/or blood of rats, mice, guinea pigs, rabbits and humans. The concentration of LtB_4 required for half-maximal chemotactic activity in most cell types is approximately 10^{-10} M which makes it the most active chemotactic agent so far discovered. At higher concentrations *in vitro* (about 10^{-8} M) LtB_4 also causes the release of lysosomal enzymes from neutrophils.

Thus LtB_4 synthesised locally at the site of an inflammatory reaction may function physiologically as a chemoattractant hastening the arrival of leucocytes to the damaged tissues. However, this same physiological mechanism, if it goes uncontrolled, may actually damage the host tissue either because leucocytes phagocytose and destroy perfectly normal host cells or because these leucocytes release lysosomal enzymes under the influence of LtB_4 which digest away the host tissue. In this way LtB_4 may contribute to the pathology of chronic inflammatory diseases like rheumatoid arthritis.

Effect of anti-inflammatory drugs on leukotriene biosynthesis
NSAID like aspirin have little or no lipoxygenase-blocking activity at concentrations which are achieved therapeutically. At first sight these drugs would appear to be of little use for proving or disproving that leukotrienes are involved in inflammation. However, this is not the case. Since leucocytes have both cyclooxygenase and lipoxygenase activities they are able to generate both prostanoids and leukotrienes. Therapeutic doses of NSAID which prevent cyclooxygenase activity would thus be expected to rechannel leucocyte arachidonic acid metabolism down the lipoxygenase pathway to increase formation of leukotrienes (as shown in Fig. 10.3). If leukotrienes are physiologically relevant modulators of leucocyte activity *in vivo* then NSAID should, at the appropriate concentration, increase leucocyte migration and accumulation at the inflamed locus.

Several authors have reported that aspirin and other NSAID may either stimulate or inhibit leucocyte migration *in vitro* depending upon the dose used. At low concentrations which selectively block cyclooxygenase activity, leucocyte migration is normally increased. At very high concentrations which prevent not only cyclooxygenase but also lipoxygenase activity, leucocyte migration is generally inhibited. Clearly results of experiments like these add weight to the possibility that leukotrienes are indeed physiological mediators of acute inflammation.

Aspirin and other NSAID are not ideal anti-inflammatory drugs. Many years of clinical experience with NSAID have revealed that these drugs reduce the symptoms of inflammation but have little effect on the underlying causes of the conditions. In other words NSAID may control the symptoms but do not cure chronic inflammatory conditions. This failure may be because NSAID do not prevent (and indeed they may even increase) leukotriene biosynthesis. Some drugs do prevent both prostanoid and leukotriene biosynthesis. Glucocorticoid steroids, for example, prevent conversion of arachidonic acid to all of its metabolites and clinically are very powerful anti-inflammatory agents which may be used successfully in clinical conditions which are resistant to NSAID therapy. An example is bronchial asthma which is benefited by glucocorticoids but is unaffected by NSAID. Unfortunately glucocorticoids have many harmful side effects which limit their clinical usefulness. A better approach would be the development of drugs which block both cyclooxygenase and lipoxygenase pathways. The major examples of dual blockers are BW755C and benoxaprofen. BW755C blocks anaphylactic contractions of lung tissue *in vitro* and decreases the number of PMNs and monocytes in inflammatory exudates from carrageenan-injected rat paw. Furthermore, BW755C

Fig. 10.3. The major classes of drugs which affect the biosynthesis of inflammatory prostanoids and leukotrienes.

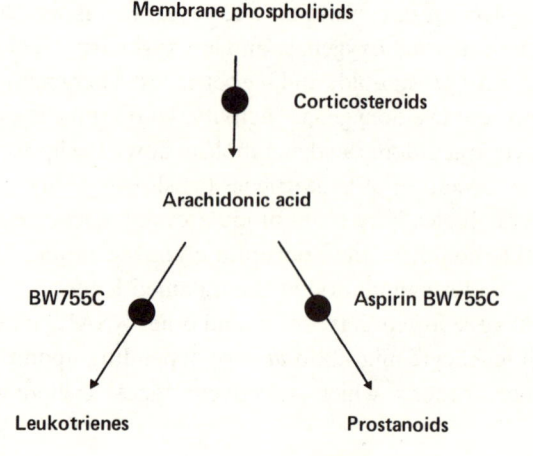

Membrane phospholipids

↓ Corticosteroids

Arachidonic acid

BW755C ↙   ↘ Aspirin BW755C

Leukotrienes   **Prostanoids**

abolishes the indomethacin-induced increase in leucocyte migration *in vivo* which, as already discussed, is probably caused by redirection of arachidonic acid to the lipoxygenase pathway to form LtB_4. The anti-inflammatory activity of such dual-blocking drugs is described in Chapter 2.

Prostanoids and cancer

The evidence that prostanoids have a part to play in the growth and spread of cancer is discussed below. More detailed reviews are provided by Jaffe & Santoro (1977) and by Bennett (1979).

Prostanoid synthesis by tumours

That solid tumours of different types affecting different organs of the body actively synthesise prostanoids is well established. Tumour homogenates and cultured tumour cells contain prostanoids (mostly E series prostaglandins) and convert exogenous arachidonic acid to prostanoids. Table 10.4 shows the range of tumour cells/organs which have been shown to exhibit prostanoid-synthesising capacity. Although some tumours do synthesise small amounts of PGI_2, TxA_2 and PGD_2 the major prostanoids formed are PGE_2 and $PGF_{2\alpha}$.

Some authors have attempted to demonstrate that tumours also synthesise and release prostanoids *in vivo*. This has been achieved by measuring prostanoid levels either in circulating blood or in blood removed from a vein draining the tumour. The results of any experiment of this type should be interpreted cautiously since assay techniques cannot be relied upon to give accurate estimates of prostanoids in blood. However, several studies have revealed the presence of large amounts of PGE_2 and $PGF_{2\alpha}$ in blood draining from medullary thyroid tumours in man (Williams *et al.*, 1968).

Table 10.4. *Tumours which synthesise prostanoids*

Tumour homogenates	Tumour cells in culture
Breast (dimethylbenzanthracene-induced in the rat)	MC5-5 fibroblasts (transformed with methylcholanthrene, mouse)
Lung (mouse)	FS6 fibrosarcoma (mouse)
Thyroid (medullary carcinoma)	HSDM fibrosarcoma (mouse)
Adrenals (phaeochromocytoma)	HeLa (cervical cancer cells, human)
Kidney (adenocarcinoma)	Colon (adenocarcinoma cells, human)
Liver, breast, lung, pancreas, uterus (all human)	

Effect of prostaglandins on cell differentiation

The effect of prostaglandins on cell differentiation and growth is a most controversial area of research. It appears that PGE_2 and $PGF_{2\alpha}$ (other prostanoids have yet to be studied) can either inhibit or facilitate cell replication depending upon the cell system studied. Most studies with malignant or transformed cells in culture suggest that the predominant effect of E series prostaglandins is inhibitory (Bennett, 1979). In accord with this view is the observation of Santoro, Philpott & Jaffe (1976) that in mice daily treatment with the stable PGE_2 analogue, 16,16-dimethyl-PGE_2, delayed the appearance of B16 melanoma tumours induced experimentally and so decreased tumour size and prolonged the animals' survival time. It is probable (although by no means yet proven) that, in those cell lines in which PGEs cause inhibition of cell replication, the final mechanism of action is likely to be by increasing intracellular cAMP formation.

Effect of NSAID on tumour growth

From the above it is not immediately clear whether the prostaglandins which are formed by tumour cells actually restrict or enhance tumour cell growth. The finding that NSAID like indomethacin and flurbiprofen have no effect on tumour growth *in vitro* but are of therapeutic benefit in some experimental cancers in animals (see Table 10.5) would

Table 10.5. *Effects of NSAID on cancer growth*

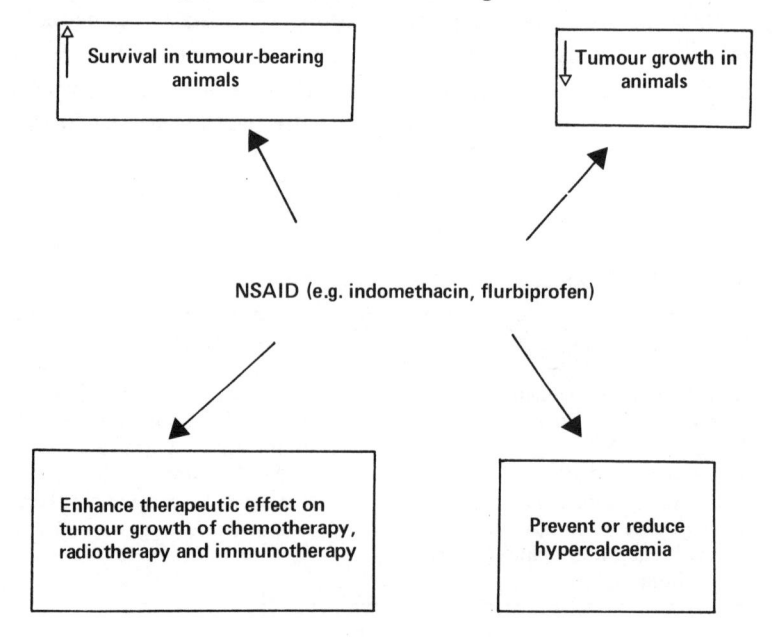

Survival in tumour-bearing animals ↑		Tumour growth in animals ↓
	NSAID (e.g. indomethacin, flurbiprofen)	
Enhance therapeutic effect on tumour growth of chemotherapy, radiotherapy and immunotherapy		Prevent or reduce hypercalcaemia

suggest that prostaglandins formed within the tumour (*in vivo* if not *in vitro*) promote cell growth.

The discrepancy between effects *in vitro* and *in vivo* of NSAID on tumour cell replication may be explained in two ways. Either NSAID block the formation by the tumour of a prostanoid which stimulates cell growth (perhaps PG endoperoxides, TxA_2 or PGI_2, which have not yet been studied) or these drugs increase the formation of lipoxygenase products of arachidonic acid *in vivo* (perhaps leukotrienes) which inhibit cell growth. Clearly, much more research is necessary before firm conclusions can be made and before NSAID (or prostaglandin analogues) can be introduced as alternative or additional anti-cancer therapy in man.

References

Abdel-Halim, M. S., Ekstedt, J. & Anggard, E. (1979). Determination of prostaglandins $F_{2\alpha}$, E_2, D_2 and 6-keto-$PGF_{1\alpha}$ in human cerebrospinal fluid. *Prostaglandins*, **17**, 405–9.

Abdel-Halim, M. S., Hamberg, M., Sjöqvist, B. & Anggard, E. (1977). Identification of prostaglandin D_2 as a major prostaglandin in homogenates of rat brain. *Prostaglandins*, **14**, 633–43.

Advenier, C., Bidet, D., Floch-Saint-Aubin, A. & Reinier, A. (1982). Contribution of prostaglandins and thromboxanes to the adenosine and ATP-induced contraction of guinea pig isolated atria. *British Journal of Pharmacology*, **77**, 39–44.

Afonso, S., Barchow, G. T. & Rouse, G. G. (1974). Indomethacin and the prostaglandin hypothesis of coronary blood flow regulation. *Journal of Physiology*, **241**, 299–308.

Ahlqvist, D. A., Duenes, J. A., Madson, T. H., Romero, J. C., Dozois, R. R. & Malagelada, J.-R. (1982). Postaglandin generation from gastroduodenal mucosa: regional and species differences. *Prostaglandins*, **24**, 115–25.

Alabaster, V. (1980). Metabolism of arachidonic acid and its endoperoxide (PGH_2) to myotropic products in guinea pig and rabbit isolated lungs. *British Journal of Pharmacology*, **69**, 479–89.

Ally, A. I., Boucher, R., Knowles, M. R. & Eling, T. E. (1982). Metabolism of prostaglandin endoperoxides by microsomes from human lung parenchyma and comparison with metabolites produced by pig, bovine, rat, mouse and guinea pig. *Prostaglandins*, **24**, 575–84.

Altsheler, P., Klahr, S., Rosenbaum, R. & Slatopolsky, E. (1978). Effects of inhibitors of prostaglandin synthesis on renal sodium excretion in normal dogs and dogs with decreased renal mass. *American Journal of Physiology*, **235**, F338–44.

Anderson, A., Haynes, P. J., Guilleband, J. & Turnbull, A. C. (1976). Reduction of menstrual blood loss by prostaglandin synthetase inhibitors. *Lancet*, **i**, 774–5.

Anggard, E., Bohman, S. O., Griffin, S. E., Larsson, C. & Maunsbach, A. B. (1972). Subcellular localisation of the prostaglandin system in the rabbit renal papilla. *Acta physiologica scandinavica*, **84**, 231–46.

Anggard, E. & Samuelsson, B. (1965). Biosynthesis of prostaglandins from arachidonic acid in guinea pig lung. *Journal of Biological Chemistry*, **240**, 3518–29.

Anturane Reinfarction Trial Research Group (1978). Sulphinpyrazone in the prevention of sudden death after myocardial infarction. *New England Journal of Medicine*, **302**, 250–6.

Armstrong, D. T. & Grimwich, D. I. (1972). Blockade of spontaneous and LH induced ovulation in rats by indomethacin, an inhibitor of prostaglandin biosynthesis. *Prostaglandins*, **1**, 21–8.

Armstrong, H. M., Blackwell, G. J., Flower, R. J., McGiff, J. C., Mullane, K. M. & Vane, J. R. (1976). Genetic hypertension in rats is accompanied by a defect in renal prostaglandin catabolism. *Nature*, **260**, 582–6.

Ashida, S. I. & Abiko, J. (1978). Effect of ticlopidine and acetylsalicylic acid on generation of prostaglandin I_2-like substance in rat arterial tissue. *Thrombosis Research*, **13**, 901–8.

Ashida, Y., Saijo, T., Kuriki, H., Makino, H., Terao, S. & Maki, Y. (1983). Pharmacological profile of AA-861, a 5'-lipoxygenase inhibitor. *Prostaglandins*, **26**, 955–72.

Aspirin Myocardial Infarction Study Research Group (AMIS) (1980). A randomised controlled trial of aspirin in persons recovered from myocardial infarction. *Journal of the American Medical Association*, **243**, 661–9.

Au, T. L. S., Collins, G. A., Harvie, C. J. & Walker, M. J. A. (1979). The actions of prostaglandin I_2 and prostaglandin E_2 on arrhythmias produced by coronary occlusion in the rat and dog. *Prostaglandins*, **18**, 707–20.

Barrett, S., Blockey, M. A. de B., Brown, J. M., Cumming, I. A., Goding, J. R. & Mole, B. J. (1971). Initiation of the oestrus cycle in the ewe by infusions of $PGF_{2\alpha}$ to the autotransplanted ovary. *Journal of Reproduction and Fertility*, **24**, 136–7.

Barrowman, J. A., Bennett, A., Hillenbrand, P., Rolles, K., Pollock, D. J. & Wright, J. T. (1975). Diarrhoea in medullary carcinoma of the thyroid: evidence for the role of prostaglandins and the therapeutic effect of nutmeg. *British Medical Journal*, **3**, 11–12.

Barrter, F. C., Pronove, P. & Gill, R. (1962). Hyperplasia of the juxtaglomerular complex with hyperaldosteronism and hypokalaemic alkalosis. *American Journal of Medicine*, **33**, 811–28.

Belch, J. J. F., Newman, P., Drury, J. K., McKenzie, F., Capell, P. H., Lieberman, P., Forbes, C. D. & Prentice, C. R. M. (1983). Intermittent epoprostenol (prostacyclin) infusion in patients with Raynaud's syndrome. *Lancet*, **i**, 313–15.

Bennett, A. (1971). Cholera and prostaglandins. *Nature*, **231**, 536.

Bennett, A. (1979). Prostaglandins and cancer. In *Practical Applications of Prostaglandins and their Synthesis Inhibitors*, ed. S. M. M. Karim, pp. 149–88. Lancaster: MTP Press.

Bennett, A. & del Tacca, P. (1975). Prostaglandins in human colonic carcinoma. *Gut*, **16**, 409.

Bennett, A., Stamford, I. F. & Unger, W. G. (1973). Prostaglandin E_2 and gastric acid secretion in man. *Journal of Physiology*, **229**, 349–60.

Berhman, H. R., Grimwich, D. I., Hichens, M. & MacDonald, G. J. (1978). Effect of hypophysectomy, prolactin and prostaglandin $F_{2\alpha}$ on gonadotrophin binding *in vivo* and *in vitro* in the corpus luteum. *Endocrinology*, **103**, 349–57.

Bergström, S., Carlsson, L. A. & Oro, L. (1964). Effect of prostaglandins on catecholamine-induced changes in the free fatty acids of plasma and on blood pressure in the dog. *Acta physiologica scandinavica*, **60**, 170–80.

Bergström, S., Dressler, F., Krabisch, L., Ryhage, R. & Sjövall, T. (1962). The isolation and structure of a smooth muscle stimulating factor in normal sheep and pig lungs. *Arkives für Kemistrie*, **20**, 63–6.

Bergström, S. & Sjövall, T. (1957). The isolation of prostaglandin. *Acta chemica scandinavica*, **11**, 1086–7.

Bergström, S. & Sjövall, T. (1960). The isolation of prostaglandin E from sheep prostate glands. *Acta chemica scandinavica*, **14**, 1701–80.

Berry, E. M., Edmonds, J. F. & Wyllie, J. H. (1971). Release of prostaglandin E_2 and unidentified factors from ventilated lungs. *British Journal of Surgery*, **58**, 189–92.

Berry, P. A. & Collier, H. O. J. (1964). Bronchoconstrictor action for antagonism of a slow reacting substance from anaphylaxis of guinea pig isolated lung. *British Journal of Pharmacology*, **23**, 201–16.

Bertele, V., Cerletti, C., Schiepatti, A., DiMinno, G. & de Gaetano, G. (1981). Inhibition of thromboxane synthetase does not necessarily prevent platelet aggregation. *Lancet*, **i**, 1057–8.

Bito, L. Z., Baroody, R. A. & Reitz, M. E. (1977). Dependence of pulmonary prostaglandin metabolism on carrier mediated transport processes. *American Journal of Physiology*, **232**, E382–7.

Bito, L. Z., Davson, H. & Hollingsworth, J. R. (1976). Facilitated transport of prostaglandins across the blood–cerebrospinal fluid and blood–brain barrier. *Journal of Physiology*, **253**, 273–85.

Blackwell, G. J. & Flower, R. J. (1983). Inhibition of phospholipase. *British Medical Bulletin*, **39**, 260–4.

Blackwell, G. J., Flower, R. J. & Vane, J. R. (1975a). Some characteristics of the prostaglandin synthesising system in rabbit kidney microsomes. *Biochimica et biophysica acta*, **398**, 178–90.

Blackwell, G. J., Flower, R. J. & Vane, J. R. (1975b). Rapid reduction of prostaglandin 15-hydroxydehydrogenase in rat tissues after treatment with protein synthesis inhibitors. *British Journal of Pharmacology*, **55**, 233–8.

Blackwell, G. J., Flower, R. J., Nijkamp, F. P. & Vane, J. R. (1978). Phospholipase A_2 activity of guinea pig isolated perfused lungs: stimulation and inhibition by anti-inflammatory steroids. *British Journal of Pharmacology*, **62**, 79–89.

Blasingham, M. C. & Nasjletti, A. (1980). Differential renal effects of cyclooxygenase inhibition in sodium replete and sodium deprived dog. *American Journal of Physiology*, **239**, F360–5.

Blasingham, M. C., Shade, R. E. & Share, L. (1980). The effect of meclofenamate on renal blood flow in the unanaesthetised dog: relation to renal prostaglandins and sodium balance. *Journal of Pharmacology and Experimental Therapeutics*, **214**, 1–4.

Bolger, P. M., Eisner, G. M. & Ramwell, P. W. (1978). Renal actions of prostacyclin. *Nature*, **271**, 467–9.

Bolton, T. (1979). Mechanism of action of transmitters and other substances on smooth muscle. *Physiological Reviews*, **59**, 606–718.

Bonne, C., Martin, B., Watado, M. & Regnault, F. (1981). The antagonism of prostaglandins I_2, E_1 and D_2 by prostaglandin E_2 in human platelets. *Thrombosis Research*, **21**, 13–22.

Borgeat, P., Hamberg, M. & Samuelsson, B. (1976). Transformation of arachidonic acid and dihomo-γ-linolenic acid by rabbit PMN leucocytes. *Journal of Biological Chemistry*, **251**, 7816–20.

Borgeat, P. & Samuelsson, B. (1979). Transformation of arachidonic acid by rabbit polymorphonuclear leucocytes: formation of a novel dihydroxyeicosatetraenoic acid. *Journal of Biological Chemistry*, **254**, 2643–6.

Botting, J. H. & Salzman, R. (1974). The effect of indomethacin on the release of prostaglandin E_2 and acetylcholine from guinea pig isolated ileum at rest and during field stimulation. *British Journal of Pharmacology*, **50**, 119–24.

Boyer, T. D., Zia, P. & Reynolds, T. B. (1979). Effect of indomethacin and prostaglandin A_1 on renal function and plasma renin activity in alcoholic liver disease. *Gastroenterology*, **72**, 215–22.
Brater, D. C. (1979). Effect of indomethacin on salt and water haemostasis. *Clinical Pharmacology and Therapeutics*, **25**, 322–30.
Breddin, K., Loew, D., Lechner, K., Uberla, K. & Walter, E. (1979). Secondary prevention of myocardial infarction: comparison of acetylsalicylic acid, phenprocoumon and placebo. A multicentre two year prospective study. *Thrombosis and Haemostasis*, **40**, 225–36.
Brodie, M. J., Hensby, C. N., Parke, A. & Gordon, D. (1980). Is prostacyclin the major pro-inflammatory prostanoid in joint fluid? *Life Sciences*, **27**, 603–8.
Bunting, S., Gryglewski, R., Moncada, S. & Vane, J. R. (1976). Arterial walls generate from prostaglandin endoperoxides a substance (prostaglandin X) which relaxes strips of mesenteric and coeliac arteries and inhibits platelet aggregation. *Prostaglandins*, **12**, 897–913.
Burka, J. P. & Eyre, P. (1974). The immunological release of slow reacting substance of anaphylaxis from bovine lung. *Canadian Journal of Physiology and Pharmacology*, **52**, 942–51.
Burnstock, G., Cockes, T., Paddle, B. & Staszowska-Barczak, J. (1975). Evidence that prostaglandin is responsible for the rebound contraction following stimulation of non-adrenergic, non-cholinergic (purinergic) inhibitory nerves. *European Journal of Pharmacology*, **31**, 360–2.
Bygdeman, M. (1979). Menstrual regulation with prostaglandins. In *Practical Applications of Prostaglandins and their Synthesis Inhibitors*, ed. S. M. M. Karim, pp. 267–82. Lancaster: MTP Press.
Canadian Co-operative Study Group (1978). A randomised trial of aspirin and sulphinpyrazone in threatened stroke. *New England Journal of Medicine*, **299**, 53–9.
Caton, M. (1972). Chemistry, structure and availability. In *The Prostaglandins: Pharmacological and Therapeutic Advances*, ed. M. F. Cuthbert, pp. 1–22. London: Heinemann Press.
Cazenave, J. P., Dejana, E., Kinlough-Rathbone, R. L., Richardson, M., Packham, M. A. & Mustard, J. F. (1979). Prostaglandins I_2 and E_1 reduce rabbit and human platelet adherence without inhibiting serotonin release from adherent platelets. *Thrombosis Research*, **15**, 273–9.
Chang, W. C., Murota, S-I. & Tsurufuji, S. (1977). Thromboxane B_2 transformed from arachidonic acid in carregeenan-induced granuloma. *Prostaglandins*, **13**, 17–24.
Chang, M. C., Saksena, S. K. & Hunt, D. M. (1974). Effect of prostaglandin $F_{2\alpha}$ on ovulation and fertilisation in the rabbit. *Prostaglandins*, **5**, 341–8.
Chapple, D. J., Dusting, G. J., Hughes, R. & Vane, J. R. (1978). A vagal reflex contributes to the hypotensive effect of prostacyclin in anaesthetised dogs. *Journal of Physiology*, **281**, 43P.
Chaudhury, M. M. & Jacobsen, E. D. (1978). Prostaglandin cytoprotection of gastric mucosa. *Gastroenterology*, **74**, 58–63.
Christ, E. J. & van Dorp, D. A. (1972). Comparative aspects of prostaglandin biosynthesis in animal tissues. *Biochimica et biophysica acta*, **270**, 537–45.
Ciabottini, G., Pugliesi, F. & Cinotti, G. A. (1980). Furosemide and prostacyclin release in man. *Clinical Research*, **28**, 442.
Claesson, H. E. & Malmsten, C. (1977). On the interrelationship between prostaglandin

G_2 and cyclic nucleotides in platelet function. *European Journal of Biochemistry*, **76**, 277–84.

Coceani, F. (1974). Prostaglandins and the central nervous system. *Annals of International Medicine*, **133**, 119–29.

Coceani, F., Dreifuss, J. J., Puglisi, L. & Wolfe, L. S. (1969). Prostaglandins and membrane function, In *Prostaglandins, Peptides and Amines*, eds P. Mantegazza & E. W. Horton, pp. 73–81. London: Academic Press.

Cohen, M. M., Cheung, G. & Lister, D. N. (1980). Prevention of aspirin induced foecal blood loss by prostaglandin E_2. *Gut*, **21**, 602–6.

Coker, S. J., Hughes, B., Parratt, J. R., Rodger, I. W. & Zeitlin, I. J. (1982). The release of prostanoids during the acute pulmonary response to *E. Coli* endotoxin in anaesthetised cats. *British Journal of Pharmacology*, **78**, 561–70.

Coker, S. J. & Parratt, J. R. (1981). The effects of prostaglandins E_2, $F_{2\alpha}$, prostacyclin, flurbiprofen and aspirin on arrhythmias resulting from coronary artery ligation in anaesthetised rats. *British Journal of Pharmacology*, **74**, 155–9.

Collier, H. O. J. (1969). A pharmacological analysis of aspirin. *Advances in Pharmacology and Chemotherapy*, **7**, 333–405.

Collier, H. O. J. (1984). The aspirin story. In *Discoveries in Pharmacology*, vol. 2, ed. M. J. Parnham & J. Bruinvels. Holland: Elsevier Press.

Collier, H. O. J. & Schneider, C. (1972). Nociceptive response to prostaglandins and analgesic action of aspirin and morphine. *Nature*, **236**, 141–3.

Collier, H. O. J. & Sweatman, P. J. (1968). Antagonism by fenamates of prostaglandin $F_{2\alpha}$ and slow reacting substance on human bronchial muscle. *Nature*, **219**, 864–5.

Cooper, M. J., Hammond, D. & Schulz, R. H. (1979). Veterinary uses of prostaglandins. In *Practical Application of Prostaglandins and their Synthesis Inhibitors*, ed. S. M. M. Karim, pp. 189–216. Lancaster: MTP Press.

Coronary Drug Project Research Group (1976). Aspirin in coronary heart disease. *Journal of Chronic Diseases*, **29**, 625–42.

Cranston, W. I., Hellon, R. F. & Mitchell, D. (1975). A dissociation between fever and prostaglandin concentration in cerebrospinal fluid. *Journal of Physiology*, **253**, 583–8.

Crutchley, D. J. & Piper, P. J. (1975). Comparative bioassay of prostaglandin E_2 and its pulmonary metabolites. *British Journal of Pharmacology*, **54**, 397–9.

Cuthbert, M. F. & Gardiner, P. J. (1981). Actions of prostaglandins and other arachidonic acid metabolites on airway function. In *The Prostaglandin System: Endoperoxides, Prostacyclin and Thromboxanes*, eds F. Berti & G. P. Velo, pp. 323–42. New York: Plenum Press.

Dahlen, S. E. (1983). Pulmonary effects of leukotrienes. *Acta physiologica scandinavica*, supplement 512.

Davenport, H. W., Warner, H. A. & Code, C. F. (1964). Functional significance of gastric mucosal barrier to sodium. *Gastroenterology*, **47**, 142–52.

Dawson, W., Boot, J. R., Cockerill, A. F., Mallen, D. & Osborne, D. J. (1976). Release of novel prostaglandins and thromboxanes after immunological challenge of guinea pig lung. *Nature*, **262**, 699–702.

Deckmyn, H., Font, L., van Hemelen, C., Carreras, L. O., Defrayn, G. & Vermylen, J. (1983). Low prostacyclin synthetase activity of fetal rat aorta: progressive increase after birth. *Life Sciences*, **33**, 117–19.

DeDeckere, E. A. M., Nugteren, D. H. & Ten Hoor, F. (1977). Prostacyclin is the major prostaglandin released from the isolated perfused rabbit and rat heart. *Nature*, **268**, 160–3.

de Gaetano, G. (1981). Platelets, prostaglandins and thrombotic disorders. In *Clinics in Haematology*, vol. 10, ed. C. R. M. Prentice, pp. 297–326. London: Saunders.
DeWitt, D. L. & Smith, W. L. (1982). Monoclonal antibodies against PGI_2 synthase: an immunoradiometric assay for quantitating the enzyme. *Methods in Enzymology*, **86**, 240–5.
Dickmann, Z. & Spilman, C. H. (1975). Prostaglandins in rabbit blastocysts. *Science*, **190**, 997–8.
DiMinno, G., Silver, M. J. & de Gaetano, G. (1979). Prostaglandins as inhibitors of human platelet aggregation. *British Journal of Haematology*, **43**, 637–47.
DiRosa, M. & Persico, P. (1979a). Latency and reversion of hydrocortisone inhibition of prostaglandin biosynthesis in rat leucocytes. *Agents and Actions*, supplement 4, 63–8.
DiRosa, M. & Persico, P. (1979b). Mechanism of inhibition of prostaglandin biosynthesis by hydrocortisone in rat leucocytes. *British Journal of Pharmacology*, **66**, 161–3.
Downie, J., Poyser, N. L. & Wunderlich, M. (1974). Level of prostaglandins in human endometrium during the normal menstrual cycle. *Journal of Physiology*, **236**, 456–72.
Drouin, J., Ferland, L., Bernard, J. & Labrie, F. (1976). Site of the *in vivo* stimulating effect of prostaglandins on LH release. *Prostaglandins*, **11**, 367–76.
Dubocovich, M. L. & Langer, S. Z. (1975). Evidence against a physiological role of prostaglandins in the regulation of noradrenaline release in the cat spleen. *Journal of Physiology*, **251**, 737–62.
Dunn, M. J. (1976). Renal prostaglandin synthesis in the spontaneously hypertensive rat. In *Spontaneous Hypertension: Its Pathogenesis and Complications*. Proceedings of an International Symposium on the SHR, pp. 359–67. National Institute of Health, USA, Publication No. 77–1179.
Dunn, M. J., Beck, T. R., Kintner, L. B. & Hassid, A. (1983). The effects of vasopressin and vasopressin analogues upon renal synthesis of prostaglandins. In *Prostaglandins and the Kidney*, ed. M. J. Dunn, C. Patrono & G. A. Cinotti, pp. 151–66. New York: Plenum Press.
Dunn, M. J., Staley, R. S. & Harrison, M. (1976). Characterisation of prostaglandin production in tissue culture of rat renal medullary cells. *Prostaglandins*, **12**, 37–49.
Dusting, G. J., Chapple, D. J., Hughes, R., Moncada, S. & Vane, J. R. (1978). Prostacyclin induced coronary vasodilatation in anaesthetised dogs. *Cardiovascular Research*, **12**, 720–30.
Dusting, G. J., Moncada, S. & Vane, J. R. (1977). Prostacyclin (PGX) is the endogenous metabolite responsible for relaxation of coronary arteries induced by arachidonic acid. *Prostaglandins*, **13**, 3–15.
Dyerberg, J., Bang, H. O., Stofferson, E., Moncada, S. & Vane, J. R. (1978). Eicosapentaenoic acid and prevention of thrombosis and atherosclerosis. *Lancet*, **ii**, 117–19.
Eckenfells, A. & Vane, J. R. (1972). Prostaglandins, oxygen tension and smooth muscle tone. *British Journal of Physiology*, **45**, 451–62.
Egg, D., Herold, M., Rumpl, E. & Gunther, R. (1980). Prostaglandin $F_{2\alpha}$ levels in human cerebrospinal fluid in normal and pathological conditions. *Journal of Neurology*, **222**, 239–48.
Ehrenpreis, S., Greenberg, J. & Belman, S. (1973). Prostaglandins reverse inhibition of electrically-induced contractions of guinea pig ileum by morphine, indomethacin and acetylsalicylic acid. *Nature*, **245**, 280–2.
Eisen, V. & Walker, D. I. (1976). Effect of ionising radiation on prostaglandin-like activity in tissues. *British Journal of Pharmacology*, **57**, 527–32.

Elwood, P. C., Cochrane, A. L., Burr, M. L., Sweetnam, P. M., Williams, G., Welsby, E., Hughes, S. J. & Renton, R. (1974). A randomised controlled trial of acetylsalicylic acid in the secondary prevention of mortality from myocardial infarction. *British Medical Journal*, **i**, 436–40.

Elwood, P. C. & Sweetnam, P. M. (1979). Aspirin and secondary mortality after myocardial infarction. *Lancet*, **ii**, 1313–15.

Eskay, R. L., Warberg, J., Mical, R. S. & Porter, J. C. (1975). Prostaglandin E_2 induced release of LHRH into hypophysial portal blood. *Endocrinology*, **97**, 816–24.

Feldberg, W. & Gupta, K. P. (1973). Pyrogen fever and prostaglandin-like activity in cerebrospinal fluid. *Journal of Physiology*, **228**, 41–53.

Feldberg, W. & Kellaway, C. H. (1938). Liberation of histamine and formation of lysolecithin-like substances by cobra venom. *Journal of Physiology*, **94**, 187–226.

Ferguson, W. W., Edmonds, A. W. & Starling, J. R. (1973). Protective effect of prostaglandin E_1 (PGE_1) on lysosomal enzyme release in serotonin induced gastric ulceration. *Annals of Surgery*, **177**, 648–53.

Ferreira, S. H., Moncada, S., & Vane, J. R. (1971). Indomethacin and aspirin abolish prostaglandin release from the spleen. *Nature*, **231**, 237–9.

Ferreira, S. H., Nakamura, M. & Abreu Castro, M. S. (1978). The hyperalgesic effects of prostacyclin and PGE_2. *Prostaglandins*, **16**, 31–8.

Ferreira, S. H. & Vane, J. R. (1974). New aspects of the mode of action of nonsteroid anti-inflammatory drugs. *Annual Review of Pharmacology*, **14**, 57–73.

Ferreira, S. H. & Vane, J. R. (1978). Mode of action of anti-inflammatory agents which are prostaglandin synthetase inhibitors. In *Anti-Inflammatory Drugs – Handbook of Experimental Pharmacology*, vol. 50/11, eds J. R. Vane & S. H. Ferreira, pp. 348–98. Berlin: Springer-Verlag.

Feuerstein, N. & Ramwell, P. W. (1981). *In vivo* and *in vitro* effects of endotoxin on prostaglandin release from rat lung. *British Journal of Pharmacology*, **73**, 511–16.

Finck, D. & Katz, P. L. (1972). Prevention of cholera-induced intestinal secretion in the cat by aspirin. *Nature*, **238**, 273–4.

Fitzgerald, G. A., Friedman, L. A., Miyamori, I., O'Grady, J. & Lewis, P. J. (1979). A double blind placebo-controlled crossover study of prostacyclin in man. *Life Sciences*, **25**, 665–72.

Fitzpatrick, P. A., Bundy, G. L., Gorman, R. R. & Honohan, T. (1978). 9,11-epoxyiminoprosta-5,13-dienoic acid is a thromboxane A_2 antagonist in human platelets. *Nature*, **275**, 764–6.

Flamembaum, W. & Kleinman, J. G. (1977). Prostaglandins and renal function, or 'A trip down the rabbit hole'. In *The Prostaglandins*, vol. 3, ed. P. W. Ramwell, pp. 267–328. New York: Plenum Press.

Flemström, G. & Garner, A. (1982). Gastroduodenal HCO_3^- transport: characteristics and proposed role in acidity regulation and mucosal protection. *American Journal of Physiology*, **242**, G183–93.

Flower, R. J. (1974). Drugs which inhibit prostaglandin biosynthesis. *Pharmacological Reviews*, **26**, 33–67.

Flower, R. J. & Blackwell, G. J. (1979). Anti-inflammatory steroids induce biosynthesis of a phospholipase A_2 inhibitor which prevents prostaglandin biosynthesis. *Nature*, **278**, 456–9.

Flower, R. J., Gryglewski, R., Herbaczynska-Cedro, K. & Vane, J. R. (1972). Effects of anti-inflammatory drugs on prostaglandin biosynthesis. *Nature*, **238**, 104–6.

Folco, G. C., Omini, C., Sautebin, L. & Berti, F. (1977). Arachidonic acid metabolites in the lung: effect of histamine and slow reacting substance of anaphylaxis. In *The Prostaglandin System: Endoperoxides, Prostacyclin and Thromboxanes*, eds F. Berti & P. Velo, pp. 343–64. New York: Plenum Press.

Forstermann, U., Herrting, G. & Neufang, B. (1984). The importance of endogenous prostaglandins other than prostacyclin for the modulation of contractility of some rabbit blood vessels. *British Journal of Pharmacology*, **81**, 623–30.

Frolich, J. C., Wilson, T. W. & Sweetman, B. J. (1975). Urinary prostaglandins: identification and origin. *Journal of Clinical Investigation*, **55**, 763–90.

Fulgraff, G. & Brandenbusch, G. (1974). Comparison of the effects of the prostaglandins A_1, E_2 and $F_{2\alpha}$ on kidney function in dogs. *Pflügers Archiv*, **349**, 9–17.

Fulgraff, G. & Meiforth, A. (1971). Effects of prostaglandin E_2 on excretion and reabsorption of sodium and fluid in rat kidneys (micropuncture studies). *Pflügers Archiv*, **330**, 243–56.

Ganguli, M., Tobian, L. & Azar, S. (1977). Evidence that prostaglandin synthesis inhibitors increase the concentration of sodium and chloride in rat renal medulla. *Circulation Research*, supplement 1, I135–9.

Gardiner, P. J. (1975). The effect of some natural prostaglandins on isolated human bronchial muscle. *Prostaglandins*, **10**, 607–16.

Gardiner, P. J. & Collier, H. O. J. (1980). Specific receptors for prostaglandins in airways. *Prostaglandins*, **19**, 819–41.

Garner, A. & Heylings, J. R. (1979). Stimulation of alkaline secretion in amphibian isolated gastric mucosa by 16,16 dimethyl PGE_2 and $PGF_{2\alpha}$. A proposed explanation for some of the cytoprotective actions of prostaglandins. *Gastroenterology*, **76**, 497–503.

Gates, W., Grimes, D. A., Haber, R. J. & Tyler, C. W. (1977). Abortion deaths associated with the use of prostaglandin $F_{2\alpha}$. *American Journal of Obstetrics and Gynaecology*, **127**, 219–25.

Gemsa, D., Barlin, E., Leser, H.-G., Deimann, W. & Seitz, M. quoted in Lewis, G. P. (1983). Immunoregulatory activity of metabolites of arachidonic acid and their role in inflammation. *British Medical Bulletin*, **39**, 243–8.

Gerber, J. G., Ellis, E. F. & Nies, A. S. (1977). The effect of a prostaglandin endoperoxide analogue on renal function and haemodynamics. *Federation Proceedings*, **36**, 402.

Gilbert, D. A., Field, A. D., Silverstein, F. E., Weinberg, C. & Saunders, D. R. (1981). 15-R-15 methyl prostaglandin E_2 cytoprotection in aspirin-induced gastric mucosal injury – an endoscopic study. *Gastroenterology*, **80**, 1155.

Gillis, C. N. & Roth, J. A. (1976). Pulmonary disposition of circulating vasoactive hormones. *Biochemical Pharmacology*, **25**, 2547–53.

Gilmore, N., Vane, J. R. & Wyllie, M. G. (1968). Prostaglandins released by the spleen. *Nature*, **218**, 1135–40.

Gimson, A. E. S., Hughes, R. D., Mellon, P. J., Woods, H. F., Langley, P. G., Canalese, J., Williams, R. & Weston, M. J. (1980). Prostacyclin to prevent platelet aggregation during charcoal haemoperfusion in fulminant hepatic failure. *Lancet*, **i**, 173–5.

Glatt, M., Kalin, H., Wagner, K. & Brune, B. (1977). Prostaglandin release from macrophages: an assay system for anti-inflammatory drugs *in vitro*. *Agents and Actions*, **7**, 321–6.

Glazer, G. & Bennett, A. (1974). Elevation of prostaglandin-like activity in the blood and peritoneal exudate of dogs with acute pancreatitis. *British Journal of Surgery*, **61**, 922–4.

Goehlert, U. G., Ng Ying Kin, N. M. K. & Wolfe, L. S. (1981). Biosynthesis of prostacyclin in rat cerebral microvessels and the choroid plexus. *Journal of Neurochemistry*, **36**, 1192–201.

Goldblatt, M. W. (1933). A depressor substance in seminal fluid. *Journal of the Society for Chemistry and Industry*, **52**, 1056–7.

Gordon, D., Bray, M. A. & Morley, J. (1976). Control of lymphokine secretion by prostaglandins. *Nature*, **262**, 401–2.

Gorman, R. R. (1982). Mechanism of action of prostacyclin and thromboxane A_2. In *Prostaglandins in Clinical Medicine*, eds. K. K. Wu & E. C. Rossi, pp. 21–34. Chicago: Year Book Medical Publishers.

Gorman, R. R., Bundy, G. L., Peterson, D. C., Sun, F. F., Miller, O. V. & Fitzpatrick, F. A. (1977a). Inhibition of human platelet thromboxane synthetase by 9,11-azoprost-5,13-dienoic acid. *Proceedings of the National Academy of Sciences, U.S.A.*, **74**, 4007–11.

Gorman, R. R., Bunting, S. & Moncada, S. (1977b). Modulation of human platelet adenylate cyclase by prostacyclin (PGX). *Prostaglandins*, **13**, 377–88.

Granström, E. (1981). Metabolism of prostaglandins and thromboxanes. In *The Prostaglandin System: Endoperoxides, Prostacyclin and Thromboxanes*, eds F. Berti, G. P. Velo, pp. 39–48. New York: Plenum Press.

Granström, E. & Kindahl, H. (1978). Radioimmunoassay of prostaglandins and thromboxanes. *Advances in Prostaglandin and Thromboxane Research*, **5**, 119–32.

Greenberg, R., Osman, G. H. Jnr, O'Keefe, E. H. & Antonaccio, M. J. (1979). The effects of captopril (SQ 14,225) on bradykinin-induced bronchoconstriction in the anaesthetised guinea pig. *European Journal of Pharmacology*, **57**, 287–94.

Grenier, F. C., Rollins, T. E. & Smith, W. L. (1981). Bradykinin-induced prostaglandin biosynthesis by renal papillary collecting tubule cells in culture. *American Journal of Physiology*, **241**, F94–104.

Grenier, F. C. & Smith, W. L. (1978). Formation of 6-keto $PGF_{1\alpha}$ by collecting tubule cells isolated from rabbit renal papilla. *Prostaglandins*, **16**, 759–72.

Griffiths, R. J. & Moore, P. K. (1983). Effect of 6-keto prostaglandin E_1 on sympathetic neurotransmission in the vas deferens. *Journal of Pharmacy and Pharmacology*, **35**, 184–6.

Gross, J. B. & Barrter, F. C. (1973). Effects of prostaglandins E_1, A_1 and $F_{2\alpha}$ on renal handling of salt and water. *American Journal of Physiology*, **225**, 218–24.

Gruber, U. F., Fuser, P., Frick, J., Loosli, J., Matt, E. & Segesser, D. (1977). Sulphinpyrazone and post-operative deep vein thrombosis. *European Surgery Research*, **9**, 303–10.

Gryglewski, R., Panczenko, B., Korbut, R., Grodzinska, L. & Ocetkiewicz, A. (1975). Corticosteroids inhibit prostaglandin release from perfused mesenteric blood vessels of rabbit and from perfused lungs of sensitised guinea pigs. *Prostaglandins*, **10**, 343–55.

Gryglewski, R., Salmon, J. A., Ubatuba, F. B., Weatherley, B. C., Moncada, S. & Vane, J. R. (1979). Effects of all-cis-5,8,11,14,17 eicosapentaenoic acid and PGH_3 on platelet aggregation. *Prostaglandins*, **18**, 453–78.

Gustafsson, L., Hedqvist, P. & Lundgren, G. (1980). Pre- and post-junctional effects of prostaglandin E_2, prostaglandin synthetase inhibitors and atropine on cholinergic transmission in guinea pig ileum and bovine iris. *Acta physiologica scandinavica*, **10**, 401–11.

Gutknecht, G. D., Duncan, G. W. & Wyngarden, L. J. (1972). Inhibition of prostaglandin

$F_{2\alpha}$ or LH induced luteolysis in the pseudopregnant rabbit by 17β estradiol. *Proceedings of the Society for Experimental Biology and Medicine*, **139**, 406–19.

Ham, E. A., Cirillo, V. J., Zanetti, M. E., Shen, T. Y. & Kuehl, F. A. Jnr (1972). Studies on the mode of action of anti-inflammatory drugs. In *Prostaglandins in Cellular Biology*, eds P. W. Ramwell & B. B. Pharriss, pp. 345–52. New York: Plenum Press.

Hamberg, M. (1972). Inhibition of prostaglandin synthesis in man. *Biochemical and Biophysical Research Communications*, **49**, 720–6.

Hamberg, M. & Fredholm, B. B. (1976). Isomerisation of prostaglandin H_2 to prostaglandin D_2 in the presence of serum albumen. *Biochimica et biophysica acta*, **431**, 189–93.

Hamberg, M. & Samuelsson, B. (1971). On the metabolism of prostaglandins E_1 and E_2 in man. *Journal of Biological Chemistry*, **246**, 6713–21.

Hamberg, M. & Samuelsson, B. (1973). Detection and isolation of an endoperoxide intermediate in prostaglandin biosynthesis. *Proceedings of the National Academy of Sciences, U.S.A.*, **70**, 899–903.

Hamberg, M., Svensson, J. & Samuelsson, B. (1975). Thromboxanes: a new group of biologically active compounds derived from prostaglandin endoperoxides. *Proceedings of the National Academy of Sciences, U.S.A.*, **72**, 2994–8.

Hamberg, M., Svensson, J., Wakabayashi, T. & Samuelsson, B. (1974). Isolation and structure of two prostaglandin endoperoxides that cause platelet aggregation. *Proceedings of the National Academy of Sciences, U.S.A.*, **71**, 345–9.

Hamberg, M. & Wilson, M. (1973). Structures of new metabolites of prostaglandin E_2 in man. *Advances in the Biosciences*, **9**, 39–51.

Hanley, S. P., Cockbill, S. R., Bevan, J. & Heptinstall, S. (1981). Differential inhibition by low dose aspirin of human venous prostacyclin synthesis and platelet thromboxane synthesis. *Lancet*, **i**, 969–71.

Hansen, H. (1976). 15-hydroxyprostaglandin dehydrogenase – a review. *Prostaglandins*, **12**, 647–79.

Harris, W. H., Salzman, E. W., Athanasoulis, C. A., Waltman, A. C. & De Sanctis, R. W. (1977). Aspirin prophylaxis of venous thromboembolism after total hip replacement. *New England Journal of Medicine*, **297**, 1246–9.

Hassid, A., Shebusky, A. & Dunn, M. J. (1983). Metabolism of prostaglandins by human renal enzymes: presence of 9-hydroxyprostaglandin dehydrogenase activity in human kidney. *Advances in Prostaglandin, Thromboxane and Leukotriene Research*, **11**, 499–504.

Hawkey, K. (1982). Evidence that prednisolone is inhibitory to the cyclooxygenase activity of human rectal mucosa. *Prostaglandins*, **23**, 397–409.

Hedqvist, P. (1969). Modulating effect of prostaglandin E_2 on noradrenaline release from the isolated cat spleen. *Acta physiologica scandinavica*, **75**, 511–12.

Hedqvist, P. (1970). Studies on the effects of prostaglandins E_1 and E_2 on the sympathetic neuromuscular transmission in some animal tissues. *Acta physiologica scandinavica*, supplement 345.

Hedqvist, P. (1976). Prostaglandin action on transmitter release at adrenergic neuroeffector junctions. *Advances in Prostaglandin and Thromboxane Research*, **1**, 357–63.

Hedqvist, P., Dahlen, S. H. & Björk, J. (1982). Leukotrienes and lipoxygenase products. *Advances in Prostaglandin, Thromboxane and Leukotriene Research*, **11**, 187–200.

Hedqvist, P. & Wennmalm, A. (1971). Comparison of the effect of prostaglandins E_1, E_2

and $F_{2\alpha}$ on the sympathetically stimulated rabbit heart. *Acta physiologica scandinavica*, **83**, 156–62.

Henderson, K. M. & McNatty, K. P. (1977). A possible interrelationship between corticotrophin stimulation and prostaglandin $F_{2\alpha}$ inhibition of steroidogenesis by granulosa luteal cells *in vitro*. *Journal of Endocrinology*, **73**, 71–8.

Heptinstall, S., Bevan, J., Cockbill, H., Hanley, S. P. & Parry, M. J. (1980). Effects of a selective inhibitor of thromboxane synthetase on human blood platelet behaviour. *Thrombosis Research*, **20**, 219–30.

Herman, A. G. & Vane, J. R. (1975). Release of renal prostaglandins during endotoxin-induced hypotension. *European Journal of Pharmacology*, **39**, 79–90.

Heyman, M. A., Rudolph, A. M. & Silverman, W. H. (1976). Closure of the ductus arteriosus in premature infants by inhibition of prostaglandin synthesis. *New England Journal of Medicine*, **295**, 530–3.

Higashihara, C., DuBose, T. D. & Kokko, J. P. (1978). Direct examination of chloride transport across papillary collecting duct of the rat. *American Journal of Physiology*, **235**, F219–26.

Higgs, G. A., Cardinal, D. C., Moncada, S. & Vane, J. R. (1979*a*). Microcirculatory effects of prostacyclin (PGI_2) in the hamster cheek pouch. *Microvascular Research*, **18**, 245–54.

Higgs, G. A., Flower, R. J. & Vane, J. R. (1979*b*). A new approach to anti-inflammatory drugs. *Biochemical Pharmacology*, **28**, 1959–61.

Higgs, G. A., Moncada, S. & Vane, J. R. (1978*a*). Inflammatory effects of prostacyclin (PGI_2) and 6-oxo $PGF_{1\alpha}$ in the rat paw. *Prostaglandins*, **16**, 153–62.

Higgs, G. A., Moncada, S. & Vane, J. R. (1978*b*). Prostacyclin (PGI_2) reduces the number of 'slow moving' leucocytes in hamster cheek pouch. *Journal of Physiology*, **280**, 55P.

Higgs, G. A. & Vane, J. R. (1983). Inhibition of cyclooxygenase and lipoxygenase. *British Medical Bulletin*, **39**, 265–70.

Hinsdale, J. G., Engel, J. J. & Wilson, D. E. (1974). Prostaglandin E in peptic ulcer disease. *Prostaglandins*, **6**, 495–500.

Hirsch, P. D., Campbell, W. B., Willerson, J. T. & Hillis, L. D. (1981). Prostaglandins and ischaemic heart disease. *American Journal of Medicine*, **71**, 1009–28.

Holmes, L. & Horton, E. W. (1968). Prostaglandins in the central nervous system. In *Prostaglandins: Symposium of the Worcester Foundation for Experimental Biology*, pp. 21–38. New York: Interscience.

Holyrode, M. C., Altounyan, E. E. C., Cole, M., Dixon, M. & Elliot, E. V. (1981). Bronchoconstriction in man produced by leukotrienes C and D. *Lancet*, **ii**, 17–18.

Horton, E. W., Main, I. M. M., Thompson, C. J. & Wright, P. M. (1968). Effect of orally administered prostaglandin E_1 on gastric secretion and gastrointestinal motility in man. *Gut*, **9**, 655–8.

Horton, E. W. & Poyser, N. L. (1976). Uterine luteolytic hormone: a physiological role for prostaglandin $F_{2\alpha}$. *Physiological Reviews*, **56**, 595–651.

Hoult, J. R. S. & Moore, P. K. (1977). Pathways of $PGF_{2\alpha}$ metabolism in mammalian kidneys. *British Journal of Pharmacology*, **60**, 627–32.

Hoult, J. R. S. & Moore, P. K. (1978). Sulphasalazine is a potent inhibitor of prostaglandin 15-hydroxydehydrogenase: possible basis for therapeutic action in ulcerative colitis. *British Journal of Pharmacology*, **64**, 6–8.

Humes, J. L., Bonney, R. J., Pelus, L., Dahlgren, M. E., Sadowski, S. & Kuehl, F. A. Jnr (1977). Macrophages synthesise and release prostaglandins in response to inflammatory stimuli. *Nature*, **269**, 149–50.

Hyman, A. L., Mathe, A. A., Lippton, H. L. & Kadowitz, P. J. (1981). Prostaglandins and the lung. *Medical Clinics of North America*, **65**, 789–808.
Itskovitz, H. D., Terragno, N. A. & McGiff, J. C. (1974). Effect of a renal prostaglandin on the distribution of blood flow in the isolated kidney. *Circulation Research*, **34**, 270–6.
Jaffe, B. M. (1974). Prostaglandins and cancer – an update. *Prostaglandins*, **6**, 453–61.
Jaffe, B. M. & Santoro, M. G. (1977). Prostaglandins and cancer. In *The Prostaglandins*, vol. 3, ed. P. W. Ramwell, pp. 329–51. New York: Plenum Press.
Johansson, C., Aly, A., Nilsson, E. & Flemstrom, G. (1982). Stimulation of gastric bicarbonate secretion by E_2 prostaglandin in man. *Advances in Prostaglandin, Thromboxane and Leukotriene Research*, **12**, 395–401.
Johansson, C., Kollberg, B., Nordeman, R., Samuelsson, R. & Bergström, S. (1980). Protective effect of prostaglandin E_2 in the gastrointestinal tract during indomethacin treatment of rheumatic disease. *Gastroenterology*, **78**, 479–83.
Johnston, H. H., Herzog, J. P. & Lauler, D. P. (1967). Effect of prostaglandin E_1 on renal haemodynamics, sodium and water excretion. *American Journal of Physiology*, **213**, 939–46.
Jorgenson, K. A. & Pedersen, R. S. (1981). Familial deficiency of prostacyclin production stimulating factor in the haemolytic uraemic syndrome in children. *Thrombosis Research*, **21**, 311–15.
Kam, S. T., Portoghese, P. S., Dunham, E. W. & Garrard, J. M. (1979). 9,11-azo-13-oxa-15-hydroxyprostanoic acid: a potent thromboxane synthetase inhibitor and a PGH_2/TxA_2 receptor antagonist. *Prostaglandins and Medicine*, **3**, 279–90.
Kamikawa, Y., Serizawa, K. & Shimo, Y. (1977). Some possibilities for prostaglandin mediation in the contractile response to ADP of the guinea pig digestive tract. *European Journal of Pharmacology*, **45**, 199–203.
Karim, S. M. M. (1975). *Prostaglandins and Reproduction*. Lancaster: MTP Press.
Karim, S. M. M. (1979). *Practical Applications of Prostaglandins and their Synthesis Inhibitors*. Lancaster: MTP Press.
Karim, S. M. M., Carter, D. C., Bhana, D. & Ganeson, P. A. (1973). Effect of orally administered prostaglandin E_2 and its 15 methyl analogues on gastric secretion. *British Medical Journal*, **i**, 143–6.
Karim, S. M. M. & Rao, B. (1975). General Introduction. In *Prostaglandins and Reproduction*, ed. S. M. M. Karim, pp. 1–16. Lancaster: MTP Press.
Katzen, D. R., Pong, S. S. & Levine, L. (1975). Distribution of prostaglandin E 9-ketoreductase and NAD^+ dependent and $NADP^+$ dependent 15-hydroxy-prostaglandin dehydrogenase in the renal medulla and cortex of various species. *Research Communications in Chemistry, Pathology and Pharmacology*, **12**, 781–7.
Kaupilla, A. & Ylikorkala, O. (1977). Indomethacin and tolfenamic acid in primary dysmenorrhoea. *European Journal of Obstetrics, Gynaecology and Reproductive Biology*, **7/2**, 59.
Kenimer, J. G., Goldberg, V. & Blecher, M. (1977).The endocrine system: Interaction of prostaglandins with adenylyl cyclase:cyclic AMP systems. In *The Prostaglandins*, vol. 3, ed. P. W. Ramwell, pp. 77–108. New York: Plenum Press.
Kennedy, I., Coleman, R. A., Humphrey, P. P. A., Levy, G. P. & Lumley, P. (1982). Studies on the characterisation of prostanoid receptors: a proposed classification. *Prostaglandins*, **24**, 667–89.

Kloeze, J. (1967). Influence of prostaglandins on platelet adhesion and platelet aggregation. In *Prostaglandins, Proceedings of the 2nd Nobel Symposium*, eds S. Bergström & B. Samuelsson, pp. 241–52. New York: Interscience.

Konzett, H. & Rossler, R. (1940). Versuchsanordnungen zu untersuchungen an der bronchial-muskulatur. *Naunyn Schmiederbergs Archives of Pathology and Pharmacology*, **195**, 71–4.

Kramer, H. J., Dusing, R. & Stinnisbeck, B. (1980). Interaction of conventional and antikaliuretic diuretics with the renal prostaglandin system. *Clinical Science*, **59**, 67–70.

Kuriyama, H. & Makita, Y. (1982). Modulation of neuromuscular transmission by endogenous and exogenous prostaglandins in the guinea pig mesenteric artery. *Journal of Physiology*, **327**, 431–48.

Kurzrok, R. & Lieb, C. (1930). Biochemical studies of human semen: action of semen on the human uterus. *Proceedings of the Society for Experimental Biology and Medicine*, **28**, 268–72.

Labhsetwar, A. P. (1975). Prostaglandins and studies related to reproduction in laboratory animals. In *Prostaglandins and Reproduction*, ed. S. M. M. Karim, pp. 241–70. Lancaster: MTP Press.

Lahav, M., freud, A. & Lindner, H. R. (1976). Abrogation by prostaglandin $F_{2\alpha}$ of LH stimulated cyclic AMP accumulation in isolated rat corpora lutea of pregnancy. *Biochemical and Biophysical Research Communications*, **68**, 1294–1300.

Lands, W. E. M. (1979). The biosynthesis and metabolism of prostaglandins. *Annual Review of Physiology*, **41**, 633–52.

Langer, S. Z. (1981). Presynaptic regulation of the release of noradrenaline. *Pharmacological Reviews*, **32**, 337–59.

Larsson, C., Weber, P. & Anggard, E. (1974). Arachidonic acid increases and indomethacin decreases plasma renin activity in the rabbit. *European Journal of Pharmacology*, **28**, 391–4.

LeDuc, L. E. & Needleman, P. (1979). Regional localisation of prostaglandin and thromboxane synthesis in the dog stomach and intestinal tract. *Journal of Pharmacology and Experimental Therapeutics*, **211**, 181–8.

Lee, S.-C. & Levine, L. (1974). Prostaglandin metabolism. 1. Cytoplasmic reduced NAD dependent and microsomal reduced NADP dependent prostaglandin E 9-ketoreductase in monkey and pigeon tissues. *Journal of Biological Chemistry*, **249**, 1369–75.

Lee, S.-C., Pong, S-S., Katzen, D., Wu, K-Y. & Levine, L. (1975). Distribution of prostaglandin 9-ketoreductase and type I and II 15-hydroxyprostaglandin dehydrogenase in swine kidney medulla and cortex. *Biochemistry*, **14**, 142–5.

Legge, M. (1979). Prostaglandins and angiography. In *Practical Applications of Prostaglandins and their Synthesis Inhibitors.* ed. S. M. M. Karim, pp. 77–88. Lancaster: MTP Press.

Lewis, R. A., Austen, K. F., Drazen, J. F., Clarke, D. A., Marfat, A. & Corey, E. J. (1980). Slow reacting substance of anaphylaxis: Identification of leukotrienes C1 and D from human and rat sources. *Proceedings of the National Academy of Sciences, U.S.A.*, **77**, 3710–14.

Lewy, R. I., Wiener, L., Smith, J. B., Wallinsky, P., Silver, M. J. & Saia, J. (1979). Comparison of plasma concentrations of thromboxane B_2 in Prinzmetals variant angina and classical angina pectoris. *Clinical Cardiology*, **2**, 404–6.

Liggins, G. C. (1978). Ripening of the cervix. In *Seminars in Peritanology*, vol. **11**, no. 3, eds. T. K. Oliver & T. H. Kirschbaum, pp. 261–94. New York: Grune & Stratton.

Lim, R. K. S., Guzman, F., Rodgers, D. W., Goto, K., Braun, C., Dickerson, G. D. & Eagle, R. J. (1964). Site of action of narcotic and non-narcotic analgesics determined by blocking bradykinin-evoked visceral pain. *Archives internationales de pharmacodynamie et de thérapie*, **152**, 25–58.

Livio, M., Villa, S. & de Gaetano, G. (1980). Long-lasting inhibition of platelet prostaglandin but normal vascular prostacyclin generation following sulphinpyrazone administration to rats. *Journal of Pharmacy and Pharmacology*, **32**, 718–19.

Longmore, D. B., Bennett, G., Gueirrara, D., Smith, M., Bunting, S., Reed, P., Moncada, S., Read, N. G. & Vane, J. R. (1981). Prostacyclin: a solution to some problems of extracorporeal circulation. *Lancet*, **i**, 1002–5.

Lonigro, A. J., Itskovitz, H. D., Crowshaw, K. & McGiff, J. C. (1973). Dependency of renal blood flow on prostaglandin synthesis in the dog. *Circulation Research*, **32**, 712–17.

Lulich, K. M. & Paterson, J. W. (1980). An *in vitro* study of various drugs on central and peripheral airways of the rat. A comparison with human airways. *British Journal of Pharmacology*, **68**, 633–6.

McCracken, J. A., Carlsson, J. C., Glen, M. E., Goding, J. R., Baird, D. T., Green, K. & Samuelsson, B. (1972a). Prostaglandin $F_{2\alpha}$ identified as a luteolytic hormone in the sheep. *Nature*, **238**, 129–34.

McCracken, J. A., Baird, D. T. & Goding, J. R. (1972b). Factors affecting the secretion of steroids from the transplanted ovary in the sheep. *Recent Progress in Hormone Research*, **27**, 537–82.

McGiff, J. C. (1981). Prostaglandins, prostacyclin and thromboxanes. *Annual Review of Pharmacology*, **21**, 478–509.

McGiff, J. C., Spokas, E. G. & Wong, P. Y.-K. (1982). Stimulation of renin release by 6-oxo-prostaglandin E_1 and prostacyclin. *British Journal of Pharmacology*, **75**, 137–44.

MacGlashan, D. W. Jnr, Schleimer, R. P., Peters, S. P., Schulman, G. E. S., Adams, K. G. III, Newball, H. M. & Lichtenstein, L. M. (1982). Generation of leukotrienes by purified human lung mast cells. *Journal of Clinical Investigation*, **70**, 747–51.

Machin, S. J., Carreras, L. O., Chamone, D. A. G., DeFreyn, G., Darden, M. & Vermylen, J. (1981). Familial deficiency of thromboxane synthetase. *British Journal of Haematology*, **47**, 629–35.

Machin, S. J., Defreyn, G., Chamone, D. A. F. & Vermylen, J. (1980). Plasma 6 keto $PGF_{1\alpha}$ levels after plasma exchange in thrombotic thrombocytopaenic purpura. *Lancet*, **i**, 661.

MacIntyre, D. E. (1977). Selective inhibition of platelet response to bisenoic prostaglandins. *British Journal of Pharmacology*, **60**, 293P.

Mackenzie, I. Z. & Embrey, M. P. (1978). The influence of vaginal prostaglandin E_2 gel on the subsequent labour. *British Journal of Obstetrics and Gynaecology*, **85**, 567–71.

Malmsten, C. (1979). Prostaglandins, thromboxanes and platelets. *British Journal of Haematology*, **41**, 453–8.

Manabe, T. & Steer, M. L. (1980). Protective effects of PGE_2 on diet-induced acute pancreatitis in mice. *Gastroenterology*, **78**, 777–81.

Mathe, A. (1977). Prostaglandins and the lung. In *The Prostaglandins*, vol. 3, ed. P. W. Ramwell, pp. 169–224. New York: Plenum Press.

Mathe, A., Hedqvist, P., Holmgren, A. & Svanborg, N. (1973). Bronchial hyperreactivity to prostaglandin $F_{2\alpha}$ and histamine in patients with asthma. *British Medical Journal*, **i**, 193.

Mathe, A. & Levine, L. (1973). Release of prostaglandins and metabolites from guinea pig lungs: inhibition by catecholamines. *Prostaglandins*, **4**, 877–90.
Matuchansky, C. & Bernier, J. J. (1971). Effects of prostaglandin E_1 on net and unidirectional movement of water and electrolytes across jejunal mucosa in man. *Gut*, **12**, 854–60.
Mennie, A. T. & Daley, V. (1973). Aspirin in radiation-induced diarrhoea. *Lancet*, **i**, 1131.
Messina, E. J., Weiner, R. & Kaley, G. (1976). Prostaglandins and local circulatory control. *Federation Proceedings*, **35**, 2367–75.
Messina, E. J., Weiner, R. & Kaley, G. (1977). Arteriolar reactive hyperaemia: modification by inhibitors of prostaglandin synthesis. *American Journal of Physiology*, **232**, H571–5.
Miller, O. V., Aiken, J. W., Shebuski, R. J. & Gorman, R. R. (1981). 6 keto prostaglandin E_1 is not equipotent to prostacyclin (PGI_2) as an anti-aggregatory agent. *Prostaglandins*, **20**, 391–400.
Milton, A. S. (1982). Prostaglandins in fever and the mode of action of anti-pyretics. In *Handbook of Experimental Pharmacology*, **60**, 257–304. Berlin: Springer-Verlag.
Milton, A. S. & Wendlandt, S. (1971). Effects on body temperature of prostaglandins of the A, E and F series on injection into the third ventricle of unanaesthetised cats and rabbits. *Journal of Physiology*, **218**, 325–36.
Mishima, K. & Kuriyama, H. (1976). Effect of prostaglandins on electrical and mechanical activities of the guinea pig stomach. *Japanese Journal of Pharmacology*, **26**, 537–48.
Misiewicz, J. J., Waller, S. L., Kiley, N. & Horton, E. W. (1969). Effect of oral prostaglandin E_1 on intestinal transit in man. *Lancet*, **i**, 648.
Miyamoto, T., Ogino, N., Yamamoto, S. & Hayaishi, O. (1976). Purification of prostaglandin endoperoxide synthase from bovine vesicular gland microsomes. *Journal of Biological Chemistry*, **251**, 2629–36.
Moncada, S. (1980). Prostacyclin and thromboxane A_2 in the regulation of platelet vascular interactions. In *Hemostasis, Prostaglandins and Renal Disease*, eds G. Remuzzi, G. Mecca & G. de Gaetano, pp. 175–88. New York: Raven Press.
Moncada, S., Ferreira, S. H. & Vane, J. R. (1978). Bioassay of prostaglandins and biologically active substances derived from arachidonic acid. *Advances in Prostaglandin and Thromboxane Research*, **5**, 211–36.
Moncada, S., Ferreira, S. H. & Vane, J. R. (1979). Pain and inflammatory mediators. In *Handbook of Experimental Pharmacology*, **50**, 588–608. Berlin: Springer-Verlag.
Moncada, S., Gryglewski, R., Bunting, S. & Vane, J. R. (1976). An enzyme isolated from arteries transforms prostaglandin endoperoxides to an unstable substance that inhibits platelet aggregation. *Nature*, **263**, 663–5.
Moncada, S., Herman, A. G., Higgs, G. A. & Vane, J. R. (1977). Differential formation of prostacyclin (PGX or PGI_2) by layers of the arterial wall: an explanation for the anti-thrombotic properties of vascular endothelium. *Thrombosis Research*, **11**, 323–44.
Moncada, S. & Korbut, R. (1978). Dipyridamole and other phosphodiesterase inhibitors act as anti-thrombotic agents by potentiating endogenous prostacyclin. *Lancet*, **i**, 1286–9.
Moncada, S. & Vane, J. R. (1979). Arachidonic acid metabolites and the interactions between platelets and blood vessel walls. *New England Journal of Medicine*, **300**, 1142–7.

Moore, P. K. & Griffiths, R. J. (1983a). 6-keto-prostaglandin E_1. *Prostaglandins*, **26**, 509–17.

Moore, P. K. & Griffiths, R. J. (1983b). Formation of 6-keto-prostaglandin E_1 in mammalian kidneys. *British Journal of Pharmacology*, **79**, 149–55.

Moore, P. K. & Hoult, J. R. S. (1978). Prostaglandin metabolism in rabbit kidney: identification and properties of a prostaglandin 9-hydroxydehydrogenase. *Biochimica et biophysica acta*, **528**, 267–87.

Moore, P. K. & Hoult, J. R. S. (1980). Pathophysiological states modify levels in rat plasma of factors which inhibit synthesis and enhance breakdown of PG. *Nature*, **288**, 271–3.

Morgan, R. O. & Pek, S. B. (1982). Glucose evokes dose-related increases in prostaglandins and insulin secretion from incubated rat islets. *Diabetes*, **31**, supplement 97A.

Morley, J., Hanson, J. M. & Rumjanek, U. M. (1984). Lymphokines. In *Textbook of Immunopharmacology*, eds. M. M. Dale & J. C. Foreman, pp. 170–86. Oxford: Blackwell.

Morris, H. R., Taylor, G. W., Piper, P. J. & Tippins, J. R. (1980). Structure of slow reacting substance of anaphylaxis from guinea pig lung. *Nature*, **285**, 104–6.

Muirhead, E. E., Germain, G., Leach, B. E., Pitcock, J. A., Stephenson, P., Brooks, B., Brosius, W. L., Daniels, E. G. & Kinman, W. (1972). Production of renomedullary prostaglandins by renomedullary interstitial cells grown in tissue culture. *Circulation Research*, **30**, supplement 11, 161–71.

Mullane, K. & Moncada, S. (1980). Prostacyclin release and the modulation of some vasoactive hormones. *Prostaglandins*, **20**, 25–49.

Mustard, J. F. & Kinlough-Rathbone, R. L. (1980). Prostaglandins and platelets. *Annual Review of Medicine*, **31**, 89–96.

Nakano, J. & McCurdy, J. R. (1967). Cardiovascular effects of prostaglandin E_1. *Journal of Pharmacology and Experimental Therapeutics*, **156**, 538–47.

Nasjletti, A. & Malik, U. (1982). Interrelations between prostaglandins and vasoconstrictor hormones: contribution to blood pressure regulation. *Federation Proceedings*, **41**, 2394–9.

Needleman, P., Douglas, J. R., Jakschik, B., Stoeklein, A. S. & Johnson, E. M. Jnr (1974). Release of renal prostaglandins by catecholamines: relationship to renal endocrine function. *Journal of Pharmacology and Experimental Therapeutics*, **188**, 453–60.

Needleman, P., Moncada, S., Bunting, S., Vane, J. R., Hamberg, M. & Samuelsson, B. (1976). Identification of an enzyme in platelet microsomes which generates thromboxane A_2 from prostaglandin endoperoxides. *Nature*, **261**, 558–62.

Needleman, P., Raz, A., Ferrendelli, J. A. & Minkes, M. (1977). Applications of imidazole as a selective inhibitor of thromboxane synthetase in human platelets. *Proceedings of the National Academy of Sciences, U.S.A.*, **74**, 1716–20.

Nissen, H. M. (1968). On lipid droplets in renal interstitial cells. 11. A histological study on the number of droplets in salt depletion and acute salt repletion. *Zeitschrift für Zellforschung und mikrokopische Anatomie*, **85**, 483–91.

Nowak, J. & Wennmalm, A. (1978). Influence of indomethacin and of prostaglandin E_1 on total and regional blood flow in man. *Acta physiologica scandinavica*, **102**, 484–91.

Nompleggi, D., Myers, L., Castell, D. O. & Dubois, A. (1980). Effect of prostaglandin E_2 analogue on gastric emptying and secretion in Rhesus monkeys. *Journal of Pharmacology and Experimental Therapeutics*, **212**, 491–5.

Nugteren, D. H. & Hazelhof, E. (1973). Isolation and properties of intermediates in prostaglandin biosynthesis. *Biochimica et biophysica acta*, **326**, 448–61.

Nylander, B., Robert, A. & Andersson, S. (1974). Gastric secretory inhibition by certain methyl analogues of prostaglandin E_2 following intestinal administration in man. *Scandinavian Journal of Gastroenterology*, **9**, 759–65.

O'Brien, J. R. (1968). Effect of salicylate on human platelets. *Lancet*, **i**, 779–81.

Ogino, N., Miyamoto, T., Yamamoto, S. & Hayaishi, O. (1977). Prostaglandin endoperoxide E isomerase from bovine vesicular gland microsomes, a glutathione-requiring enzyme. *Journal of Biological Chemistry*, **252**, 890–5.

Okahara, T., Imanishi, M. & Yamamoto, K. (1983). Biosynthesis of prostaglandins and thromboxanes in the dog kidney. In *Prostaglandins and the Kidney*, eds M. J. Dunn, C. Patrono & G. A. Cinotti, pp. 53–60. New York: Plenum Press.

Olley, P. M., Coceani, F. & Bodach, E. (1976). E type prostaglandins: A new emerging therapy for certain cyanotic congenital heart malformations. *Circulation*, **53**, 728–31.

Olsen, U. B. (1976). Clonidine-induced increase of renal prostaglandin activity and water diuresis in conscious dogs. *European Journal of Pharmacology*, **36**, 95–101.

Olsson, A. G. (1980). Intravenous prostacyclin for ischaemic ulcers in peripheral arterial disease. *Lancet*, **ii**, 1076.

Omini, C., Moncada, S. & Vane, J. R. (1977). The effects of prostacyclin (PGI_2) on tissues which detect prostaglandins. *Prostaglandins*, **14**, 625–32.

Orloff, J., Handler, J. & Bergström, S. (1965). Effect of PGE_1 on the permeability response of toad bladder to vasopressin, theophylline and adenosine 3'5'-monophosphate. *Nature*, **205**, 397–8.

Orohek, J., Douglas, J. S., Lewis, A. J. & Bouhuys, A. (1973). Prostaglandin regulation of airways smooth muscle. *Nature*, **245**, 84–9.

Pace-Asciak, C. (1975). Activity profiles of prostaglandin 15- and 9-hydroxydehydrogenase and Δ-13-reductase in the developing rat kidney. *Journal of Biological Chemistry*, **250**, 2795–800.

Pace-Asciak, C. (1976). Decreased renal prostaglandin catabolism precedes the onset of hypertension in the developing rat kidney. *Nature*, **263**, 510–12.

Pace-Asciak, C., Carrara, M. C., Rangaraj, G. & Nicolau, K. C. (1978). Enhanced formation of PGI_2, a potent hypotensive substance, by aortic rings and homogenates of the spontaneously hypertensive rat. *Prostaglandins*, **15**, 1005–12.

Pace-Asciak, C. & Rosenthale, A. R. (1982). Constriction of the isolated intact rat kidney by vasopressin is strongly opposed by PGE_1 and PGE_2. *Progress in Lipid Research*, **20**, 605–8.

Pace-Asciak, C. & Wolfe, L. S. (1970). Biosynthesis of prostaglandins E_2 and $F_{2\alpha}$ from tritium labelled arachidonic acid by rat stomach homogenate. *Biochimica et biophysica acta*, **218**, 539–42.

Pace-Asciak, C. & Wolfe, L. S. (1971). A novel prostaglandin derivative formed from arachidonic acid by rat stomach homogenate. *Biochemistry*, **10**, 3657–64.

Patrono, C., Wennmalm, A. & Ciabottini, G. (1979). Evidence for an extra-renal origin of urinary prostaglandin E_2 in healthy men. *Prostaglandins*, **18**, 623–9.

Peatfield, A. C., Piper, P. J. & Richardson, P. S. (1982). The effect of leukotriene C_4 on mucin release into the cat trachea *in vitro* and *in vivo*. *British Journal of Pharmacology*, **72**, 391–3.

Persantine–Aspirin Reinfarction Study Research Group (PARIS) (1980). Persantine and aspirin in coronary heart disease. *Circulation*, **62**, 449–61.

Peskar, B. M., Weiler, H., Kroner, E. E. & Peskar, B. A. (1981). Release of prostaglandins from small intestinal tissue in human and rat *in vitro* and the effect of endotoxin in rat *in vivo*. *Prostaglandins*, **21**, supplement, 9–14.

Pharriss, B. B., Cornette, J. C. & Gutknecht, G. D. (1970). Vascular control of luteal steroidogenesis. *Journal of Reproduction and Fertility*, **10**, supplement, 97–103.

Pickles, V. R. (1957). A plain muscle stimulant in the menstrual fluid. *Nature*, **180**, 1198–9.

Piper, P. J. & Samhoun, M. N. (1982). Stimulation of arachidonic acid metabolism and generation of thromboxane A_2 by leukotrienes B_4, C_4 and D_4 in guinea pig lung *in vitro*. *British Journal of Pharmacology*, **77**, 391–3.

Piper, P. J. & Vane, J. R. (1969). Release of additional factors in anaphylaxis and its antagonism by anti-inflammatory drugs. *Nature*, **223**, 29–35.

Piper, P. J. & Vane, J. R. (1971). The release of prostaglandins from lung and other tissues. *Annals of the New York Academy of Science*, **180**, 363–85.

Pong, S. S. & Levine, L. (1976). Biosynthesis of prostaglandins in rabbit renal cortex. *Research Communications in Chemistry, Pathology and Pharmacology*, **13**, 115–23.

Potts, W. J. & East, P. F. (1972). Effect of prostaglandin E_2 on the body temperature of conscious rat and cat. *Archives internationales de pharmacodynamie et de thérapie*, **197**, 31–6.

Powell, W. S., Hammarström, S., Samuelsson, B. & Sjoberg, B. (1974). Prostaglandin $F_{2\alpha}$ receptor in human corpora lutea. *Lancet*, **i**, 1120.

Poyser, N. L. (1981). *Prostaglandins in Reproduction*. New York: Research Studies Press.

Ratner, A., Wilson, M. C. & Peake, G. T. (1973). Antagonism of prostaglandin promoted pituitary cAMP accumulation and growth hormone secretion *in vitro* by 7 oxa 13 prostynoic acid. *Prostaglandins*, **3**, 413–19.

Raz, A., Minkes, M. S. & Needleman, P. (1977). Endoperoxides and thromboxanes. Structural determinants for platelet aggregation and vasoconstriction. *Biochimica et biophysica acta*, **488**, 305–11.

Richardson, P. S., Phopps, R. J., Balfre, K. & Hall, R. L. (1978). Respiratory tract mucus. *Ciba Foundation Symposium*, **54**, pp. 111–31. Amsterdam: Elsevier Press.

Rittenhouse-Simmons, S. (1979). Production of diglyceride from phosphatidylinositol in activated human platelets. *Journal of Clinical Investigation*, **63**, 580–7.

Robert, A. (1975). An intestinal disease produced experimentally by a prostaglandin deficiency. *Gastroenterology*, **69**, 1045–7.

Robert, A. (1977). Prostaglandins and the gastrointestinal tract. In *The Prostaglandins*, vol. 3, ed. P. W. Ramwell, pp. 225–66. New York: Plenum Press.

Robert, A. (1981). Prostaglandins and the gastrointestinal tract. In *Physiology of the Gastrointestinal Tract*, ed. L. R. Johnson, pp. 1407–34. New York: Raven Press.

Robert, A., Nezamis, J. E., Hanchar, A. J., Lancaster, C. & Klappin, M. S. (1976a). The enteropooling assay to test for diarrhoea due to prostaglandins. *Federation Proceedings*, **35**, 457–9.

Robert, A., Nezamis, J. E., Lancaster, C. & Hanchar, A. J. (1979). Cytoprotection by prostaglandins in rats: prevention of gastric necrosis produced by alcohol, HCl, NaOH, hypertonic NaCl and thermal injury. *Gastroenterology*, **77**, 433–43.

Robert, A., Schultz, J. R., Nezamis, J. E. & Lancaster, C. (1976b). Gastric anti-secretory and anti-ulcer properties of PGE_2, 15-methyl PGE_2 and 16,16-dimethyl PGE_2: intravenous, oral and intrajejunal administration. *Gastroenterology*, **70**, 359–70.

Roberts, L. J., Lewis, R. D., Oates, J. A. & Austen, W. F. (1979). Prostaglandins and

12-hydroxy-5,8,10,14-eicosatetraenoic acid production by ionophore-stimulated rat serosal mast cells. *Biochimica et biophysica acta*, **575**, 185–92.

Robertson, R. P., & Chen, M. (1977). A role for prostaglandin E in defective insulin secretion and carbohydrate intolerance in diabetes mellitus. *Journal of Clinical Investigation*, **60**, 1045–7.

Robertson, R. P., Robertson, D., Roberts, J., Maas, R. L., Fitzgerald, G. A., Friesinger, G. C. & Oates, J. A. (1981). Thromboxane A_2 in vasotonic angina pectoris: evidence from direct measurements and inhibitor trials. *New England Journal of Medicine*, **304**, 998–1003.

Rose, J. C., Kot, P. A., Ramwell, P. W., Dykos, M. & O'Neill, W. P. (1976). Cardiovascular response to prostaglandin endoperoxides in the dog. *Proceedings of the Society for Experimental Biology and Medicine*, **153**, 209–12.

Rosendorff, C. & Cranston, W. I. (1968). Effects of salicylate on human temperature regulation. *Clinical Science*, **35**, 81–91.

Rosenkranz, R. P., Thayer, R. & Killam, K. F. (1980). Effects of prostaglandin E_2 on brain levels of GABA in mice. *Proceedings of the Western Pharmacological Society, U.S.A.*, **23**, 147–50.

Rosenthale, M. E., Dervinis, A. & Kassarich, J. (1974). Bronchodilator activity of the prostaglandins E_1 and E_2. *Journal of Pharmacology and Experimental Therapeutics*, **178**, 541–8.

Roth, G. J. & Majerus, P. W. (1975). The mechanism of the effect of aspirin on human platelets. 1. Acetylation of a particulate protein fraction. *Journal of Clinical Investigation*, **56**, 624–32.

Rubin, R. P., Kelly, K. L., Halenda, S. P. & Laychock, S. G. (1979). Arachidonic acid metabolism in rat acinar pancreatic cells: calcium mediated stimulation of the lipoxygenase system. *Prostaglandins*, **24**, 179–94.

Rudick, J., Gonde, M., Dreiling, D. A. & Janowitz, J. (1971). Effects of prostaglandin E_1 on pancreatic exocrine function. *Gastroenterology*, **60**, 272–8.

Ruwart, M. J., Klepper, M. S. & Rush, B. D. (1980). Prostaglandin stimulation of gastrointestinal tract in post-operative ileus rats. *Prostaglandins*, **19**, 415–26.

Ruwart, M. J., Bush, B. D., Friedle, N. M., Piper, R. C. & Kolaya, G. J. (1981). Protective effects of 16,16-dimethyl PGE_2 on the liver and kidney. *Prostaglandins*, **21**, supplement, 97–102.

Saeed, S. A., McDonald-Gibson, W. J., Cuthbert, J., Copas, J. L., Schneider, C., Gardiner, P. J., Butt, N. M. & Collier, H. O. J. (1977). Endogenous inhibitors of prostaglandin synthetase. *Nature*, **270**, 32–6.

Saikh, A. A. & Klaiber, A. L. (1974). Effect of sequential treatment with oestradiol and $PGF_{2\alpha}$ on the length of the primate menstrual cycle. *Prostaglandins*, **6**, 253–62.

Salmon, J. A. & Amy, J.-J. (1973). Levels of prostaglandin $F_{2\alpha}$ in amniotic fluid during pregnancy and labour. *Prostaglandins*, **4**, 523–33.

Salmon, J. A. & Flower, R. J. (1979). Prostaglandins and related compounds. In *Hormones in Blood*, vol. 2, eds C. H. Gray & V. H. T. James, pp. 237–319. London: Academic Press.

Samuelsson, B. (1965). On the incorporation of oxygen in the conversion of 8,11,14 eicosatrienoic acid into PGE_1. *Journal of the American Chemical Society*, **87**, 3011–13.

Samuelsson, B., Goldyne, M., Granström, E., Hamberg, M., Hammarström, S. & Malmsten, C. (1978). Prostaglandins and thromboxanes. *Annual Review of Biochemistry*, **42**, 997–1029.

Sandler, M., Karim, S. M. M. & Williams, E. D. (1968). Prostaglandins in amine-peptide secreting tumours. *Lancet*, **ii**, 1053–4.

Santoro, M. G., Philpott, G. W. & Jaffe, B. M. (1976). Inhibition of tumour growth *in vivo* and *in vitro* by prostaglandin E. *Nature*, **263**, 777–9.

Sato, J., Jyujo, T., Hirono, M. & Iesaka, T. (1975). Effects of indomethacin, an inhibitor of prostaglandin synthesis, in rats. *Journal of Endocrinology*, **64**, 395–6.

Sautebin, L., Spagnuolo, C., Galli, C. & Galli, G. (1978). A mass fragmentographic procedure for the simultaneous determination of HETE and $PGF_{2\alpha}$ in the central nervous system. *Prostaglandins*, **16**, 985–8.

Schiessel, R., Allison, J. G., Barzilai, A., Fleisher, L. A., Matthews, J. B., Merhav, A. & Solen, W. (1980). Failure of 16,16-dimethyl PGE_2 to stimulate alkaline secretion in the isolated amphibian gastric mucosa. *Gastroenterology*, **78**, 1513–17.

Schulman, E. S., Adkinson, N. F. Jnr & Newball, H. H. (1982). Cyclooxygenase metabolites in human lung anaphylaxis: airways vs parenchyma. *Journal of Applied Physiology*, **53**, 589–95.

Shea-Donohue, P. T., Myers, L., Castell, D. O. & Dubois, A. (1980). Effect of prostacyclin on gastric emptying and secretion in Rhesus monkeys. *Gastroenterology*, **78**, 1476–9.

Shebuski, R. J. & Aiken, J. W. (1980). Angiotensin II stimulation of renal prostaglandin synthesis elevates circulating prostacyclin in the dog. *Journal of Cardiovascular Pharmacology*, **2**, 667–77.

Shibouta, Y., Inada, Y. & Terashita, Z. (1979). Angiotensin II stimulated release of thromboxane A_2 and prostacyclin in isolated, perfused kidneys of spontaneously hypertensive rats. *Biochemical Pharmacology*, **28**, 3601–9.

Silberbauer, K. & Sinzinger, H. (1978). Cortex and medulla of rat kidney generate different amounts of PGI_2-like activity. *Thrombosis Research*, **13**, 1111–18.

Silver, M. J., Smith, J. B., Ingerman, C. M. & Kocsis, C. (1973). Arachidonic acid induced human platelet aggregation and prostaglandin formation. *Prostaglandins*, **4**, 863–75.

Silver, M. J., Smith, J. B., McKean, M. L. & Bills, C. (1980). Prostaglandins and thromboxanes – current concepts. In *Haemostasis, Prostaglandins and Renal Disease*, eds G. Remuzzi, G. Mecca & G. de Gaetano, pp. 159–73. New York: Raven Press.

Sinzinger, H., Clopath, P., Silberbauer, K. & Winter, M. (1980). Is the variability in the susceptibility of various species to atherosclerosis due to inborn deficiency in prostacyclin (PGI_2) formation? *Experientia*, **36**, 321–3.

Smith, J. B., Ingerman, C. M. & Silver, M. J. (1976). Formation of prostaglandin D_2 during endoperoxide-induced platelet aggregation. *Thrombosis Research*, **9**, 413–18.

Smith, J. B. & Willis, A. L. (1971). Aspirin selectively inhibits prostaglandin production in human platelets. *Nature*, **231**, 235–7.

Smith, W. L., Grenier, F. C., Dewitt, D. L., Garcia-Perez, A. & Bell, T. G. (1983). Cellular compartmentalisation of the biosynthesis and function of PGE_2 and PGI_2 in the renal medulla. In *Prostaglandins and the Kidney*, eds. M. J. Dunn, C. Patrono & G. A. Cinotti, pp. 27–39. New York: Plenum Press.

Soll, A. H. (1980). Specific inhibition by prostaglandins E_2 and I_2 of histamine stimulated [^{14}C]aminopyrine accumulation and cyclic adenosine monophosphate generation by isolated canine parietal cells. *Journal of Clinical Investigation*, **65**, 1222–9.

Soll, A. H. & Whittle, B. J. R. (1981). Interaction between prostaglandins and cyclic AMP in the gastric mucosa. *Prostaglandins*, **21**, supplement, 39–45.

Spies, H. G. & Norman, R. L. (1973). Luteinising hormone release and ovulation induced by the intraventricular infusion of prostaglandin E_1 into pentobarbitone-blocked rats. *Prostaglandins*, **4**, 131–41.

Starjne, L. (1973). Dual alpha adrenoceptor mediated control of secretion of sympathetic neurotransmitter: one mechanism dependent and one independent of prostaglandin E. *Prostaglandins*, **3**, 111–16.

Starjne, L. (1979). Role of prostaglandins and cyclic adenosine monophosphate in release. In *Release of Catecholamines from Adrenergic Neurones*, ed. D. M. Paton, pp. 111–42. London: Pergamon Press.

Starke, K. (1977). Regulation of noradrenaline release by pre-synaptic receptor systems. *Reviews in Physiology, Biochemistry and Pharmacology*, **77**, 1–116.

Stokes, J. B. (1979). Effect of prostaglandin E_2 on chloride transport across the thick ascending limb of Henle. Selective inhibition of the medullary portion. *Journal of Clinical Investigation*, **64**, 495–502.

Stokes, J. B. (1981). Integrated actions of renal medullary prostaglandins in the control of water excretion. *American Journal of Physiology*, **240**, F471–80.

Stone, K. J. & Hart, M. (1975). Prostaglandin E_2 9-ketoreductase in rabbit kidney. *Prostaglandins*, **10**, 273–88.

Stone, K. J. & Hart, M. (1976). Inhibition of renal PGE_2 9-ketoreductase by diuretics. *Prostaglandins*, **12**, 197–207.

Susic, H., Nasjletti, A. & Malik, K. U. (1981). Inhibition by bradykinin of the vascular action of angiotensin II in the dog kidney. *Journal of Pharmacology and Experimental Therapeutics*, **218**, 102–7.

Svensson, J., Strandberg, K., Tuvemo, T. & Hamberg, M. (1977). Thromboxane A_2 effect on airway and vascular smooth muscle. *Prostaglandins*, **14**, 425–36.

Sweatman, W. J. F. & Collier, H. O. J. (1968). Effects of prostaglandins on human bronchial muscle. *Nature*, **217**, 69.

Szczeklik, A., Gryglewski, R. J., Nizankowski, E., Nizankowski, R. & Musia, J. (1978a). Pulmonary and anti-platelets effect of intravenous and inhaled prostacyclin in man. *Prostaglandins*, **16**, 651–660.

Szczeklik, A., Gryglewski, R. J., Musia, J., Grodzinska, L., Serwonska, M. & Marcinkiewicz, E. (1978b). Thromboxane generation and platelet aggregation in survivors of myocardial infarction. *Thrombosis and Haemostasis*, **40**, 66–74.

Szczeklik, A., Nizankowski, R., Skawinski, S., Szczeklik, J., Gluszko, P. & Gryglewski, R. J. (1979). Successful therapy of advanced arteriosclerosis obliterans with prostacyclin. *Lancet*, **i**, 1111–14.

Szczeklik, A., Szczeklik, J. & Nizankowski, R. (1980). Haemodynamic changes induced by prostacyclin in man. *British Heart Journal*, **44**, 254–8.

Tan, S. Y., Sweet, P. & Mulrow, R. J. (1977). Impaired renal production of prostaglandin E_2: a newly identified lesion in human essential hypertension. *Prostaglandins*, **15**, 139–49.

Tomlinson, H. M., Ringold, H. J., Qureshi, M. C. & Forchielli, E. (1972). Relationship between inhibition of prostaglandin synthesis and drug efficacy: support for the current theory on mode of action of aspirin-like drugs. *Biochemical and Biophysical Research Communications*, **46**, 552–9.

Tyler, H. M., Parry, M. J., Wood, B. A. & Saxton, C. A. P. D. (1981). Administration to man of a new inhibitor of thromboxane synthesis. *Lancet*, **i**, 629–32.

van Dorp, D. A., Beerthuis, R. K., Nugteren, D. H. & Vonkemann, H. (1964). Enzymatic

conversion of all-*cis*-polyunsaturated fatty acids into prostaglandins. *Nature*, **203**, 839–41.
van Ouderas, F. J. G. & Buytenhek, M. (1982). Purification of PGH synthase from sheep vesicular gland. *Methods in Enzymology*, **80**, 60–8.
Vane, J. R. (1969). The release and fate of vasoactive hormones in the circulation. *British Journal of Pharmacology*, **35**, 209–42.
Vane, J. R. (1971). Inhibition of prostaglandin synthesis as a mechanism of action for aspirin-like drugs. *Nature*, **231**, 232–5.
Vane, J. R. (1978). The mode of action of aspirin-like drugs. *Agents and Actions*, **8**, 430–43.
Vargaftig, B. B., Chignard, M. & Benveniste, J. (1981). Present concepts on the mechanisms of platelet aggregation. *Biochemical Pharmacology*, **30**, 263–71.
Vargaftig, B. B. & Dao, N. H. (1972). Selective inhibition by mepacrine of the release of rabbit aorta contracting substance evoked by administration of bradykinin. *Journal of Pharmacy and Pharmacology*, **24**, 159–61.
Vargaftig, B. B. & Zirinis, P. (1973). Platelet aggregation induced by arachidonic acid is accompanied by release of potential inflammatory mediators distinct from PGE_2 and $PGF_{2\alpha}$. *Nature*, **244**, 114–16.
Veale, W. L., Cooper, K. E. & Pittman, Q. J. (1977). Role of prostaglandins in fever and temperature regulation. In *The Prostaglandins*, vol. 3, ed. P. W. Ramwell, pp. 145–67. New York: Plenum Press.
Vermylen, J., Chamone, D. A. F. & Verstraete, M. (1979). Stimulation of prostacyclin release from vessel wall by BAY g6575, an antithrombotic compound. *Lancet*, **i**, 518–20.
Voelkel, N. F., Worthen, S., Reeve, J. T., Henson, P. M. & Murphy, R. C. (1982). Non-immunological production of leukotrienes induced by platelet activating factor. *Science*, **218**, 286–9.
von Euler, U. S. (1934). Zur kenntnis der pharmakalogischen wirkungen von nativsekreten und extrekten mannlicher accessorsicher geschlechtsdrusen. *Naunyn Schmiedebergs Archives of Experimental Pathology and Pharmacology*, **175**, 78–84.
von Euler, U. S. & Eliasson, R. (1967). *Prostaglandins* (Monographs in Medicinal Chemistry). New York: Academic Press.
Walker, J. L. (1973). The regulatory function of prostaglandins in the release of histamine and SRS-A from passively sensitised human lung tissue. *Advances in the Biosciences*, **9**, 235–9.
Walker, J. R., Smith, M. J. H. & Ford-Hutchinson, A. W. (1976). Anti-inflammatory drugs, prostaglandins and leucocyte migration. *Agents and Actions*, **6**, 602–6.
Watkins, W. D., Petersen, M. B., Crone, R. K., Shannon, D. C. & Levine, L. (1980). Prostacyclin and prostaglandin E_1 for severe pulmonary artery hypertension. *Lancet*, **i**, 1083.
Weber, P. C., Larsson, C., Anggard, E., Hamberg, M., Corey, E. J., Nicolaou, K. & Samuelsson, B. (1976). Stimulation of renin release from rabbit kidney cortex by arachidonic acid and prostaglandin endoperoxides. *Circulation Research*, **39**, 868–74.
Weber, P. C., Scherer, B. & Larsson, C. (1977). Increase of free arachidonic acid by furosemide in man as the cause of prostaglandin and renin release. *European Journal of Pharmacology*, **41**, 329–32.
Weichmann, B. M. & Tucker, S. S. (1982). Contraction of guinea pig uterus by synthetic leukotrienes. *Prostaglandins*, **24**, 245–53.

Weksler, B. B., Marcus, A. J. & Jaffe, E. A. (1977). Synthesis of prostaglandin I_2 (prostacyclin) by cultured human and bovine endothelial cells. *Proceedings of the National Academy of Sciences, U.S.A.*, **74**, 3922–6.

Wennmalm, A. (1971). Studies on mechanisms controlling the secretion of neurotransmitters in the rabbit heart, *Acta physiologica scandinavica*, **82**, supplement 365.

Wennmalm, A. (1979). Prostaglandins and cardiovascular function: some biochemical and physiological effects. *Scandinavian Journal of Clinical Laboratory Investigation*, **39**, 399–405.

Westfall, T. C. (1977). Local regulation of adrenergic neurotransmission. *Physiological Reviews*, **57**, 659–94.

Whorton, A. R., Smigel, P. & Oates, J. A. (1978). Regional differences in prostacyclin formation by the kidney, prostacyclin is a major prostaglandin of renal cortex. *Biochimica et biophysica acta*, **529**, 176–80.

Whittle, B. J. R. (1977). Mechanisms underlying gastric mucosal damage induced by indomethacin and bile salts and the actions of prostaglandins. *British Journal of Pharmacology*, **60**, 455–60.

Whittle, B. J. R., Moncada, S. & Vane, J. R. (1978a). Formation of prostacyclin by the gastric mucosa and its action on gastric function. *Prostaglandins*, **15**, 704–5.

Whittle, B. J. R., Moncada, S. & Vane, J. R. (1978b). Comparison of the effects of prostacyclin (PGI_2), prostaglandin E_1 and D_2 on platelet aggregation in different species. *Prostaglandins*, **16**, 373–88.

Whittle, B. J. R., Mugridge, K. G. & Moncada, S. (1979). Use of the rabbit transverse stomach strip to identify and assay prostacyclin, PGA_2, PGD_2 and other prostaglandins. *European Journal of Pharmacology*, **53**, 167–72.

Williams, E. D., Karim, S. M. M. & Sandler, M. (1968). Prostaglandin secretion by medullary carcinoma of the thyroid: a possible cause of the associated diarrhoea. *Lancet*, **i**, 22–5.

Williams, K. I. & El Tahir, K. E. H. (1980). Effect of uterine stimulants on prostacyclin production by the pregnant rat myometrium. 1. Oxytocin, bradykinin and $PGF_{2\alpha}$. *Prostaglandins*, **19**, 31–8.

Williams, K. I., El Tahir, K. E. H. & Marienkiewicz, E. (1979). Dual action of prostacyclin (PGI_2) on the rat pregnant uterus. *Prostaglandins*, **17**, 667–72.

Williams, T. (1983). Interactions between prostaglandins, leukotrienes and other mediators of inflammation. *British Medical Bulletin*, **39**, 239–42.

Williams, T. & Piper, P. J. (1980). The action of chemically pure SRS-A on the microcirculation *in vivo*. *Prostaglandins*, **19**, 779–89.

Willis, A. L., Vane, F. M., Kuhn, D. C., Scott, C. A. & Petrin, M. (1974). An endoperoxide aggregator (LASS) formed in platelets in response to thrombotic stimuli: purification, identification and unique biological significance. *Prostaglandins*, **8**, 453–507.

Wilson, D. E. & Levine, R. A. (1969). Decreased canine gastric mucosal blood flow induced by prostaglandin E_1: a mechanism for its inhibitory effect on gastric secretion. *Gastroenterology*, **56**, 1268.

Wilson, D. E., Winman, G., Quatermass, J. & Tao, P. (1975). Effects of an orally administered prostaglandin analogue (16,16-dimethyl prostaglandin E_2) on human gastric secretion. *Gastroenterology*, **69**, 607–11.

Wolfe, L. S., Pappius, H. M. & Marion, J. (1976). The biosynthesis of prostaglandins by brain tissue in vitro. *Advances in Prostaglandin and Thromboxane Research*, **1**, 345–55.

Wong, PY-K., Terragno, D. A., Terragno, N. A. & McGiff, J. C. (1977). Dual effects of bradykinin on prostaglandin metabolism: relationship to the dissimilar vascular actions of kinins. *Prostaglandins*, **13**, 1113–25.

Wright, J. P., Young, G. O., Klaff, L. J., Weers, L. A., Price, S. K. & Marks, I. N. (1982). Gastric mucosal prostaglandin E levels in patients with gastric ulcer disease and carcinoma. *Gastroenterology*, **82**, 263–7.

Yen, S.-S., Mathe, A. & Dugan, J. J. (1976). Release of prostaglandins from healthy and sensitised guinea pig lung and trachea by histamine. *Prostaglandins*, **11**, 227–34.

Yoshimoto, T. & Yamamoto, S. (1982). Partial purification and assay of thromboxane A synthase from bovine platelets. *Methods in Enzymology*, **86**, 106–9.

Zambraski, E. J. & Dunn, M. J. (1981). Prostaglandins and renal function in chronic bile duct ligated-cirrhotic dogs. *Kidney International*, **19**, 218–25.

Zenser, T. V., Herman, C. A. & Gorman, R. R. (1977). Metabolism and action of the prostaglandin endoperoxide PGH_2 in rat kidney. *Biochemical and Biophysical Research Communications*, **79**, 357–63.

Ziboh, V. A., Maruta, H., Lord, J., Cagle, W. D. & Lucky, W. (1979). Increased biosynthesis of thromboxane A_2 by diabetic patients. *European Journal of Clinical Investigation*, **9**, 223–8.

Zipser, R. D., Radvan, G., Duke, E. & Little, T. E. (1982). Increased urinary thromboxane B_2 and reduced prostaglandin E_2 in hepatorenal syndrome. *Clinical Research*, **30**, 468.

Zusmann, R. M. (1981). Prostaglandins and water excretion. *Annual Review of Medicine*, **32**, 369–74.

Zusman, T. V. & Keiser, H. R. (1977). Prostaglandin E_2 biosynthesis by rabbit renomedullary interstitial cells in tissue culture. *Journal of Biological Chemistry*, **252**, 2069–71.

Index

abortion 117
adrenoceptors 215–17
AH 19437 35, 161
aldosterone 193
algesia 58, 224–5
anaphylactic shock 39, 153, 155
angina pectoris 75–6
angiography 147
arachidonic acid 5
arrhythmias 76
arthritis 222, 230
aspirin 51–62, 95–7, 102–3, 136, 195
asthma 171–2
ATP 141

Barrter's syndrome 197
benoxaprofen 63–4
bioassay 27
blood pressure 71
blood vessels 70–3, 133, 170, 182–3, 223, 230
brain 200–2
bronchial secretion 171
bronchial smooth muscle 159–65
BW 755C 63–4, 232

calcium 81–3, 87, 130–1, 211–13
cancer 143, 145, 233–5
capillary permeability 223–4
carbenoxolone 41
catabolism 20–3, 123–6, 83, 155–9, 180–2
central effects
    anti-convulsant 203
    behaviour 202
    temperature control 203–4
    water and food intake 207–9
cerebrospinal fluid (CSF) 201–2
cervix 116, 118
chemokinesis 226, 231
chemotaxis 226, 231

cholera 134, 143
clonidine 196, 216
corpus luteum 34, 100
corticosteroids 43–51, 228, 232
cyclic AMP 33, 81–3, 86–9, 106–7, 155, 194, 208, 234
cyclic GMP 203
cyclooxygenase 17–18
cytoprotection
    colon 138
    gastric 135
    intestinal 136
    liver and pancreas 138

dazoxiben 24–7
dialysis 98
diarrhoea 127, 143, 146
diet 93
dihomolinolenic acid 5–6, 9
distribution 24–7
diuretic drugs 195–6
ductus arteriosus 119–20

eicosatetraynoic acid (ETYA) 37, 215
eicosatrienoic acid 5–6, 9
eicosapentaenoic acid 5–6, 93–4
endogenous inhibitors of prostanoid formation 67, 93, 124
endothelium 69, 149, 89–91, 223–7
endotoxins 143, 204–7
enteropooling 134

fertilisation 113–14
fertility (male) 121
fluoroindomethacin 55
forced expiratory volume (FEV) 161
FPL 55712 63, 169, 171

gas chromatography–mass spectrometry 2, 28, 29

gastric acid secretion 131–4
glutathione 39, 124, 174
gonadotrophins 102–5, 208
Graafian follicle 100
gut motility 126–31

heart 73–4
heart rate 71
hepatorenal disease 198
hydroxyeicosatetraenoic acids (HETEs) 38, 61, 201, 229
hydroxyperoxyeicosatetraenoic acids (HPETEs) 38, 61, 229
hyperalgesia 224
hypertension 74–5
hypothalamus 58, 205–9

ibuprofen 55
imidazole 41, 65
indomethacin 51, 55, 58, 60, 169, 183, 197, 214
inflammatory mediators 221–33
intestinal secretion 134
irradiation 143
isotopic oxygen 41, 65

kallikrein–kinin system 185
kidney
    blood flow 182, 192
    collecting tubules 176
    countercurrent exchange 192–3
    glomerulus 176
    interstitial cells 177
    urine formation 186–95
Konzett–Rossler preparation 162, 164–5, 170

labile aggregation stimulatory substances (LASS) 84
labour 118, 120
lactation 117
leucocytes 38, 49, 62, 222, 225–227
leukotrienes 35–40, 112, 127, 168–71, 229–31
linoleic acid 6
lipoxygenase 62–4, 35
lungs, PG release from 46–8, 150–5
luteolysis 100, 105–11
lymphokines 227
lysosomes 141

macrocortin 43–51
macrophages 227
malondialdehyde 11
meclofenamic acid 183
menstruation 111–13, 120, 145
mepacrine 42
microcirculation 223, 230
mucosal 'barrier' 139

mucus 139, 171
myocardial infarction 76–7, 92, 95–7

naproxen 55, 60
nephron 174–8
neurotransmission
    ganglionic 219
    parasympathetic 165, 217–18
    sympathetic 209–15
nitroglycerin 99
NO 164 33, 65
non-steroid anti-inflammatory drugs (NSAID)
    effects of 51–62, 116, 120, 227–9
    mechanism of action 53–6

oedema 152, 223–4
ovulation 105
6 oxo PGE_1 23, 71, 87, 163, 171, 193
6 oxo $PGF_{1\alpha}$ 17, 71, 149, 171, 201

pain 224–5
pancreas 135, 137
pancreatitis 146
para-bromophenacyl bromide (*p*BPB) 42
paracetamol 58, 62, 206
paralytic ileus 146
parenchyma 150, 159–61, 169
parietal cell 134
parturition 115–17
phagocytosis 222
phenylbutazone 55
phospholipase enzymes 6–9
phospholipids 8
placenta 100, 114–15
platelets
    adhesion and aggregation 81–91
    anti-platelet drugs 95–9
    in disease 91–3
    PGI_2 : TxA_2 balance 89–91
    structure 80
Plexaura homomalla 26
prostacyclin synthetase 20
prostaglandin A_1 (PGA_1) 134
prostaglandin A_2 (PGA_2) 4, 71, 73
prostaglandin analogues 111, 117, 129, 133, 136, 168, 234
prostaglandin B_2 (PGB_2) 4
prostaglandin C_2 (PGC_2) 4
prostaglandin D_1 (PGD_1) 4
prostaglandin D_2 (PGD_2) 4, 33, 70, 71, 123, 149, 165, 201, 222, 226, 233
prostaglandin E_1 (PGE_1) 4, 6, 33, 34, 128, 129, 202, 207, 210
prostaglandin E_2 (PGE_2) 3, 6, 33, 34, 70, 71, 101, 123, 146, 148, 163, 189, 193, 200, 207, 210, 222, 223, 226
prostaglandin endoperoxide D isomerase 19

Index

prostaglandin endoperoxide E isomerase 18–19
prostaglandin endoperoxide synthase 17–18
prostaglandin $F_{1\alpha}$ ($PGF_{1\alpha}$) 3, 107, 202
prostaglandin $F_{2\alpha}$ ($PGF_{2\alpha}$) 3, 6, 70, 71, 101, 106, 123, 148, 163, 193, 200, 202, 207, 211, 221, 226
prostaglandin $F_{2\beta}$ ($PGF_{2\beta}$) 3, 165, 166
prostaglandin G_2 (PGG_2) 11, 12, 71, 83–6, 167, 174
prostaglandin H_2 (PGH_2) 11, 12, 71, 167, 174
prostaglandin I_2 (PGI_2) 15–17, 23, 58, 66, 69, 71, 87, 98–9, 101, 123, 129, 131, 149, 174, 193, 201, 207, 210, 221, 223, 226, 233
prostaglandin I_3 (PGI_3) 93
prostaglandin metabolites 21–4, 71
prostate gland 1

rabbit aorta contracting substance (RCS) 10–12
rabbit aorta contracting substance releasing factor (RCS-RF) 44
receptors 32–5, 106
renin-angiotensin system 177–8, 183–7, 193–5, 198, 209

SC 19220 35, 59, 130, 161, 207
semen 2, 3, 101, 121
seminal vesicle 2
slow reacting substance of anaphylaxis (SRS-A) 37, 53
slow reacting substance from cobra venom (SRS-C) 37
sodium salicylate 61
specific airways conductance (S_{GAW}) 161
spectrophotometry 28, 31, 38
sperm 121
sulphasalazine 121

thermoregulation 203–207
thromboxane A_2 (TxA_2),
thromboxane B_2 (TxB_2) 12–15, 53, 65, 71, 87, 149, 168, 191, 201, 207, 221, 226, 233
thromboxane synthetase 19–20
tumour 233–5

U46619 85, 129, 131, 171
ulcerative colitis 145
ulcers 142–3
uterus 100, 115, 116

vasopressin 190–1, 209